Systems Engineering and Architecting

Creating Formal Requirements

Systems Engineering and Architecting

Creating Formal Requirements

Laurence Bellagamba

CRC Press
Taylor & Francis Group
Boca Raton London New York

CRC Press is an imprint of the
Taylor & Francis Group, an **informa** business

CRC Press
Taylor & Francis Group
6000 Broken Sound Parkway NW, Suite 300
Boca Raton, FL 33487-2742

First issued in paperback 2017

© 2012 by Taylor & Francis Group, LLC
CRC Press is an imprint of Taylor & Francis Group, an Informa business

No claim to original U.S. Government works
Version Date: 20120221

ISBN 13: 978-1-138-07775-1 (pbk)
ISBN 13: 978-1-4398-8140-8 (hbk)

Visit the Taylor & Francis Web site at
http://www.taylorandfrancis.com

and the CRC Press Web site at
http://www.crcpress.com

Dedication

To Julie and especially Michelle, who helped

Contents

Preface

This book was written to take a step to fulfill a goal that George Friedman stated in his president's keynote address in 1994 at just the second meeting of the International Council on Systems Engineering. George asked his audience to provide a mathematical basis for doing systems engineering. Such a basis is now called *formal requirements*, which are explicit, executable instructions to do something that can be verified by logic or examination. Since George asked, substantial advances have been gradually made in our ability to provide formal requirements for doing many aspects of software engineering and embedded systems. These successful efforts provide the insights needed to start the process for systems engineering. Also in the years since, the need to rationally control the interactions of families of systems has developed into a major concern. So we now need formal methods to do architecting as well.

The book describes a set of formal requirements and shows examples of their use. The actual formal requirements themselves are written in *Mathematica*® and are available online at http://www.wolfram.com. In retrospect, formulating the formal requirements is actually much easier than inventing how to accomplish systems engineering and architecting tasks in the first place. The job to make formal requirements is more illumination than invention, so embellishing and adding to the set of formal requirements are best done by many people rather than a few individuals. Therefore, all my colleagues are encouraged to get the set and recommend improvements or additions. My hope is that over time talented individuals will collectively achieve George's goal.

The book was designed to enable systems engineers or architects to learn exactly how to accomplish many key activities of their discipline, as well as to caution that many of the activities remain more art than science and are not yet amenable to formal requirements.

Using the formal requirements as presented enables the book to serve as a text for an introductory course in systems engineering and architecting. In such a setting, the emphasis is on learning what to do. Each formal requirement can be taken as is and adhered to with minimum introspection. The book also supports a graduate-level course where the focus is to find better ways for us to accomplish our job. In that setting, the formal requirements themselves are the focus and the educational objective is to comprehend them, creatively improve them, and develop new formal requirements to accomplish more activities.

Mathematica® and the *Mathematica* logo are registered trademarks of Wolfram Research, Inc. (WRI – www.wolfram.com http://www.wolfram.com/) and are used herein with WRI's permission. WRI did not participate in the creation of this work beyond the inclusion of the accompanying software, and offers it no endorsement beyond the inclusion of the accompanying software.

For MATLAB® and Simulink® product information, please contact:
The MathWorks, Inc.
3 Apple Hill Drive
Natick, MA, 01760-2098 USA
Tel: 508-647-7000
Fax: 508-647-7001
E-mail: info@mathworks.com
Web: www.mathworks.com

Acknowledgments

Thanks to Generals Haywood and Helms for their unwavering support to enable the discipline of architecting to be discovered. Special thanks to the two key government employees who made it all happen, Roberta Ewart and James Przybysz. As contractors, we can be only as good as our customer allows, and these two allowed us to do our best. And most importantly, thanks to the men and women who constituted the architecting team. It was their collective efforts that enabled a new and important discipline to be born. These individuals include Ross Balestreri, Dennis Cajili, Carol Carlson, Phil Dibb, Donna Drew, Mike Gabor, Michael Gregoire, Larry Halek, Barbara Hemmerich, Mark Mason, Lance Newhart, Courtney Oh, David Rudolph, Leigh Sorgen, Quena Stropty, Robert Webb, Robin Weathers. Also special thanks to senior editor Jonathan Plant for making this book possible and project editor Iris Fahrer for accomplishing the difficult task of formulating this book.

About the Author

Laurence Bellagamba was director of engineering for Rockwell International, Northrop Grumman, and TASC. He has led systems engineering or architecting efforts for the Space Shuttle Program, Shuttle-C, Global Positioning System, Anti-satellite System, Ground Based Missile Defense, Airborne Laser, Space Based Laser, and Space Control Architecture Development. He was awarded the NASA Certificate of Appreciation. He received BSAE, MS, and PhD degrees from the School of Aeronautics and Astronautics at Purdue University.

1 Motivation, Objective, Definitions, and Approach

1.1 MOTIVATIONS AND OBJECTIVE

Systems engineering and architecting are very young sciences. The practice of each is more an art than a science. Many excellent books are available to help a practitioner increase the impact of the art or comprehend the science.[1-8]

This book aims to provide tools to do aspects of systems engineering and architecting better, faster, or cheaper. To accomplish systems engineering or architecting, one must do four things well:

1. Model system or architecture behavior
2. Make rational decisions
3. Establish natural language requirements
4. Improve systems engineering and architecting processes and products

The objective of this book is to provide *formal requirements* to do each of these four activities a variety of ways. Formal requirements are just explicit instructions— you can think of them as a computer program if you like. Though at first such a rigid structure may seem a hindrance, it actually provides two huge benefits. First, explicit instructions can be executed, that is, produce a product. Many systems engineering and architecting activities cannot yet be described as formal requirements. For example, how do you explicitly tell someone to come up with the best concept that fulfills customer requirements? The best you can do is give general guidance on how to identify candidate solutions and synthesize something new and innovative that will be appealing. The actual step-by-step process remains a mystery of the inner workings of an individual's brain. Some people are extraordinarily good at concept formulation. The extraordinary good people may be able to explain how they found their innovation, but even when they can, their explanation is insufficiently detailed for someone else to accomplish the same feat. On the other hand, you can explicitly tell someone how to find a mathematical minimum or maximum; indeed, there are many ways to do so, and some are better suited for some problems than others. Should the search for the solution concept be translatable to a mathematical optimization problem, those explicit instructions can be quite helpful. Even if you cannot make such a translation, you can also instruct someone on the many ways to compare optional solutions to determine which is best. The second benefit of formal requirements is that you can improve them and capture the improvements. If someone thinks they have a better way to do the job, they rewrite the formal requirements, execute them, and compare the results to how they were previously accomplished.

If better results are obtained, or the same results are obtained quicker or less expensively, then the new version of the formal requirements should replace the old.

The subsequent chapters provide formal requirements in *Mathematica*® to perform each of the four key activities efficiently and expertly. *Mathematica* is a convenient means to record the formal requirements because it provides a huge vocabulary of verified constructs to use to state formal requirements that are also executable. Since *Mathematica* is an interpreter-based language, the execution may not be the most efficient implementation of the formal requirements. Suitably skilled readers can translate the *Mathematica* code into whatever code they find more advantageous to use. I presume the reader has basic familiarity with *Mathematica*; for a brief introduction, see the Appendix at the end of this chapter, and for details see Reference 9. Herein, variables are denoted in bold, while mathematical functions and *Mathematica* routines are denoted in italics.

To try to achieve a quantum leap in our collective ability to do systems engineering and architecting, the author asks his colleagues to help improve and develop the formal requirements in an open source environment. A copy of all the currently available formal requirements can be found at library.wolfram.com, in the Mathematica Technology section, as the Formal Requirements for Systems Engineering and Architecting entry. The reader is encouraged to provide modified or additional formal requirements. The method to do so is explained at the website.

1.2 DEFINITIONS

Figure 1.1 identifies the key words that need to be defined to fully explain the objective. This section defines words as used in this book. Please check to ensure that how a term is used in this book is indeed what you mean when using the same word or phrase. Otherwise, you and I may use the same word to mean very different things.

Formal requirements are explicit instructions that are *unambiguous, executable,* and *correct.* Formal requirements document what needs to be done in such a way that there is only one possible interpretation, and they produce the result consistent with the documentation. A convenient way to record formal requirements is as a

formal requirement > unambiguous > executable > correct > verified

architecture > mission > system > customer > operator > user

architecting > systems engineering > optimal solution > functional requirement > performance requirement

function > develop > design > manufacture > deploy > train > operate > maintain > dispose

performance > availability > coverage > timeliness > quality > quantity

behavior > feedback > linear > exponential > threshold > oscillation > S shaped > collapse

resonance > damping > overshoot > steady-state error

FIGURE 1.1   Words to define as used in this book.

mathematical or logical formula in the most general possible sense. All programming languages record formulas in a manner certain to be sufficiently unambiguous such that they will execute on a computer in one particular way. But no programming language can by itself achieve complete correctness. Fortunately, for most programming languages, if a program is proven correct once, it will stay correct every time it is used. This simplifies the effort to prove the correctness of formal requirements that are written in terms of other formal requirements.

Here is an example of the simplest possible formal requirement written in *Mathematica*:

```
In[5]:= ProbTrueGivenPosTest[PA_, Pfalsepos_, Pfalseneg_] :=
        (* OUTPUT: the probability a condition exists given a test indicates it does *)
        (* INPUT: *)
        (* PA = the probability the condition exists *)
        (* Pfalsepos = the probability for a false positive *)
        (* Pfalseneg = the probability for a false negative *)
        (* AUTHOR: Laurence Bellagamba at bellagamba@me.com *)
        (* COPYRIGHT © 2011 by Laurence Bellagamba *)
        (* SOURCE: self *)
        PA * (1 - Pfalseneg) / (PA * (1 - Pfalseneg) + (1 - PA) * Pfalsepos)
```

ProbTrueGivenPosTest is the name given the formal requirement. The inputs are **PA_**, **Pfalsepos_**, and **Pfalseneg_**. Between the (* and *) markings is explanatory material. When provided symbolic inputs, the formal requirements provide an abstract result:

```
In[6]:= ProbTrueGivenPosTest[pa, pf, pn]
```

$$Out[6]= \frac{pa\,(1-pn)}{(1-pa)\,pf + pa\,(1-pn)}$$

When provided numerical inputs, the formal requirements provide a numerical result:

```
In[7]:= ProbTrueGivenPosTest[.5, .1, .2]
```

Out[7]= 0.888889

Unambiguous means the requirement statement can be interpreted in only one possible way. Please note that the tricky part is to achieve the intended interpretation.

Executable means that given correct and necessary inputs, the formal requirements produce an output consistent with the unambiguous stipulation. The execution could transform a set of measurements or information into other information, or simulate the behavior of a system or architecture.[10,11] The execution could produce real-world entities in hardware or software.[12,13]

Correct means that executing the formal requirements achieves the intent for which they were written. To prove the formal requirements are correct requires they be verified.

Verified means there is proof that the formal requirements are correct. The proof requires examining the formal requirements mathematically or logically, or executing the formal requirements to obtain results that can be used as evidence to show the intent was satisfied. For formal requirements with a small finite list of possible inputs and simple rules for transforming the inputs to an output, such proofs are usually easy to obtain. For very large or infinite input variations, or very complicated or conditional rules for transforming the inputs to outputs, it is possible that the formal requirements may only be verified for stipulated conditions.

Mission:
Detect in the shortest time, an attack against any satellite operated by the United States, France or
United Kingdom

Constituent systems:
System 1: Existing ground based radars in the continental United States operated by the US
Air Force
System 2: To be developed ground based telescope to acquired by British Army, located in Australia,
operated by the Australian Air Force
System 3: To be developed ground based radars to be acquired by French Army, with fielding
locations to be determine, and operated by the French Army
System 4: Existing ground based telescopes located in Norway, operated by US Air Force
System 5: Space bases surveillance satellite to be acquired by the US Air Force, operated by the US
Air Force
System 6: Optical sensors placed on commercial satellites. The satellites are operated by private entities.
The sensor data is transformed into products by contractors working for the US Air Force.
System 7: Command and control systems jointly operated by the US Air Force, British Air Force and
French Air Force

FIGURE 1.2   An architecture consists of multiple systems, which perform a mission even if
one system is deleted.

System is a combination of people, machines, raw materials, methods, and environment that produces a product or service. Should any portion of the system be deleted, the product or service would not be produced.

Customer is the name given to the person or persons acquiring the system.

Operator is the name given to the person or persons maintaining the system.

User is the name given to the person or persons utilizing the system's products or services. One or more of the customers, operators, or users may be the same person or persons, or they may all be different.

Architecture consists of multiple systems that redundantly perform at least one *mission*. Should a system be deleted, the mission would still be accomplished, but perhaps not as well. The constituent systems may have different customers, operators, or users.

Mission is an attempt to achieve a quantifiable objective, usually subject to constraints. Figure 1.2 shows an example mission and a corresponding architecture.

Systems engineering is the process of determining, documenting, and verifying the optimal functional and performance requirements for a system.

Architecting is the process of determining the *optimal* systems for the architecture.

Optimal is defined as the minimum value of an index of performance (IP) that is a function of independent variables, subject to constraints involving the independent variables. Should IP need to be maximized, minimizing $-IP$, or $1/IP$, achieves the desired result.

Optimal solution is the set of values for the independent variables that produce the optimal index of performance.

Function is a task or activity to be performed by a system. There are eight primary functions: develop, design, manufacture, deploy, train, operate, maintain, and dispose.

Develop functions determine what system will best achieve the customer's, operator's, and user's needs.

Design functions determine the instructions for creating the real-world items that will constitute the system.

Manufacture functions create the real-world items.

Deploy functions place the system in its operating environment.

Train functions prepare the operators to operate the system and the users to use the system.

Operate functions are the activities performed by the system in an associated environment for a user.

Maintain functions are the activities performed by operators to keep the system operating.

Dispose functions address what to do with the system after it ceases to operate.

A *functional requirement* states a task or activity the system is to perform.

Performance is an observable and verifiable result during or after completing a functional requirement.

A *performance requirement* is an indication of how, or how well, the function is to be accomplished. Specifying at least one of the following can almost always denote performance requirements: availability, coverage, timeliness, quality, or quantity. Alternatively, a performance requirement may stipulate methodologies to use to perform the function, or give explicit instruction as to how to accomplish the function.

Availability addresses under what circumstances, or the fraction of the time in which, the function is to be provided.

Coverage addresses where geographically or spatially the function is to be provided.

Timeliness addresses when the function is to be provided.

Quality addresses a user's measure of goodness that the function is to provide.

Quantity addresses how much of something associated with the function is to be provided.

Figure 1.3 shows an example of functional and performance requirements.

Whether or not formal requirements for systems engineering are also adequate for architecting depends on how *architecting* is defined. For some, *architecture* is the term for how a system is partitioned so that design and manufacture can be modularized. For example, an automobile manufacturer may create a vehicle architecture that consists of one or more basic structural platforms on which different

---

Operate functional requirement:
– Lift chair above floor.

Corresponding performance requirements:
– Distance measured from floor to any point on chair is at least 0.5 m, plus or minus 0.1 m.
– Acceleration is less than 0.2 g, plus or minus 0.05 g.
– Vector normal to seat surface deviates less than 1 degree from vector normal to floor.
– Time to achieve is 5 seconds, plus or minus 0.1 second.

FIGURE 1.3 Functional requirements state an activity to accomplish; performance requirements state how, or how well.

engines and transmission, bodies, and interiors can be mixed and matched to achieve vehicles that appeal to buyers based on features and price. The software community uses the terms architecture and architecting this way as well. The aim of this type of architecting is to establish the most versatile modules from which the final whole system will be constructed. Formal requirements for systems engineering enable this type of architecting.

For people in the command, control, communications, intelligence, surveillance, and reconnaissance communities, the word architecture is used to address the nature, structure, and flow of information between constituent systems. The fundamental aim of this architecting is to create opportunities to efficiently reuse information across systems. What the constituent systems should be is not relevant to the challenge. Formal requirements for systems engineering also enable this type of architecting.

Architecting as defined herein is the task to determining which constituent systems would best achieve one or more missions, with command and control of those systems but one consideration. Formal requirements for doing systems engineering are helpful for this type of architecting, but additional formal requirements are also needed.

When operating, systems and architectures exhibit *behavior*, which is at least one performance measure varying with an independent variable, such as time. All behavior is the result of the performance measures changing directly as a result of independent inputs (referred to as *open loop dynamics*) or as the result of the current state or a change in state of the performance measures (referred to as *closed loop dynamics*, or, more simply, *feedback*). Fundamental behaviors are linear, exponential, threshold, oscillation, S shaped, and collapse. All behaviors are combinations of these fundamental manifestations.

A convenient method for describing dynamic behavior is with differential equations that define the rate of change of the performance measures in terms of the performance measures and other features of the system or architecture, that is:

$$dy/dt = f(\mathbf{y}, \mathbf{t})$$

where **y** are the performance measures, **t** is the independent variable, and *f* the functional relationship incorporating the features. Such equations quantify the consequences of feedback to the system or architecture.

The *Mathematica* routine *DSolve* solves differential equations symbolically, while the routine *NDSolve* solves the equations numerically. The *Mathematica* routine *Plot* enables us to visualize the solutions. Let's use these routines to define and visualize the fundamental behaviors.

Linear behavior occurs when the change in performance is constant, that is:

$$dy/dt = \mathbf{constant}$$

so the performance grows or shrinks proportionally with the independent variable. The differential equation and solution are

```
In[8]:= sol = DSolve[{y'[t] == constant, y[0] == y0}, y[t], t]

Out[8]= {{y[t] → constant t + y0}}
```

For **y0** = 0 and **constant** = 1, or −1, the results are

```
In[9]:= y0 = 0;
        constant = 1;
        Plot[Evaluate[ y[t] /. sol], {t, 0, 10},
          AxesLabel -> {"independent variable", "performance"},
          PlotLabel → "Linear Behavior with constant = 1"]
        constant = -1;
        Plot[Evaluate[ y[t] /. sol], {t, 0, 10},
          AxesLabel -> {"independent variable", "performance"},
          PlotLabel → "Linear Behavior with constant = -1"]
```

Out[11]=

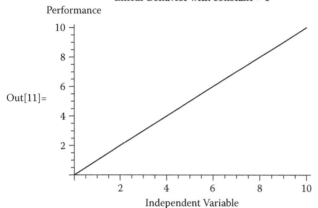

Out[13]=

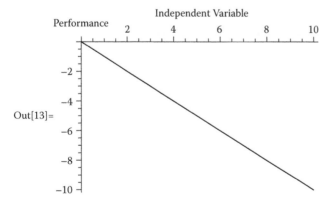

Exponential behavior occurs when the performance changes in proportion to a growth rate times the current performance level, which can be modeled as the following differential equation, with the indicated solution:

```
In[14]:= Clear[y0];
        sol = DSolve[{y'[t] == growthrate * y[t], y[0] == y0}, y[t], t]
```

$$Out[15]= \left\{\left\{y[t] \rightarrow e^{growthrate\ t}\ y0\right\}\right\}$$

For **growthrate** and **y0** both equal to 1, the behavior is

```
In[16]:= growthrate = 1.;
         y0 = 1;
         Plot[Evaluate[y[t] /. sol], {t, 0, 10},
           AxesLabel → {"independent variable", "performance"}, PlotLabel → "Exponential Behavior"]
```

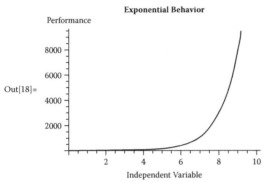

Out[18]=

When the performance of a system or architecture is the combination of many oscillations at different fundamental frequencies, a second kind of exponential growth, called *resonance*, occurs when an input matches one of these fundamental frequencies of the system or architecture and there is little or no natural damping. The magnitude of the performance at the common frequency (called the *resonant peak*) as a function of the *damping ratio* is

For **gain** = 0.001, **resource** = 10 and **y0** = 1, the behavior is:

```
In[61]:= gain = .001;
         resource = 10;
         y0 = 1;
         Plot[Evaluate[y[t] /. sol], {t, 0, 1000},
           AxesLabel → {"independent variable", "performance"}, PlotLabel → "S Shaped Behavior"]
```

Out[64]=

Exponential growth's relentless upward curve fools us into believing the behavior will continue unabated forever. Yet this is never the case. Either the resources that enable the growth eventually deplete, or an as yet unperceived feedback grows stronger in proportion to the growth. Once such counters stop the growth, the result is a partial or complete collapse. In the case of resonance, the resulting huge outputs can literally destroy the system or architecture. As will be shown later in this section, we can purposefully attempt to control exponential growth by introducing counter forces in proportion to the performance and the rate of change of the performance, bringing the behavior to a desired level.

Threshold behavior occurs when the performance grows in proportion to the difference between a desired goal and the current performance, which is modeled as the following differential equation, with the indicated solution:

```
In[21]:= Clear[y0, gain];
        sol = DSolve[{y'[t] == gain * (goal - y[t]), y[0] == y0}, y[t], t]
Out[22]= {{y[t] → e^(-gain t) (-goal + e^(gain t) goal + y0)}}
```

For **gain** −0.5, **goal** = 10 and y0 = 0, the behavior is:

```
In[19]:= gain = .5;
        goal = 0;
        y0 = 10;
        Plot[Evaluate[y[t] /. sol], {t, 0, 10},
          AxesLabel → {"independent variable", "performance"},
          PlotLabel → "Decreasing Threshold Behavior"]
```

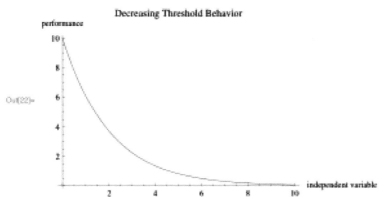

Out[22]=

Notice how the performance decreases at a slower and slower rate. This increase in time or effort for the output to continue to decrease is modeled by the variable **gain**. Such behavior can also be exhibited for thresholds that grow, as illustrated here:

```
In[27]:= gain = .5;
        goal = 10;
        y0 = 0;
        Plot[Evaluate[y[t] /. sol], {t, 0, 10},
         AxesLabel → {"independent variable", "performance"},
         PlotLabel → "Increasing Threshold Behavior"]
```

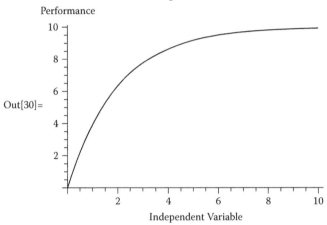

Out[30]=

Many systems and architectures exhibit oscillations in which the performance varies up and down with the same or different amplitudes at the same, multiple, or varying frequencies. Oscillation behavior can occur when the rate of change of the rate of change is proportional to the performance of the system, such as in a pendulum, which is modeled as

```
In[31]:= Clear[sol, y0, yp0];
        sol = DSolve[{y''[t] == -frequency^2 * y[t], y[0] == y0, y'[0] == yp0}, y[t], t]
```

$$\text{Out[32]}= \left\{\left\{y[t] \rightarrow \frac{\text{frequency } y0 \, \text{Cos}[\text{frequency } t] + yp0 \, \text{Sin}[\text{frequency } t]}{\text{frequency}}\right\}\right\}$$

When **frequency** = **y0** = 1 and **yp0** = 0, the behavior is

```
In[33]:= frequency = 1;
        y0 = 1;
        yp0 = 0;
        Plot[Evaluate[y[t] /. sol], {t, 0, 100},
         AxesLabel → {"time", "angle from vertical"}, PlotLabel → "Pure Oscillations"]
```

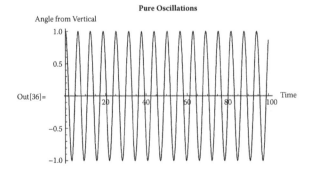

Out[36]=

Oscillations can also occur whenever at least two entities compete, for example, the population of **victims** and **predators** may oscillate for certain values of birthrates, natural death rates, and how often the predators kill the victims. In such situations, two differential equations are needed, one for each aspect of the system. Here is an example:

```
In[37]:= victimgrowthrate = .5;
        victimpredation = .1;
        predatordeathrate = .1;
        predatorpredation = .05;
        predators0 = 20;
        victims0 = 10 000;
        sol = NDSolve[{victims'[t] == victimgrowthrate * victims[t] -
              victimpredation * victims[t] * predators[t], predators'[t] ==
              predatorpredation * victims[t] * predators[t] - predatordeathrate * predators[t],
              victims[0] == victims0, predators[0] == predators0}, {victims, predators}, {t, 500}];
        Plot[Evaluate[victims[t] /. sol], {t, 0, 500}, PlotRange → {0, 100},
          PlotLabel → "Victim's Population", AxesLabel → {"time", "population"}]
        Plot[Evaluate[predators[t] /. sol], {t, 0, 500}, PlotRange → {0, 100},
          PlotLabel → "Predator's Population", AxesLabel → {"time", "population"}]
```

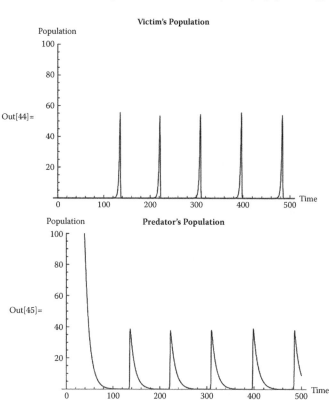

Oscillations can also occur when corrective actions to achieve a goal are delayed relative to the perception of the performance, as modeled here:

```
In[101]:= Clear[y, t, y0, yp0];
    gain = .5;
    goal = 10;
    delay = 3;
    y0 = 0;
    yp0 = 1;
    sol = NDSolve[{y'[t] == gain * (goal - y[t - delay]), y[0] == y0}, y, {t, 0, 200}]
    Plot[Evaluate[y[t] /. sol], {t, 0, 200},
      AxesLabel → {"independent variable", "performance"},
      PlotLabel → "Goal Seeking with Measurement Delay Induced Oscillations"]
```

NDSolve::ihist : Conditions given at t = 0.` will be interpreted as initial history functions for t /; t ≤ 0.. ≫

Out[107]= {{y → InterpolatingFunction[{{0., 200.}}, <>]}}

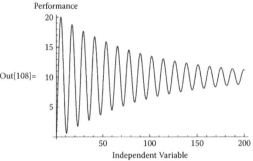

These oscillations also exhibit *overshoot* (*Abs*[**y**[t] – **goal**] varies from the goal), and notice the peak values are gradually decreasing to what is called the *steady-state error*. Selecting values for **gain** and **delay** changes how much the performance overshoots the goal and how quickly the oscillations are reduced to a specified fraction of the goal.

Overshoot can also be exhibited when the rate of change of rate of change of the performance is proportional to both the current performance and the rate of change of the performance:

```
In[54]:= Clear[frequency, y0, yp0, damping, ygain, y, t]
    sol =
      DSolve[{y''[t] == - (2 * damping * frequency + ygain) * y'[t] - (frequency^2 + ygain) * y[t],
        y[0] == y0, y'[0] == yp0}, y[t], t]
```

Out[55]= $\left\{\left\{y[t] \rightarrow \right.\right.$

$$\left(-2 \text{ damping } e^{\frac{1}{2} t \left(-2 \text{ damping frequency-ygain}-\sqrt{-4 \left(\text{frequency}^2+\text{ygain}\right)+(2 \text{ damping frequency+ygain})^2}\right)} \text{ frequency y0 } + \right.$$

$$2 \text{ damping } e^{\frac{1}{2} t \left(-2 \text{ damping frequency-ygain}+\sqrt{-4 \left(\text{frequency}^2+\text{ygain}\right)+(2 \text{ damping frequency+ygain})^2}\right)} \text{ frequency y0 } -$$

$$2 e^{\frac{1}{2} t \left(-2 \text{ damping frequency-ygain}-\sqrt{-4 \left(\text{frequency}^2+\text{ygain}\right)+(2 \text{ damping frequency+ygain})^2}\right)} \text{ yp0 } +$$

$$2 e^{\frac{1}{2} t \left(-2 \text{ damping frequency-ygain}+\sqrt{-4 \left(\text{frequency}^2+\text{ygain}\right)+(2 \text{ damping frequency+ygain})^2}\right)} \text{ yp0 } -$$

$$e^{\frac{1}{2} t \left(-2 \text{ damping frequency-ygain}-\sqrt{-4 \left(\text{frequency}^2+\text{ygain}\right)+(2 \text{ damping frequency+ygain})^2}\right)} \text{ y0 ygain } +$$

$$e^{\frac{1}{2} t \left(-2 \text{ damping frequency-ygain}+\sqrt{-4 \left(\text{frequency}^2+\text{ygain}\right)+(2 \text{ damping frequency+ygain})^2}\right)} \text{ y0 ygain } +$$

$$e^{\frac{1}{2} t \left(-2 \text{ damping frequency-ygain}-\sqrt{-4 \left(\text{frequency}^2+\text{ygain}\right)+(2 \text{ damping frequency+ygain})^2}\right)}$$

$$y0 \sqrt{-4 \left(\text{frequency}^2+\text{ygain}\right)+(2 \text{ damping frequency+ygain})^2} +$$

$$e^{\frac{1}{2} t \left(-2 \text{ damping frequency-ygain}+\sqrt{-4 \left(\text{frequency}^2+\text{ygain}\right)+(2 \text{ damping frequency+ygain})^2}\right)} y0$$

$$\left.\sqrt{-4 \left(\text{frequency}^2+\text{ygain}\right)+(2 \text{ damping frequency+ygain})^2}\right) \Big/$$

$$\left.\left.\left(2 \sqrt{-4 \left(\text{frequency}^2+\text{ygain}\right)+(2 \text{ damping frequency+ygain})^2}\right)\right\}\right\}$$

For **damping** = 0.1, **frequency** = **yp0** = 1, and **ypgain** = **ygain** = **y0** = 0, this results in

```
In[56]:= damping = .1;
         frequency = 1;
         ypgain = 0;
         ygain = 0;
         y0 = 0;
         yp0 = 1;
         Plot[Evaluate[y[t] /. sol], {t, 0, 30},
           AxesLabel → {"independent variable", "performance"},
           PlotLabel → "Damping or State and Derivative Feedback Oscillations"]
```

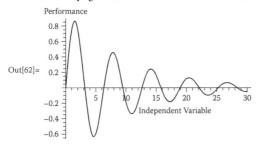

Damping or State and Derivative Feedback Oscillations

Out[62]=

This behavior could be a natural result of the system or architecture, or we can cause it to happen if we can measure the current performance value or the rate of change of the performance measure and use that information to influence the rate of change in performance. One way to do this is to select nonzero values **ypgain** and **ygain**. Again, the peak values of the oscillations will eventually threshold to a steady-state error.

S-shaped behavior occurs when the performance changes are proportional to both the state of the system and the extent to which a capacity was achieved. This can be modeled in many ways; if the capacity is the difference between a resource and the performance measure, then the equations and solution are

```
In[63]:= Clear[gain, y0, resource, y, t];
         sol = DSolve[{y'[t] == gain * y[t] * (resource - y[t]), y[0] == y0}, y[t], t]
```

Solve::ifun : Inverse functions are being used by Solve, so
              some solutions may not be found; use Reduce for complete solution information. »

$$\text{Out[64]= } \left\{ \left\{ y[t] \to \frac{e^{\text{gain resource } t} \text{ resource } y0}{\text{resource} - y0 + e^{\text{gain resource } t} y0} \right\} \right\}$$

For **gain** = 0.001, **resource** = 10 and **y0** = 1, the behavior is:

```
In[61]:= gain = .001;
         resource = 10;
         y0 = 1;
         Plot[Evaluate[y[t] /. sol], {t, 0, 1000},
           AxesLabel → {"independent variable", "performance"}, PlotLabel → "S Shaped Behavior"]
```

S Shaped Behavior

Out[65]=

Collapse behavior occurs when the resources needed to achieve the performance decrease in proportion to the growth in performance. Here is an example where the resource is S shaped, starting large and then ending small:

```
In[69]:= Clear[resource, gain, y0];
        gain = 1;
        resource0 = 100.;
        consumption = -1;
        y0 = 10;
        sol = NDSolve[{y'[t] == gain * y[t] - (1 - resource[t] / resource0) * y[t], resource'[t] ==
             consumption * y[t], y[0] == y0, resource[0] == resource0}, {y, resource}, {t, 0, 10}]
        Plot[Evaluate[y[t] /. sol], {t, .1, 10}, PlotRange → All,
          AxesLabel → {"independent variable", "performance"}, PlotLabel → "Collapse Behavior"]
Out[74]= {{y → InterpolatingFunction[{{0., 10.}}, <>],
          resource → InterpolatingFunction[{{0., 10.}}, <>]}}
```

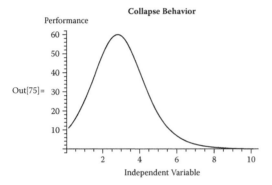

The performance first peaks, then collapses, because it is dependent on **resource**, which is exhibiting the following behavior:

```
In[76]:= Plot[Evaluate[resource[t] /. sol], {t, 0, 10},
          AxesLabel → {"independent varibale", "resouce quantity"},
          PlotLabel → "Resource Available"]
```

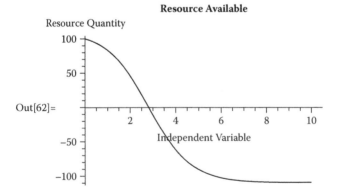

Collapse behavior can also occur very suddenly, such as when a structural column buckles. Under increasing load the column compresses more and more without noticeably bending, and then at a critical load, in the blink of an eye, the column bends to the point of breaking. Systems and architectures can also appear to be performing fine, until a single point failure occurs or a set of coupled failures occur, immediately causing the system or architecture to cease to function as desired.

Good system engineers and architects must discover the possible sources of catastrophic collapse and prudently avoid their manifestation, and must comprehend the natural feedback loops inherent in the system or architecture. Once we've arrived at such an understanding, we may elect to introduce purposeful feedback to influence the behavior. But we must do so with extreme caution, as *unintended consequences* are the result of unrecognized feedback loops taking effect. Such consequences can negate all the good we attempted to achieve. Systems and architectures fail because the systems engineers or architects:

1. Misconstrue an internal or external interface.
2. Miss an influential feedback.
3. Improperly prepare for a single point or coupled failure.

That is why systems engineering and architecting is so challenging. Our failures are the result of omission rather than commission. We fail because we miss something we do not even know is there. Humans are naturally much better at checking for what is expected than identifying what is unexpected. That is why the successful practice of systems engineering and architecting is so dependent on learning lessons from prior mistakes. At a minimum, we can avoid those. Then it is up to us to do the best we can to avoid new mistakes when we expand the discipline into new endeavors and, at the very least, enable others to learn from our mistakes. To help us both avoid and learn from the mistakes we are about to make, we need the skills of modeling, good decision making, specifying requirements, and finding ways to improve what we do.

1.3   APPROACH

Chapter 2 addresses modeling system or architecture behavior. The focus is to achieve executable models of the proposed system or architecture so that both behavior and cost can be anticipated. Sources for mathematical models for various domains are identified. Processes to model systems and architectures are described. A key modeling consideration is how to address uncertainty. Five methods to model uncertainty are examined: regression, the Monte Carlo method, the fuzzy logic method, agent-based models, and fractals. Regression fits mathematical functions to data. Monte Carlo methods rely on probability and statistics. Fuzzy logic methods treat uncertainty as a "degree of belief" rather than probability density functions. Agent-based models address uncertainty by simulating an environment and the interactions of many entities, each of which can perceive and perform in accordance with stipulated rules. The value of fractal models is declared along with references for more information. The chapter concludes with formal requirements to monitor the key technical performance measures on a program.

All models imperfectly represent reality. Some models are useful for making decisions. The ideal practitioner of systems engineering and architecting is sufficiently knowledgeable and experienced to quickly and inexpensively implement a useful model to obtain the information needed to make a decision with the least risk of error.

Chapter 3 addresses decision making. First, principles to make a good decision, particularly a good decision in the presence of ambiguity in selection criteria and options, are presented. Then formal requirements provide for several techniques for making different types of decisions. Specifically, the following types of decisions are addressed:

1. Assess the goodness of a diagnosis.
2. Make a sequence of decisions to optimize a return.
3. Provide people a fair method to choose from a static, finite set of options.
4. Allocate a static, finite set of resources.
5. Evolve options that best satisfy multiple criteria.
6. Select from a static, finite set of options that one which best achieves multiple, dissimilar criteria.
7. Select from a static, finite set of options those which maximize benefits while minimizing detriments.
8. Do the optional endeavor with the best uncertain financial return.
9. Select a portfolio of investments that maximize return and minimize risk.
10. Choose the independent variable values that optimize an index of performance subject to constraints when the index of performance is inexpensive to determine.
11. Choose the independent variable values that optimize an index of performance subject to constraints when the index of performance is expensive to determine.
12. Define a dynamic control law to optimize a dynamic index of performance.
13. Determine an optimal configuration.
14. Determine the best strategy relative to a competitor.

The chapter ends with explaining how to apply the modeling and decision-making formal requirements to identify, assess, and manage risks. The ideal practitioner of systems engineering and architecting is sufficiently knowledgeable and experienced to quickly implement the most useful decision-making process to resolve the issues confronting the endeavor.

Chapter 4 addresses natural language requirements. First, the types of requirements needed are itemized, and then what constitutes a good natural language requirement is explained. Formal requirements to avoid ambiguities in natural language requirement statements are demonstrated. The distinction between determining and documenting requirements is discussed. Then, the following techniques to determine and document functional and performance requirements are described, along with relative strengths, weaknesses, and recommended uses:

1. Reuse from prior program
2. Interpret customer-provided documents
3. Surveys
4. Witness
5. Focus groups
6. Assess product defects
7. Concept of operations

8. Formal diagramming techniques
9. Quality matrices
10. Prototypes
11. Incremental build

The thorny topic of how to verify that one has obtained the correct requirements is discussed, as well as how to ensure the people who need to implement the requirements interpret the requirements as intended. Formal requirements are presented to create and maintain a database of requirements, along with the means to group the appropriate requirements into specifications to be provided to the item's designers, manufactures, customers, users, and operators. The differences between a system specification and an architecture specification are illustrated. Then how to plan verification efforts and capture the resulting verification evidence is discussed. Finally, formal requirements are presented to predict when the requirement generation and verification processes will be completed.

The ideal practitioner of system engineering and architecting is sufficiently knowledgeable and experienced to quickly and inexpensively implement the most cost-efficient methods to obtain a minimal but complete set of natural language requirements, write the requirements in the least ambiguous manner possible, prepare a specification that is comprehended, plan the minimal required verification efforts, and obtain the information needed to determine if each requirement is achieved.

Chapter 5 addresses how to incrementally improve the practice of systems engineering and architecting in a statistically meaningful way. First, the need for improving both process and products is argued. Then, surveys are shown that can be used by the team engaged in the effort to assess their current level of process maturity and product quality, and identify the most important areas of improvement.

The ideal practitioner of systems engineering and architecting is sufficiently knowledgeable and experienced to quickly determine what aspect of the effort most urgently needs improvement and quickly define and implement the means to improve.

Each chapter begins with introductory remarks, followed by case studies based on real events to provide lesson-learned guidance, and then the formal requirements are presented and used in illustrative examples. Each chapter ends with a list of heuristics for performing the effort.

To put the subject matter of this book in the context of an industry systems engineering standard, Table 1.1 relates the chapters to the subprocesses documented in

TABLE 1.1

Formal Requirements for Some of the Subprocesses Defined in ANSI/EIA-632-1999

Chapter	Provides Formal Requirements for EIA 632 Subprocess
2	17, 22
3	23, 24
4	10, 14, 15, 16, 19, 25, 26, 27, 28, 29
5	4

the Government Electronics and Information Technology Association Standard, *Process for Engineering a System*, ANSI/EIA-632-1999.[14] The ANSI/EIA-632-1999 standard does not address architecting as defined herein.

1.4   SUMMARY

- The objective of this book is to provide formal requirements for modeling systems and architecture behavior, making decisions, establishing natural language requirements, and improving the practice of systems engineering or architecting.
- Formal requirements are explicit instructions that are unambiguous, executable, and correct.
- A system is a combination of people, machines, raw materials, methods, or environments that produce a product or service. Should any portion of the system be deleted, the product or service would not be produced.
- Systems engineering is the process of determining, documenting, and verifying the optimal functional and performance requirements for a system.
- Architecture consists of multiple systems, which redundantly perform at least one mission. Should a system be deleted from the architecture, the mission would still be accomplished, but perhaps not as well. The constituent systems may have different customers, operators, or users.
- Architecting is the process of determining the optimal systems to accomplish a mission.
- All system and architecture behaviors are combinations of the following fundamental behaviors, each of which is a consequence of feedback:
  - Linear
  - Exponential
  - Threshold
  - Oscillation
  - S shaped
  - Collapse
- Good system engineers and architects must comprehend the natural feedback inherent in the system or architecture they are addressing.
- Systems and architectures fail because the systems engineers or architects:
  1. Misconstrue an internal or external interface.
  2. Miss an influential feedback.
  3. Improperly prepare for a single point or coupled failure.
- The ideal practitioner of systems engineering and architecting is sufficiently knowledgeable and experienced to quickly and inexpensively:
  1. Implement a useful model to obtain the information needed to make a decision with the least risk of error.
  2. Implement the most useful decision-making process to resolve the issues confronting the endeavor.

3. Obtain the minimal but complete set of natural language requirements, write the requirements in the least ambiguous manner possible, prepare a specification that is comprehended, plan the minimal necessary verification efforts, and obtain the information needed to determine if each requirement is achieved.

4. Determine what aspect of the systems engineering or architecting effort most urgently needs improvement and quickly define and implement the means to improve.

APPENDIX 1: *MATHEMATICA* IN BRIEF

Upon initiation, *Mathematica* creates a notebook to use and retain the results. Saved notebooks can be opened and modified. The basic working entity in the notebook is a cell. Every time the Enter key is pressed, a new cell is made. The contents of the cell can be text, mathematical expressions that are to be evaluated, or programs given definition to be executed later.

Mathematica is an interpreter-based language, which is geek speak for saying the contents for any or all selected cells are immediately executed by *Mathematica* upon selecting the cells and typing the Shift-Enter keys simultaneously. Between executions, *Mathematica* remembers all defined variables and programs (which I will call *routines*), so sometimes it is prudent to use the *Mathematica* function *Clear* to erase these from memory before use again. *Mathematica* denotes what it took as an input by

$$\text{In}[\mathbf{n}]:=$$

and *Mathematica* denotes the result of its processing as

$$\text{Out}[\mathbf{n}]=$$

where **n** will be the integer count of the total number of inputs and outputs accomplished so far.

A variable is given a name and value by typing it and using the "=" sign as follows:

```
In[77]:= variable = 1
Out[77]= 1
```

The double equal sign "==" is a logical test of whether or not the left-hand side of an expression is identical to the right-hand side. Here are two examples:

```
In[78]:= 1 == 2
Out[78]= False
In[79]:= variable == 1
Out[79]= True
```

In *Mathematica*, the "=" sign is best interpreted as "is defined to be," while the double equal sign, "==", is best interpreted to be "is the same as?" and the colon equal sign ":=" is best interpreted to be "will execute" the subsequent set of commands when provided all the information on the left-hand side of the expression.

Output is suppressed by ending the command line with a semicolon:

In[80]:= **anothervariable = 2;**

A variable can be given a symbolic name:

In[81]:= **variable = symbol**

Out[81]= symbol

Notice even though, as far as *Mathematica* is concerned, **symbol** has neither a numerical nor text association, it is still perceived to be the same as **variable**:

In[82]:= **variable == symbol**

Out[82]= True

Lists of text symbols are denoted between quote marks, for example:

In[83]:= **symbol = "a text string";**

Now **symbol** is defined and since it was previously associated to be defined as **variable**, **variable**'s value is no longer "1" but the same text string associated with **symbol**:

In[84]:= **symbol**
        variable
        variable == symbol

Out[84]= a text string

Out[85]= a text string

Out[86]= True

The user can define a function with variable inputs, such as

$$y = a*t^2 + b*t + c$$

as follows:

In[87]:= **y[t_, a_, b_, c_] := a*t^2 + b*t + c**

Mathematica returns specific values when provided specific inputs, for example:

In[88]:= **y[1, 1, 1, 1]**

Out[88]= 3

Or specific values when provided defined variables, as follows:

In[89]:= **a = b = c = t = 1;**
       y[t, a, b, c]

Out[90]= **3**

Or symbolic values when provided symbolic inputs, as follows:

In[91]:= **y[x, a1, a2, a3]**

Out[91]= **a3 + a2 x + a1 x^2**

Mathematica treats variables as global entities. That is, once defined, that definition is used whenever that variable name is invoked within the notebook.

Functions can be used with other functions provided by *Mathematica*, such as *Plot*:

In[92]:= **Plot[y[t, a, b, c], {t, 0, 10}]**

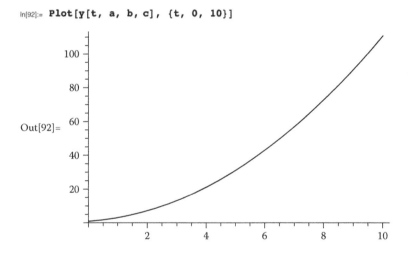

Out[92]=

Users can also define a more complicated construct, called a *Module*, which can include internal logic and documentation which are denoted as text between "(*" and "*)" symbols. A *Module* can use global variables, input values denoted in the name of the *Module*, or can use local variables, that is, variable names can be defined that have meaning "only within" the context of the *Module*, but not the global environment. Here is an example in which the quadratic equation is redefined as a *Module* that expects one input variable (**t**) and contains three local variables (**c1, c2, c3**):

```
In[93]:= QuadraticEquation[t_] := Module[{c1, c2, c3},
         (* This routine returns the value of the quadratic equation c1*t^2 + c2*t + c3 *)
         (* Input: t_ which can be a number, symbol, or an array of numbers or symbols *)
         (* Output: c1*t^2 + c2*t + c3, which will be in the same form as the input *)
         (* Note: c1, c2, and c3 are local variables
          with defined meaning only within the context of the Module *)
         c1 = c2 = c3 = 1;
         Return[c1 * t^2 + c2 * t + c3]];
```

First, note the local variables have no value assignment outside of the module:

In[94]:= **c1**

Out[94]= c1

In[95]:= **c2**

Out[95]= c2

In[96]:= **c3**

Out[96]= c3

If *QuadraticEquation* is proved an array of numbers for input, it returns the corresponding array of quadratic outputs:

In[97]:= **QuadraticEquation[{1, 2, 3}]**

Out[97]= {3, 7, 13}

QuadraticEquation is mathematically identical to the function $y[t\_, a\_, b\_, c\_]$ we defined earlier, provided $a = b = c = 1$, as follows:

In[98]:= **Plot[QuadraticEquation[s], {s, 0, 1}]**

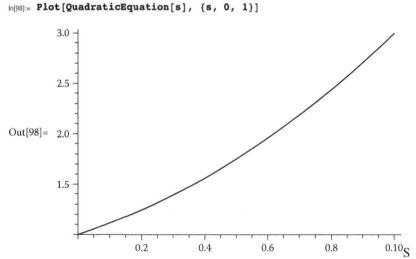

Mathematica can handle both symbolic and numerical representations. For example, if we want to solve for when a quadratic equation is 0, we can do it symbolically using the *Mathematica* routine *Solve*:

In[99]:= **Solve[y[x, c1, c2, c3] == 0, x]**

$$\text{Out[99]=}\ \left\{ \left\{ x \rightarrow \frac{-c2 - \sqrt{c2^2 - 4\,c1\,c3}}{2\,c1} \right\}, \left\{ x \rightarrow \frac{-c2 + \sqrt{c2^2 - 4\,c1\,c3}}{2\,c1} \right\} \right\}$$

Or we can do it numerically, using the *Mathematica* routine *NSolve*:

```
In[100]:= NSolve[y[x, 1, 1, 1] == 0, x]

Out[100]= {{x → -0.5 - 0.866025 i}, {x → -0.5 + 0.866025 i}}
```

REFERENCES

1. Blanchard, Benjamin S., and Fabrycky, Wolter J., *Systems Engineering and Analysis*, 3rd ed., Upper Saddle River, NJ: Prentice Hall, 1998.
2. Chapman, William L., Bahill, A. Terry, and Wymore, A. Wayne, *Engineering Modeling and Design*, New York: CRC Press, 1992.
3. Gharajeddaghi, Jamshid, *Systems Thinking: Managing Chaos and Complexity*, 2nd ed., London: Elsevier, 2006.
4. Hall, Arthur D., *A Methodology for Systems Engineering*, New York: Van Nostrand, 1962.
5. Maier, Mark W., and Rechtin, Earnhardt, *The Art of Systems Architecting*, 2nd ed., Boca Raton, FL: CRC Press, 2002.
6. Sage, Andrew P., *Systems Engineering*, New York: John Wiley & Sons, 1992.
7. Sterman, John D., *Business Dynamics: Systems Thinking and Modeling of a Complex World*, New York: McGraw Hill, 2000.
8. Wymore, A. Wayne, *Model-Based Systems Engineering*, New York: CRC Press, 1993.
9. See http://www.wolfram.com.
10. Abrial, Jean-Raymond, Borfer, Egona, and Landmacck, Hans, *Formal Methods for Industrial Application*, Berlin: Springer, 1990.
11. Potter, Ben, Sinclair, Jane, and Till, David, *An Introduction to Formal Specification and Z*, 2nd ed., Upper Saddle River, NJ: Prentice Hall, 1996.
12. Gajski, Daniel D., Vahid, Frank, Narayan, Sanjiv, and Gong, Jie, *Specification and Design of Embedded Systems*, Englewood Cliffs, NJ: Prentice Hall, 1994.
13. Berzins, Valdid, and Luqi, Lucia, *Software Engineering with Abstractions*, Reading, MA: Addison-Wesley, 1991.
14. Government Electronics and Information Technology Association (GEIA), *Standard Process for Engineering a System*, ANSI/EAI-632-1999, Arlington, VA: GEIA, 1999.

2 Model Systems and Architecture Behavior

Models enable the prediction of system or architecture behavior in order to provide information that people will use to determine requirements and to select the implementation approach. Behavior is how the system or architecture transforms a set of inputs, which may be varying with time, the environment, or the state of the system or architecture, into a set of outputs. Analysis allows for the examination of the intricacies of the input-to-output transformation. Synthesis enables the determination of the functional and performance requirements and the implementation approach that will optimize the desired transformation. Models are valuable in proportion to how quickly and inexpensively they provide trusted results to make decisions. The model may be a diagram, or a miniature physical representation of the system or architecture. The model may be the system itself operated under controlled conditions. The model may be a single input, a single-output algebraic function, or a set of simultaneous nonlinear stochastic differential equations. To develop mathematical models it is necessary to understand the system or architecture at a deep level. The modeler's adage is "All models are wrong; some models are useful." All models are wrong because aspects of reality are excluded by conscious decision or error.

The modeler's job is to find the least complicated model that can be used to obtain the information needed to make the decision at hand. For example, if one wants to determine the best architecture to maintain a timely catalog of objects in space orbit, it is sufficient to model the approximate orbits of a representative sample of the objects and the extent they can be perceived by the sensors being considered for use, as well as the location, availability, sensitivity, and capacity of the various sensors. If there are inherent time delays in directing a sensor to search for an object or in turning the sensed information into useful orbit information, those delays are important to model so one can determine the best ways to command and control the architecture. While the three-dimensional shape of each object could be modeled, it is not relevant, nor are the higher order perturbations on the orbits. However, if the need is instead to characterize the objects in orbit, then imaging of the object is necessary and the model must include the three-dimensional shape of the objects, their rotational motion, and how each object reflects or absorbs electromagnetic energy.

Determining the appropriate level of model fidelity remains more an art than a definable process. The easier mistake to avoid is having too much fidelity. This is because it usually takes more work to add more fidelity, and a good model development heuristic is to incrementally add complexity. Yet, the fear of having too little fidelity often results in models being constructed with too much fidelity. (Plus modelers get paid to model, and as more fidelity means more work, there is a natural tendency to add fidelity.) Missed fidelity errors are examples of the notorious unknown unknowns. Usually, these missing aspects are discovered during the model verification process when it is noticed the model output does not match the real world. While it is then obvious that something is wrong with the model, what exactly is wrong is often initially a mystery. After first ensuring there are no errors in the mathematics or coding, only then is it possible to determine if the mismatch between the model's results and the observed results is due to errors in the observed results, or because some aspect of reality is missing in the model's mathematics. Unfortunately, if the model verification effort was limited in scope, the missing model aspect is often discovered after the system or architecture solution is substantially developed, perhaps even in operation. Then these missing model features become lessons learned for the next similar modeling attempt. So another model development heuristic is to build on models used to produce successfully operating systems or architecture. By doing so, one absorbs the lessons learned from prior developers.

Two case studies follow, one associated with modeling systems and one associated with modeling an architecture.

Case Study 2.1: Using Models to Keep the Space Shuttle Wings from Falling Off

BACKGROUND

The Space Shuttle during ascent flight consists of three elements: the two solid rocket boosters (SRBs), the external tank (ET), and the Orbiter. The SRBs fire from liftoff until the vehicle has risen above the atmosphere, then separate from the ET and parachute into the Atlantic Ocean, landing relatively close to the launch site. The ET holds the propellant stored as liquid hydrogen and oxygen for the three Space Shuttle main engines (SSMEs), which fire from just before liftoff to the main engine cutoff (MECO), which leaves the Orbiter–ET combination just short of orbit velocity. The winged Orbiter holds the crew and payload as well as the three SSMEs. The ET is jettisoned shortly after MECO. Since the ET velocity is just below that needed to achieve orbit, the ET doesn't make it all the way around the Earth and ends up in the mid-Pacific Ocean. To achieve the final desired orbit, the Orbiter uses engines called the orbital maneuvering system (OMS), which are also used to slow the Orbiter down when it is time to leave orbit and reenter the atmosphere. The Orbiter glides to a landing.

In the mid- to late 1980s, the two potential failures of paramount concern for threatening the safe flight of the Shuttle were (1) premature loss of an SSME, and (2) exceeding the limit loads in one of the Orbiter's wings during ascent.

Here, the loss of an SSME means the loss of the thrust of SSME, not a malfunction resulting in damage. Should the thrust of any SSME be lost prior to achieving the desired orbit velocity, the Shuttle would fall back to Earth just as a motorcyclist trying to hurdle numerous buses parked side to side would fail to clear them if he fails to achieve an adequate velocity at the top of the ramp. To save the vehicle and crew in the event of the loss of a SSME, extensive analyses were conducted to determine contingency trajectories, depending on when the thrust was lost from one or more SSME. In brief, alternative trajectories were planned to abort to orbit (ATO), abort once around (AOA), find an alternative landing site (ALS), or return to launch site (RTLS). The simplest of the options is ATO, used when the Shuttle had achieved enough velocity so that it could rely on the OMS to reach at least a lower altitude orbit than that of the nominal mission. The planned mission is abandoned, but the crew can return to Earth later by performing a near-normal reentry burn. All the other abort options require blending an ascent trajectory with an entry trajectory because to get the Shuttle from point A to point B, all energy the vehicle has at point A plus any energy added from continued engine firing must be expended in order for the Orbiter to come to a safe stop at point B. If the energy needed is underestimated, then the Orbiter fails to reach point B; if energy needed is overestimated, then the Orbiter cannot come to a stop at point B. The AOA alternative was for the condition in which enough velocity could be achieved to get the Orbiter to a landing site either on the west coast of the United States, or near the launch site at Cape Canaveral in Florida. The ALS locations were on the west coasts of Europe and Africa, so if enough velocity was achieved to at least cross the Atlantic, there was a place for the Orbiter to land. Finally, if an SSME failed very early so there was insufficient energy to cross the Atlantic, the Orbiter would perform a rocket-powered loop, changing its velocity vector from heading basically due east to due west in order to fly back to the launch site; jettisoning first the SRB, then the ET, so neither impact land; and finally gliding to a landing near the launch site in Florida. Needless to say, finding, preparing for, and training the crew and mission support team for all these contingencies consumed a significant fraction of the resources available to plan Shuttle missions. But one effort consumed even more resources, and that was to ensure the atmospheric portion of the Shuttle's ascent did not cause the wings to fail. This case examines how models were made, used, and abused over a period of years to address a very uncertain situation for which the decisions had life-and-death consequences.

What Happened

The complete process is illustrated in Figure 2.1. The models used are itemized in Figure 2.2.

Predicting the structural loads on a wing traveling at high velocity in the atmosphere and therefore being severely heated involves the disciplines of aerodynamics, thermodynamics, structures, materials, and dynamics. The loads on the wing are proportional to

$$q * alpha$$

where **q** is the dynamic pressure, defined to be $1/2 * rho * V^2$, where **rho** is the atmospheric density and **V** is the vehicle's velocity relative to the wind, and **alpha** is the angle of attack of the wing (in radians). The angle of attack is measured in the plane that contains the vehicle's velocity vector and the net lift force, and is the angle between the velocity vector and the chord line of the wing. The chord line is from the front-most edge of the wing to the back-most edge. The larger the **q** or **alpha**, the higher the loads on the wing, but the exact stress and strain

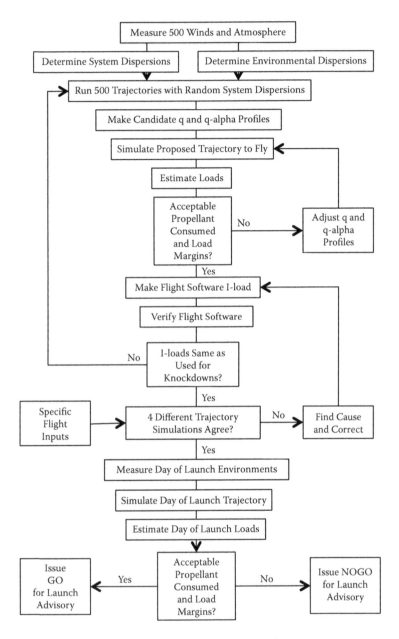

FIGURE 2.1   Space Shuttle ascent trajectory design process flow.

in the wings parts are a complicated consequence of the structural forms and material used. Back in the 1980s, how to make models that could predict stresses and strains at various points in the wing was well known, but with the computers available at the time, it could take months to complete an analysis. This effort was termed a *load cycle*. Each load cycle assessed the result of a pressure profile

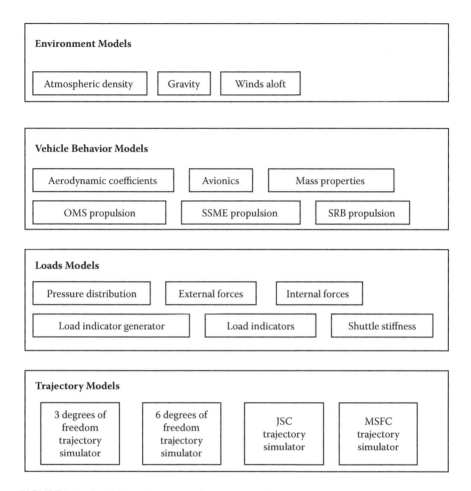

FIGURE 2.2   Models used to determine how to safely fly Shuttle ascent trajectories.

on the surface of the wing, along with the forces due to the dynamic interactions between the Orbiter, ET, and SRB, as well as the extreme temperatures the structure was exposed to since the Shuttle achieved hypersonic velocity while still in the atmosphere, resulting in a great deal of heat being generated where the structure rammed into air molecules at high velocity. What was most difficult to model with certainty was the pressure profile on the wing's surface that changes with speed, angle of attack, any yawing to left or right, the angles of the aerodynamic surfaces that can be moved, and the state of the atmosphere.

The primary means to have complete faith in the accuracy of mathematical models is to make sure the calculated predictions agree with real-world data. For regular airplanes, this is done three ways. First, scale models of the plane are placed in wind tunnels that simulate the fluid flow that would be achieved on the real full-sized wing. While in the tunnel, the net forces on the model as well as the pressure on parts of the model can be measured. These data are used to adjust the mathematical models to ensure they agree with the wind tunnel data, as well as to determine what speed and angles of attack are most critical. Second, an instrumented test article is built; machines are used to impart measured loads on the structure; and the actual strain, stress, and deflections are compared to the model predictions. Third, the first airplane built is instrumented to measure strains on its structural members and possibly pressure on parts of the wing, and the plane is flown on numerous test flights at gradually higher speeds and steeper angles of attack and in more turbulent wind environments. If the measurements show unacceptable loads are occurring in the wing, either the structure is made stronger (with the consequence of making the plane heavier, hence reducing its top speed or range) or a safe flight envelope is established that will become part of the pilot's operating instructions. Wind tunnel tests using scaled models of the Shuttle were extensively undertaken. But the Shuttle flew so fast and so high, that in order to mimic the environment one essentially had to fire a small rocket to get the gases moving fast enough past the dime-sized model in the tunnel to mimic the environment. So the wind tunnel data were useful but also fraught with uncertainties. The Shuttle wings were instrumented, but extensive flight tests of the Shuttle were not economically feasible, as each Shuttle mission cost approximately a billion dollars. (*Note*: The actual cost per mission was a notoriously difficult thing to determine. The "billion-dollar" cost estimate is roughly the annual budget divided by the four flights per year. Since a large portion of the shuttle program costs were fixed, the more missions flown, the less costly the per-flight rate. To my knowledge, no one ever determined the variable costs per mission or successfully comprehended the fixed costs sufficiently to appropriately reduce them.) As the Shuttle was to fly every few months (indeed, in the mid-1980s the desire was to fly the Shuttle every two weeks), a safe method of updating the mathematical models used to determine the best way to fly was needed in the interim, while the flight data that had been obtained to proceed through the many-month load cycle. The approach taken was to use whatever load cycle data were available to generate load indicators, which were numerical curve fits of predicted load in various parts of the wing as a function of trajectory parameters. As more instrumented flights were accomplished, the new data would then be used to conduct a new load cycle, and thereafter the load indicators would be modified if necessary to better match the newly obtained data.

Getting mass to orbit takes a tremendous amount of energy. Each pound of mass needs to be brought to approximately 17,500 miles per hour. The Shuttle was useful only if a useful payload got to orbit, so it was necessary to fly a trajectory that used all the propellant available to get the maximum payload mass to orbit. The equations of motion for a vehicle in a vacuum are simple enough that calculus of variations can be used to derive a closed-form solution for the optimal trajectory. But the equations of motion in the atmosphere, particularly for a winged vehicle like the Orbiter, add just enough complexity to prevent a calculus of variations from providing a closed-form solution. What can be discerned is that if dynamic pressure, or the angle of attack, must be constrained to achieve acceptable loads, then to minimize the propellant required, the angle of the applied thrust with respect to the horizon must vary with velocity in a manner that is dependent on the total profile of the atmospheric density that will be felt as well as the total profile of the wind that will be felt. That is, any local change to either the density or wind causes the entire thrust profile to be changed, rather than just in the vicinity close to the density or wind change. Of course, it is impossible to change a rocket's thrust angle history after it has already flown, and it is very difficult to know what the density or wind will be before it gets to that part of the sky. So the best that can be achieved is an approximate optimal trajectory during atmospheric flight. Very roughly, this approximate optimal trajectory can be accomplished by determining the velocity to rotate from the vertical motion the Shuttle starts with to clear the tower at the launch site, to an angle to be specified to achieve some horizontal velocity, then to achieve the maximum allowed dynamic pressure as soon as practical and hold it as long as practical. Of course, these conditions are completely contradictory to our criteria to minimize the loads on the wings. So, the modeler's job was to find a trajectory that balanced the need to deliver a useful payload while keeping the uncertain loads on the wing safe. Any mistake in the model that predicted the loads would be less than the wing's structural limit and could result in the destruction of the Orbiter and the death of the crew. Any mistake in overpredicting the loads would result in a reduced payload capability and thus reduced usefulness of the Shuttle.

It is one thing to determine what the trajectory should be and another for the vehicle to actually fly it. For atmospheric flight, the Shuttle flew an open-loop profile. That is, it commanded thrust levels and angles using constants at specified velocities called I-loads since they were inputs loaded into the flight software (that is, software that operated in real time using the on-board computers). The flight software made real-time adjustments to keep moments about the three primary axes of the Shuttle balanced, so the Shuttle smoothly pitched, yawed, or rolled. After the SRBs were jettisoned and flight was in vacuum, the avionics subsystem switched to a different flight software that implemented a near-optimal trajectory by adjusting the thrust angles every two seconds to minimize the propellant needed to achieve the desired MECO condition. The Shuttle's avionics subsystem had five computers operating during launch: three ran a primary set of software to execute the mission, and two ran an independently derived, coded, and verified back-up software. Prior to actually commanding anything, the five results were compared. If the five results were all about the same, it was presumed they were all correct, and the result was then communicated to the mechanical devices (called actuators) that made the needed adjustments to thrust or change the direction of the Shuttle. If one or more of the computers came up with different results, it was presumed something was wrong: some of the redundant sensor sets feeding information to one of the computers had failed, or a computer had failed, or something was wrong with the software. A complicated set of logic was undertaken to determine which of the computer outputs to trust and which to ignore. So prior to each mission, the software, including the I-loads, was run through numerous tests, some utilizing computer simulations and some utilizing copies of the real hardware-in-the-loop simulations, to verify that the flight software worked as intended. Once all this work was accomplished, people were loath to change anything about the flight software for fear of creating an unverified state. As will be seen, this had the result of increasing the workload and paradoxically of flying the Shuttle less safely than possible.

Since the atmospheric flight was open loop, any differences between how the system really performed and how it was predicted, based on simulations of the vehicle and environment behavior, resulted in the real trajectory and therefore the wing loads being different than anticipated. So to be safe, these potential dispersions between the simulated behavior and the real behavior had to be accounted for.

Every machine performs a little differently every time it is turned on. The sources of these variations are called *system dispersions*. For the sake of safety, allowances must be made for these system dispersions. For the Shuttle, on any given launch, the performance of the pumps and turbines determines if the SSME would deliver a little more or a little less thrust with a little more or little less efficiency. Dynamic and thermally induced distortion of the nozzles' shapes also impacts both the actual thrust achieved and the efficiency of the rockets. Deficiencies in the ability of the actuators to achieve and hold the commanded nozzle angles alter the true angles from those desired. The precise chemical nature and even the shape of the aluminum particles suspending in the SRB solid propellant mix have a huge impact on SRB thrust profile and burn time. The propellant in the SRBs is molded to create a void with a star-like pattern that runs along the long axis of the SRBs. The SRB thrust is proportional to how much surface area was burning, so the star is shaped to get a lot of area burning early to maximize thrust, then decrease the area and thrust as the Orbiter gained altitude to keep the dynamic pressure limits from being exceeded as well as to limit the acceleration felt by the crew. The mold is not perfectly shaped, nor is it possible to know for certain exactly how well the burning of the propellant would keep the desired varying shape. The accelerometers and gyroscope, used to sense motion and attitude, are imperfectly precise, so they provide imperfect measurements for the flight software to determine the vehicle's true velocity and attitude. Consequently, any command to orientate the SSME, or to set a thrust level based on this imperfect knowledge, was inevitably inappropriate in some manner. Indeed, since there is an unavoidable delay between when the physical phenomena were sensed and the time the on-board computers complete the calculations to determine the commands, this means the commands are always for a situation that literally no longer exists. The Earth's gravitational force varies with latitude and longitude, so any deviation in position from the planned trajectory exposes the vehicle to a slightly different gravitational force than anticipated. The mass of the vehicle was known partly by measuring the entire Orbiter at some time in the past and partly by measuring the things that were added or removed; as it is likely things are added or removed without an accurate mass update, the mass of the Orbiter is never known exactly. (Interestingly, like a lot of people, all the Orbiters gradually got heavier.) At liftoff, the freezing cold liquids in the ET cause a coat of ice to stick to the ET, and that mass has a small effect on the Shuttle. The Shuttle is big enough that the air it displaces has some mass, which provides some buoyancy, and that small buoyancy changed with temperature. All possible system dispersions were identified and each possible variation quantified. To gauge the impact of the system dispersions, 500 trajectories were simulated, each using the same I-load but each with different system dispersion values, randomly selected according to the probability distribution associated with each. The 500 cases created 500 trajectories, which in turn created 500 load indicator predictions. From these 500 results, the mean and variance of the potential impact on the trajectory and loads could be determined. The 3 sigma level of possible variation that made a load worse was called a knockdown. For example, the 3 sigma level of additional dynamic pressure that might occur was the dynamic pressure knockdown. Similarly, the 3 sigma possible additional loads to each indicator were the load indicator knockdowns. It turned out that these dispersions varied with the Shuttle's velocity. In plain English, these knockdowns protected the vehicle from 99.87% of the possible effects of the system dispersions, provided, of course, that the basic trajectory to the mean wind and atmosphere and system performance parameters was also safe.

Since both dynamic pressure and angle of attack depend on wind velocity and the dynamic pressure depends on the atmospheric density, any difference between the actual atmospheric density and wind and the values used to design the trajectory will cause the real trajectory to be different than designed. Similarly, the initial temperature of the propellant in the SRBs has some impact on the thrust and efficiency of the motor. These are examples of *environmental dispersions*. So to be safe, some allowances must be made for environmental dispersions. This was anticipated, so for years before the Shuttle first flew, balloons were launched every day near the launch site to measure the winds and atmospheric density versus altitude. The huge database of samplings was turned into a mean density and wind velocity versus altitude for every month of the year, along with the 95% possible variations in each. Also, 500 winds were chosen for each month to represent the statistical variation seen in that month from all the measured data. The 500 monthly winds and densities were used to estimate the impact of the monthly environmental dispersions in the same way the potential impact of the system dispersions were determined. In general, particularly for the spring and fall months at the launch site, when the daily winds varied the most, the environmental dispersions had up to four times the potential impact as the system dispersions.

Given all this, the solution to finding a trajectory for the Shuttle to safely fly that maximized possible payload was to establish a constraining dynamic pressure profile versus speed and a dynamic pressure times the angle of attack curve versus speed. The first was called the *q profile*, and the second the *q-alpha profile*. Selecting the most appropriate phenomena for the independent variable is always one of the first critical modeling decisions. Speed was used rather than time since speed is also monotonically increasing during the Shuttle ascent flight and is more representative of the forces acting on the vehicle at that instant. That is, there is less variability in the forces acting on the Shuttle with velocity than time. The design profiles included the knockdowns. For example, if the 3 sigma system and environmental dispersion knockdown for dynamic pressers was found to be 50 newton/m^2 and the maximum dynamic pressure for acceptable wing loads was 650 newton/m^2, the design to dynamic pressure was 600 newton/m^2. A three-degree-of-freedom trajectory simulator was created that designed the trajectory to follow the stipulated q and q-alpha profiles for the mean wind and atmospheric density. With the computers of the day, about an hour was needed to design this trajectory that flew the Shuttle for about two minutes. For any rocket, the dynamic pressure rapidly increases as speed increases, but eventually reaches a peak, after which q decreases rapidly even though the rocket is still accelerating. This is because as the rocket gains altitude, the density of the atmosphere quickly falls to zero. The Shuttle controls the maximum dynamic pressure achieved and the duration held by throttling the SSME to a lower thrust level, holding that lower level until the Shuttle is high enough that the atmospheric density is reduced, and then increasing the thrust to go faster to get more payload capability. On a graph, the thrust profile looks like a bucket. The thrust starts at maximum level, then decreases at a constant rate to a minimum level, holds that minimum level, then increases at the constant rate back to the maximum level, so the design became known as determining the optimal *thrust bucket*. Given the thrust profile, the q-alpha constraint enabled the determination of the thrust angles relative to the horizon versus velocity. Once the trajectory that satisfied the q and q-alpha profiles was found, it was given to the people in the loads community, which used it as an input for the load indicators. If any load indicators exceeded their limits (including allowance for load indicator knockdowns), the q or q-alpha profile was adjusted and the process repeated. If there was excess margin in the indicators, the profiles were adjusted to try to get more payload capability, and the process repeated. Once a seemingly acceptable trajectory was found, the three-degree-of-freedom trajectory simulator was used to determine I-loads that were given to the people in the avionics community who verified them for use in the flight software. If these I-loads were different from the I-loads used to establish the knockdowns, the system and environmental knockdowns were recalculated with the new I-loads to ensure the knockdowns were not underestimated. If the new I-load resulted in new knockdowns, this process was repeated with the new knockdowns. Since in the early days this process was often repeated, it did not take long for people to notice that the magnitude of the knockdowns was almost insensitive to the I-load values. The magnitude of the system dispersion knockdown is primarily driven by the SRB burn rate dispersion, and the environmental dispersion knockdowns are primarily driven by the magnitude of the deviations of the real wind from that used to design the trajectory.

Of course, every time a load cycle finished, the load indicators changed, so that a new trajectory had to be determined. Also, after about 20 flights there was sufficient data to check if the presumed system dispersions values had been modeled appropriately and the data did indicate the need for some changes. The most significant change concerned the SRB thrust. The flight data showed the SRBs sometimes ran hot (had extra thrust and burned out faster) and sometimes ran cold (had less thrust and burned out later than expected). There was no way to predict which would happen, just that one or the other would occur. This phenomenon was the predominant contributor to the dynamic pressure system dispersion knockdown. The historical flight data showed that if the Shuttle got to a target speed sooner than expected, the SRBs were almost certainly hot, or if the target speed was achieved late, the SRBs were likely to be cold. So the flight software was modified to note the time a specified speed was reached, and then based on the size of the difference between that time and the nominal time that speed was predicted to be achieved, the flight software calculated a proportional adjustment to the minimum SSME throttle level I-load. If the SRBs were hot, the SSMEs were throttled to a proportionally lower limit than the initial I-load called for, and vice versa.

Extensive efforts were undertaken to continue to take advantage of the actual flight data to find the best trajectories to fly. Meanwhile, since each trajectory was designed for a projected launch day and the associated atmospheric and wind conditions for that month, every time a launch slipped out of the month, the I-loads needed to be checked for the new launch date. This happened often. Indeed, the actual launch date was never close to the first planned launch date. For the summer and winter months, there wasn't a lot of variation in the atmospheric density and wind from month to month, but for the spring and fall months, there was significant month-to-month variation. Eventually, the process was modified to use seasonal environmental databases to minimize the work required to verify the I-loads for a changed launch date. Wind and atmospheric databases were established for three seasons—summer, winter, and transition, a combination of variation seen in the spring and fall months.

The process described above was time consuming, not only because of the limited computing power of the day, but mostly because people were figuring out how to do things for the first time. As a result, the government (who was the customer) decided that the best way to handle all this uncertainty was to begin all the trajectory design work 24 months before flight, with scheduled updates at regular interviews closer to launch. All of these intermediate flight products added to the workload, though only the final product was actually utilized since many inputs changed during the 24-month period. However, the repeated assessments did yield one huge benefit. There were four independently developed, coded, and operated trajectory simulations available for assessing Shuttle trajectories. First, there was the three-degree-of-freedom simulator that designed the trajectory and determined the I-loads. A second simulation came from the avionics community tool used to verify the guidance, navigation. and control software, which saw the Shuttle as having six degrees of freedom—three representing translational motion, and three representing rotational motion. The third and fourth were government-sponsored simulations, one hosted by the Johnson Space Flight Center and one hosted by the Marshall Space Flight Center. When the outputs from these four simulations were first compared, there would inevitably be discrepancies with respect to what the trajectory would be and how much propellant would be needed to get to the desired MECO condition. Upon careful review, the causes for these discrepancies were identified. Most often the cause was that the person running the simulation made a data entry error, as there were tens of thousands of numbers needed to represent the aerodynamics, mass properties, propulsion, or avionics portions of the vehicle. Sometimes programming errors by one or more parties were found. In contrast, the load indicators were developed, coded, and operated by one group of people. Often very similar trajectories provided to the loads community would yield very different load predictions. In such cases it was clear that either data input mistakes had been made, or there were errors in the indicator formulas. When such a problem was identified, the load indicator folks would look into it and eventually issue an update. The load indicator group was part of the structures group that did the load cycles. Since the loads were so critical to flight safety, the customer would periodically consider funding a second group to generate an independent estimate of the loads. But this never happened.

To avoid expending resources on trajectory designs that would ultimately not be used, it was suggested that the I-loads be updated only once, just before launch, for the measured environment of the day. This process was called *day of launch I-load update*. The wind and atmosphere would change in the approximate three-hour period between the measurement and launch, but coincidentally data were available to model this effect from the source data for the monthly winds and atmospheres, as the balloons to get the data were released on three-hour intervals. Incidentally, it took about three hours for the balloon to reach the necessary altitudes, process and transmit the measured density and wind data to the trajectory simulators, run the trajectory designer to determine the new I-load for the measured case, obtain the resulting load indicator model predictions, and, if found acceptable, load the new I-loads to the vehicle's flight software and confirm that the I-loads were stored with the intended values, leaving about 30 minutes to make the final decision whether or not to launch that day. This was possible because nearly all the I-loads produced using the monthly or seasonal data could continue to be used unchanged, while only the relative few numbers that stipulated the SSME thrust angles versus speed were updated. Day of launch I-load update offered considerable benefits. The trajectory was inherently safer since the only environmental dispersion left for the open-loop system to deal with was the change in environment during the three hours between the trajectory design and the actual flight, which was considerably less than the difference that could exist using the seasonal I-loads. Second, the trajectory would be more efficient, increasing payload capability, and reduced the probability of running out of propellant prior to achieving MECO. Last, ceasing the practice of making trajectories prior to launch that were ultimately never used would free up resources to tackle other issues.

Though day of launch I-load update was proposed before the Shuttle flew, it took over six years before it was implemented, partly because the group that invented the process was not the group that would implement the process on launch day, and partly due to the fear of loading an incorrect I-load at the last minute, but mostly because those responsible for making the decisions found it difficult to gauge the probabilistic benefits. For all the launches during the first six years of Shuttle operation, at least one load indicator was at or above its allowable limit based on the preflight predictions using the measured winds. The structure experts would huddle, and then announce they had been sufficiently conservative in designing their system dispersion knockdown such that the apparent problem was really within true tolerance, and they would recommend a go for launch. Then, for one mission, the predicted loads were so large and so numerous that they resulted in a no-go advisory. The next day, the government managers who were so adamantly opposed to the day of launch I-load update initiated implementation. Implementation required about 18 months, but day of launch I-load update became standard procedure for all Shuttle launches.

LESSONS LEARNED

Determining what to model is as important as determining how to make the model. Losing SSME thrust and breaking a wing were only two of a huge list of events that were identified that could cause the Shuttle crew to be killed. Each of these possible events was documented on what was called the Category 1 list. Prior to the first Shuttle launch, a great deal of effort was expended to prevent or mitigate the consequences from items on the list, particularly those for which the likelihood was judged to be high and it was clear what could be done to alleviate the risk. All this work almost certainly prevented a Shuttle crew from being killed. Once flights started, the remaining items on the Category 1 list were catastrophic events with unknown probabilities needing significant resources to mitigate. So the government adopted a political process to determine which if any of the residual Category 1 items were addressed. That is, in order for action to be taken to mitigate one of the remaining Category 1 events, some organization had to make the case that the event was likely enough that it needed to be addressed and that there was an affordable means to remove the risk. The two events that ended up killing Shuttle crews were both on the Category 1 list before they occurred, but neither had an effective advocate until after they happened. That the O-rings inside the joints connecting SRB segments were being singed by the hot internal gases (indicating that some of the gas was leaking into the joints) and that a redesign of the joint was necessary were both known years before the *Challenger* crew was lost. There was no attempt to address the issue prior to the crew being killed because the SRB community never asked that one be made. The fact that ice was accumulating on the ET during the wait before launch was observed from the first Shuttle launch, and indeed on all rockets that ever used cryogenic propellants. That the ice was causing ET insulation to fall off during ascent and that both ice and ET insulation were damaging the thermal protection tiles on the Orbiter were known from the first Shuttle flight, many years prior to the loss of the *Columbia* crew. Indeed, after every Orbiter landed, every single dent and ding in the tiles were mapped and damaged tiles were replaced. No attempt was made to minimize the ice buildup, reduce insulation delamination, use radar to see possible impacts as they occurred, or enable on-orbit inspection and repair prior to the *Columbia* crew being killed because the Orbiter community never initiated one. Once these events that had previously been perceived as unlikely actually happened, they became perceived as inevitable events, and huge amounts of resources went into addressing them. Indeed, after the *Challenger* loss, the entire Category 1 list was reexamined and several other items on the list were addressed in addition to modifying the SRB joint design. But the process of selecting the Category 1 items to address post *Challenger* remained political. Ironically, after the *Challenger* was lost, it was suggested that the loss of a tile be one of the items addressed, but the astronauts most urgently wanted a means to escape from the vehicle, so significant resources were expended in the futile search for ways for the crew to escape from the Orbiter in different flight modes, which consumed the resources that might have been used to prevent or mitigate the impact of tile loss or other possible catastrophic failures still on the Shuttle's Category 1 list.

If real-world test data are not available, the best method to verify a mathematical simulation is to independently develop, code, and operate at least two and ideally three independent simulations. The initial outputs from the four trajectory simulations used on the Shuttle program were always in disagreement. Had any one of the wrong trajectories been adopted as the correct simulation, the consequences could have been dire. Comparing the four simulations was the best way to identify and correct the errors. Each simulation was the province of a different community that happened to need a trajectory simulation. There was no conscious effort to create four separate simulations, so the costs of the efforts were not perceived as duplicated. Because only one group was responsible for loads, only one group was funded to create load indicators. The load indicator folks also made errors, at about the same rate as the trajectory community. But since there were no independent models for comparison, it took much longer to find errors and they were found only because of careful observations by the load indicator personnel and quite often by the people who produced the trajectories that were the input to the load indicators—or, to put it another way, luck. Often, those wrong results went undetected for a long time, so a lot of analyses had to be redone with the corrected indicators. A second independent group of people producing and predicting loads would certainly have doubled the simulation cost, but would have saved more in avoided rework costs. A second independent group would also have been safer, since a load indicator mistake could have been catastrophic. If the second group had been within the trajectory community, the load indicators could have been used directly to design the trajectory, saving considerable time by avoiding having to do trajectory analysis followed by a load analysis in series to find the best trajectory design.

Do only the analysis needed to make the decision needed. Literally thousands of Shuttle flights were planned that never happened because the criteria to which they were planned would inevitably change before the trajectories were needed. This was a complete waste of resources, with the possible exception that it did help train people to do the analysis.

The model maker is responsible for constructing a model that can be used by the decision makers. Over and over again on the Shuttle program, whether it was regarding choosing the trajectory to use, determining the implication of the load indicator prediction, deciding if to implement day of launch I-load update, or choosing which Category 1 items to mitigate, decisions were made that were counter to what the probabilistic data recommended. One of the more extreme examples was that it was considered unnecessary to model the potential change in the environment during the three hours from the final go/no-go assessment until launch for I-loads built to monthly or seasonal environments, but it was considered essential to do so for I-loads built to the launch minus three-hour wind, even though the potential impact was exactly the same. The reason appears to be that most people do not understand probability. Therefore, most people do not know how to make a decision if presented evidence in probabilistic terms. You could argue this is the decision maker's problem, and as such, people shouldn't be in decision-making roles if incapable of utilizing appropriate evidence to make decisions. But such an argument abdicates the modeler's ethical responsibility to provide the information needed to make the decision in a manner that enables the decision to be made correctly. If the modeler knows the decision maker misused the information in any way, the modeler is responsible to make this known and to find a way to present the information in a manner that the decision maker can comprehend.

The following presents a modeling case study for architecting.

Case Study 2.2: Best Systems to Acquire to Perform the Space Superiority Mission

BACKGROUND

In the mid-2000s, those in charge of acquisition planning for the U.S. Air Force Space Command (AFSPC) sought a method to assess all possible means to do all the AFSPC's missions, to enable selecting the few means that maximized the total mission accomplishment at least cost. Prior to adopting an approach for all of AFSPC, the acquisition planning organization conducted a pilot study in one mission area called Space Superiority. Space Superiority missions know where things are in space and what they are, defend our things in space from attack, and, if necessary, prevent an adversary from using their space assets. To implement the pilot program, multiple measures of goodness were defined for each Space Superiority mission area. Each measure of goodness was documented as a unit-less utility curve, so the individual goodness factors each had a potential score between 0 and 1. The features contributing to each goodness factor were the independent variables. Typically, these were qualitative ranges of performance. For example, being able to detect something in geosynchronous orbit that was less than 1 meter in diameter might score 0.9 to 1, while only being able to detect something 10 meters in diameter might score 0.1 to 0.2. Expert opinions were used to gauge how well existing or conceptual systems achieved the independent variables, as well as to estimate the approximate annual cost for each of the existing or conceptual systems. The individual goodness measures were summed using normalized weights. The weight values were also based on soliciting expert opinions. A program was created to search through all the system options to find those that provided the highest utility score while staying under the annual acquisition budget limits for the next 20 years. At the time, the U.S. Department of Defense (DoD) had five-year budget cycles, with the first and perhaps second year having some certainty. The 20-year time frame was thought necessary to ensure that the cost to maintain the systems, not just the cost to acquire the systems, was part of the selection process.

The approach was eminently rational, but the decision makers it was developed for did not know what to do with the results for several reasons. First, the goodness of one proposed collection of systems differed from another proposed collection of systems by a tiny fraction of a unit-less number. The decision makers could not relate to these small unit-less differences. Second, since the means for estimating the goodness of the individual systems, as well as the cost, was based on experts' opinions, and since the decision makers weren't these experts (though they had been asked to be), they doubted the scores the system combinations were given. Indeed, since each decision maker inevitably intuitively preferred one system over another, if the reported scores did not match their preconceptions, this inevitably cast doubt in their minds regarding all the scores. Also, the advocates for any conceptual systems saw them as eminently practical. If the advocate was also an expert judge, which did happen, these folks tended to score their identified solutions as superior to alternatives. Third, the decision makers were uncertain why the identified candidate systems were the only candidate systems. They would inevitably ask about other options that did not appear to be considered. This happened despite the effort by the acquisition planners to call for concepts from any and all potential sources. Fourth, the claimed system capabilities were doubted. Systems that existed might provide accurate capabilities, but didn't always. Systems that were in development could report either their required capabilities or their predicted capabilities, but rarely wanted to report either for fear of the data being used to attack the rationale for their program or their budgets. Often the proposed systems were little more than ideas, for which the practicality was hard to judge. Fifth, the process performers showed only the result, so it was difficult for the decision makers to see the pros and cons for an alternative family of systems they thought might be superior to the proposed results. Finally, since even next year's budget was often uncertain, and budgets 5–20 years in the future were virtually unknowable, the imposed financial constraints that drove the solution were not what the decision makers wanted. The decision makers wanted to know what was the most effective budget to try to get approved.

WHAT HAPPENED

An alternative methodology was developed. The process is shown in Figure 2.3. The process begins by establishing the capability definitions, which are documented in what became known as holistic view 1 (HV-1). (This product was originally called Helms' view, since at the time the then Colonel Helms, later General Helms, was the commander in charge in formulating requirements.) Figure 2.4 shows the content of an HV-1. One of the key HV-1 items is the architecture reference mission (ARM), which is created by the Office of the Secretary of Defense (OSD). Each ARM was defined in sufficient detail to enable predicting how well the mission is achievable with optional systems. The HV-1, along with Department of Defense Architecture Framework (DoDAF) products from

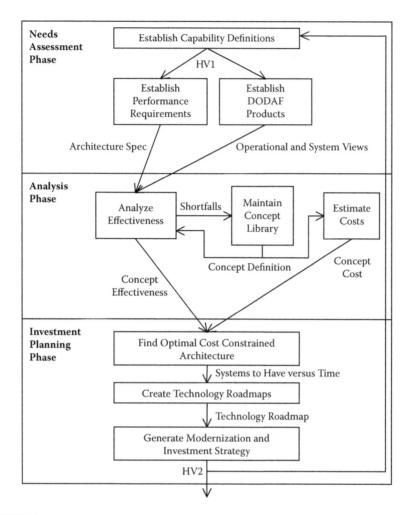

FIGURE 2.3 The process to find the optimal family of systems.

Holistic View 1: Architecture Requirements
1. Key performance parameters
2. Necessary capabilities
3. Architecture reference mission(s)
4. Performance requirements
5. DoDAF requirements

FIGURE 2.4 Holistic view 1 documents what the architecture needs to accomplish.

related programs, is used to generate the operational views (OV) and system views (SV) called for by the DoDAF process to enable identification of the most efficient means to exchange information between the Space Superiority family of systems, and those acquired and maintained by other organizations. Also, the ARM was used to establish performance requirements for each mission, not for individual systems. The results of the concept formulation stage are the key performance parameters (KPPs) and the corresponding desired values to achieve. For defending space assets, a KPP is percent of service retained. For negating space assets, a KPP is percent of an adversary service denied. For space situational awareness, a KPP is attack warning time. The analysis phase starts with an assessment of the as-is systems, which results in the identification of specific shortfalls with respect to the defined KPP. The shortfalls are used to identify trade studies to perform to find ways to use existing systems, or to define the features of proposed systems to remove the shortfall. A template was provided to record in sufficient detail the physical nature of the proposed system to gauge its technical maturity and so performance could be simulated based on physics, not just claimed based on argument. The same physical features enabled the cost to develop and maintain the proposed system to be estimated using parametric models based on actual results for similar systems in the past. Only systems with completed templates were brought forward for further consideration. Without a completed template, the concept was termed an idea, perhaps a good idea, and was documented for future reference. The analysis continued by first formulating combinations of existing and proposed systems, each called a family of systems. This was done partly by using common sense, but also by simply mathematically assembling all possible permutations in type and quantity from the available pool of existing and possible systems. Each family of systems was simulated to determine how well it would achieve the KPP and how much it might cost for the next 20 years. Each family of systems was then plotted with respect to KPP value achieved and total cost of ownership, thus enabling identifying the *sufficient frontier*, which are those candidate family of systems that provide the best KPP values at each cost value. The decision becomes to pick the one family of systems that achieves the best value consistent with the cost willing to pay. Clearly, complicated families of systems are unlikely to be completely discernible based on only one parameter, even a key parameter. For example, three families of systems may be close in cost and value, but one needs substantially less manpower to operate and another is dependent on immature technology. Though the cost of the manpower and the cost to mature the technology are estimated to determine where along the cost axis the candidate family of systems is plotted, both are clearly estimates. So of the three, it is prudent to select the option that needs the least manpower and has the most mature technology. Once a family of systems is chosen consistent with the cost constraint, the technology maturation required becomes clear. If any of the systems that make up the chosen family of systems is dependent on immature technology, then clearly those technologies need to be matured as quickly as needed to enable the system to be operational when desired. Or, if a better performing family of systems was not selected, because one or more of the member systems had immature technology, then the rationale

Holistic View 2: Integrated Roadmap
1. Family of Systems efficient frontier
2. Program timelines
3. Benefit timelines
4. Technology roadmap(s)
5. Funding profile

FIGURE 2.5 Holistic view 2 documents what architecture to implement.

for undertaking technology maturation efforts is again clear; if that technology develops, a better family of systems will be available. So technology road maps are then prepared to move the identified technologies from their current state to that needed to offer viable systems. The final step is to outline the entire time-phased family of system implementation process, which is shown in holistic view 2 (HV-2). (This product was originally called Haywood's view, because then Colonel Haywood, later General Haywood, was the commander in charge of acquisition.) Figure 2.5 shows the content of an HV-2, while Figure 2.6 shows a notional example. The chosen family of systems is cycled back to the start of the process to update the DoDAF products as necessary. The process is repeated only if the stipulations of the HV-1 change, if shortfalls close, or as technology matures.

The modeling process is explained in more detail in Figure 2.7. Rather than attempt one huge model, the modeling effort is stratified into engineering, engagement, mission, and campaign levels. The engineering-level models predict in physical terms how well the candidate systems will achieve factors relevant to the KPP chosen for the mission. The engagement models determine for each candidate system just how well the KPPs are fulfilled. The mission models mix the candidate systems with respect to type and quantity to predict how well combinations of systems will fulfill the KPP associated with the ARM. For many decision makers, this is all that is needed, and they are comfortable making decisions based on mission models. Some DoD decision makers prefer to know the military utility that the candidate family of systems will achieve. Determining military utility requires campaign models to translate the still mostly physical parameters associated with mission into net results in particular military scenarios, such as time to achieve objective, or relative loss rates. Campaign models are notoriously difficult to build and make credible because they are so difficult to verify. All the levels can be used to explicitly point out performance, sufficiency, task satisfaction. or campaign shortfalls. Each shortfall was documented with an explanatory note, an indication of what was desired, and a list of the options considered to close the shortfall. For each option that might reduce or eliminate a shortfall, physical features are defined, performance capabilities modeled, and total cost of ownership estimated. A model is used to assess how well any combination of systems achieves the mission KPP and what the total ownership cost will be, so each candidate family of systems is plotted versus KPP value

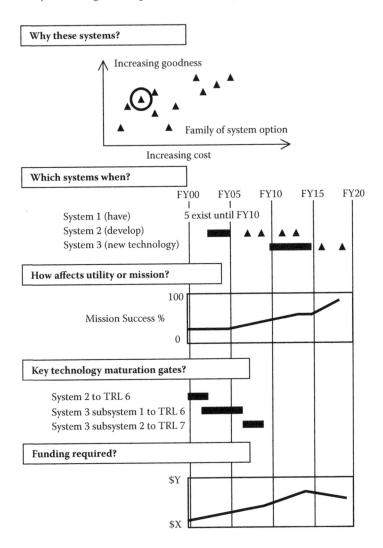

FIGURE 2.6   Notional HV2 product.

and total ownership cost. Critical secondary considerations, such as manpower required or technical maturity level, should also be easy to display for any family of systems point on the plot.

The decision makers responsible for requirements, acquisition, and operations overwhelming preferred the revised methodology. The leader of the planning organization continued to prefer the utility curve approach based on expert judgments. The appeal of the new process to those that appreciated it was mostly due to the provided information transparency. The decision makers recognized that using the OSD-provided ARM proved they were seeking systems to do exactly what the OSD had stipulated as important to be able to

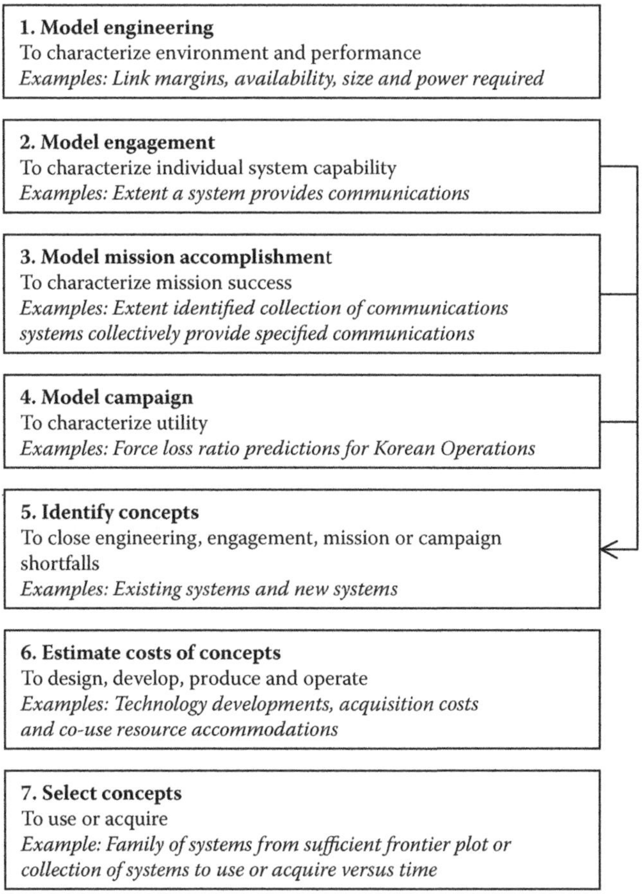

1. Model engineering
To characterize environment and performance
Examples: Link margins, availability, size and power required

2. Model engagement
To characterize individual system capability
Examples: Extent a system provides communications

3. Model mission accomplishment
To characterize mission success
*Examples: Extent identified collection of communications
systems collectively provide specified communications*

4. Model campaign
To characterize utility
Examples: Force loss ratio predictions for Korean Operations

5. Identify concepts
To close engineering, engagement, mission or campaign
shortfalls
Examples: Existing systems and new systems

6. Estimate costs of concepts
To design, develop, produce and operate
*Examples: Technology developments, acquisition costs
and co-use resource accommodations*

7. Select concepts
To use or acquire
*Example: Family of systems from sufficient frontier plot or
collection of systems to use or acquire versus time*

FIGURE 2.7   Modeling an architecture is different than modeling a system.

do, and if their contribution to achieving the ARM was superior to solutions offered by other DoD organizations, then a strong case was made to acquire the superior solution. The decision makers appreciated that they could see complete documentation on all the existing and proposed systems. They liked that the shortfalls were finite, explicit, and matched to options to close the shortfalls. More importantly, they valued the trade tree, which compared each candidate system's ability to close shortfalls, since they understand how the solution space was searched, and if they wanted, they could direct a systems inclusion or exclusion. They liked that they could see in real physical terms the potential of each alternative family of systems. They liked that they could explicitly see in terms of physical phenomena whether some combinations of systems were very close or very different in value and cost. And they liked that secondary considerations (such as manpower required or technology maturation) could be overlaid onto the sufficient frontier so they could explore the option with respect to criteria different than what might be used for the value axis. Mostly, they liked the new approach because it gave them a vocabulary to explain to each other, people trying to accomplish other missions, congressional staffers, and industry providers what they were trying to accomplish and why they thought the chosen solution was the best available.

LESSONS LEARNED

The value of the model is determined by those who use the result, not by the model makers. Utility analysis is eminently rationale, but if not used by the decision makers it is developed for, something else needs to be done. Difficult-to-verify models, such as those usually associated with military utility analysis, should generally be avoided if they detract from the credibility of the claims.

Make lots of small models that are easier to verify and thus easier to be trusted, rather than one big model. Big models are complicated, and complicated things rarely work well. Even if one gets a big model to work, for decision makers to trust the result, they will usually want an independent source to examine the model to confirm its validity. Big models are much harder to verify, even if true. For architecting in particular, what needs to be modeled and the fidelity to model are very uncertain at the start of the process. Big models are hard to change, while small models can be grouped as needed.

Architecture-level analysis needs to be a notch higher in abstraction than system-level analysis, just as system-level analysis is a notch higher in abstraction than subsystem-level analysis. Architecting as defined in this book is a new field. To find the optimal set of systems to perform a mission is not the same thing as to find the best system. But just as system engineers would be bogged down in details if they used the most detailed possible models of all the subsystems, an architect will similarly be lost if he or she relied on system-level simulations. The architect must model the accomplishment of the ARM at a level higher than that of any individual system.

In this chapter, approaches are recommended to model systems and architectures using diagrams and mathematics. Sources are identified for excellent existing models that are reasonably certain to be error free. The bulk of the chapter illustrates different means to model uncertainty—namely, by regression, Monte Carlo, fuzzy logic, agent-based programming, and fractals. The chapter concludes with recommendations on how to monitor key performance parameters during the program life cycle.

2.1   MODEL SYSTEMS AND ARCHITECTURE USING DIAGRAMS

Diagrams are useful to both comprehend and communicate aspects of systems or architectures, but rarely enable the prediction of behavior. The word *diagram* is used in the most general sense, including lists, tables, or pictorial representations.

No single diagram can communicate all aspects of the system or architecture. The following types of diagrams are usually needed:

1. *Reference mission.* This is a depiction of the representative mission for the system or architecture, denoting the primary objectives and operating environment, ideally including key constraints. A still image is traditional for easy communication, but animated presentations are now easy to make and can better depict the intention of very complicated systems or architectures.

2. *Key performance parameters.* A table that names and defines the most important results the system or architecture is to achieve, along with a quantified indication of the level sought. The level sought may stipulate a threshold (the minimum that must be achieved for the system or architecture to be acquired) or an objective (the desired goal for which cost constraints may limit the actual amount achieved).

3. *Customer, user, and operator hierarchies.* These identify customers, users, and operators with an indication of relationships. Victims, those who may lose something if the system is developed and therefore may work to prevent the system from being achieved until their losses are acceptable, are also often useful to identify.

4. *Constraints.* An itemization of any law, regulation, policy, rules, or other imposed restrictions bounding any of the primary functions or solution implementations of the system or architecture.

5. *System boundary.* A representation of the entities that are internal to the systems or architecture and entities that are external to the system or architecture, with an indication of what if any information or material crosses the boundary. Typically, the boundary is established by the customer, implicitly if not explicitly, as those aspects which they have control over versus those they must cooperate with others.

6. *State transition diagram.* For many systems, the same real-world parts need to function differently in different configurations. For example, your car has a limited set of functions at rest with the ignition turned off, another set when the ignition is set to power, a third when starting the engine, and a fourth when the transmission is engaged and the car is moving. These configurations are often referred to as *states* with subsets called *modes*. Whatever one cares to call them, the entire set of desired configurations needs to be identified along with an indication showing which configuration can transition to which configuration.

7. *Functional hierarchy.* This identifies all the functions internal to the system or architecture in a specified state. An indentured list is usually adequate to provide the necessary information.

8. *Functional flow.* This is an indication of the relative sequencing of the functions internal to the system or architecture. Ideally, it is also an indication of the material or information input or output between functions, along with controls and resources needed for each function.

9. *Functional timeline.* An indication of the durations or time limits for performing each function.

10. *Information structure.* An identification of the information that needs to be aggregated and how. This is typically shown as identified inputs, outputs, and transformations.

11. *Trade tree.* The hierarchy of implementation options considered for the system or architecture, denoting the explicit differences between the options, and the reason the chosen implementation was chosen. For systems, the options denote alternative subsystems; for architectures, the options denote alternative systems.

12. *Configured end items hierarchy.* The hierarchy of the end items chosen to implement the system or architecture functions. Each end item could be hardware, software, procedure, documentation, or whatever needs to exist for the system or architecture to exist.

13. *Functional allocation.* This is a mapping of the function hierarchy (along with associated performance requirements) to the configured end items.

14. *Interface diagram.* A naming of any information or physical relations between configured end items, usually with an indication of how the interface will be accomplished, so both parties to the interface can be confident the other will comply.

Since the diagrams created to communicate these aspects of a system or architecture are abstractions, there is no right or wrong way to create each diagram. As systems engineering and architecting disciplines have matured, prescriptions for how to develop such diagrams have also matured, and most likely will continue to do so. A prescription is desirable as it tends to standardize the content and hence meaning of the diagram. Standardization also enables reuse of the diagram, saving time, cost, and probably errors. Any particular prescription will have some strengths and weaknesses relative to any other. For some prescriptions, tools exist that can minimize the modeler's work. For example, if a modeler draws many functional flows, the tool used should be able to automatically make the corresponding functional hierarchy, as that information is completely discernible from the functional flow diagrams. Also, some prescriptions may enable the automatic translation from a functional flow depiction or an information structure depiction to the corresponding mathematical equations. Readers are encouraged to research the current diagramming prescriptions and choose the approach they find most beneficial. Options in vogue are IDEF,[1] Model Framework,[2] Department of Defense Architecture Framework,[3] SysML,[4] Unified Modeling Language,[5] colored Petri nets,[6] and Higraphs.[7]

2.2 MODEL SYSTEMS AND ARCHITECTURES USING MATHEMATICS

The number of potential domains for which mathematical models may be needed is beyond the scope of this book. To keep the cost of systems engineering or architecting effort as low as possible, and to avoid errors, it is highly desirable to reuse existing equations or software, rather than develop them from scratch. How can you be certain the reused material is error free? First, to the extent possible, use equations and code used by the largest number of people the most often. The more people who use the material for the longer period of time, the more likely someone will identify and correct errors. Second, perform exactly the same verification effort for reused equations and code as you would do with any you develop. Create test cases for which solutions are known by alternative means, and compare the reused material output to the known result. Four very useful sources for reusable mathematical models are as follows:

1. Wolfram Research website[8]
2. MATLAB® website[9]
3. National standards website[10]
4. Cambridge numerical recipes[11]

Though there are similarities in how systems and architectures should be modeled, there are great differences too. The following sections define the approach for each.

2.2.1 Mathematically Model Systems

Establishing models for systems is more art than the execution of a defined process. It is often an extraordinary difficult task to make the first appropriate fidelity model of a complex real-world situation. Once the basic model is derived for a situation, it can be reused with a fraction of the effort associated with its development. The three things to do to mathematically model a system are the following:

1. Define the functions that constitute the systems. The dividing boundary constitutes external interfaces which must be modeled to the extent an external entity (a) commands or controls the system, (b) provides material or information needed by the systems, (c) obtains material or information from the system (which is perhaps the reason for the system), and (d) influences the behavior of the system in some manner.
2. Establish at least one representative design reference mission (DRM) for the system to accomplish, such that if it is achieved, all other potential uses of the system are acceptable. This scenario illustrates the key inputs and desired outputs that are to be optimized and the key constraints on the operation of the system. Care should be taken to comprehend the input and output ranges, the uncertainty, and at what frequency or period of time updates to the inputs or knowledge of the outputs must be known. If the system is primarily mechanical, that is, not dependent on information to operate, then data flow between components of the system is not modeled. If the system is primarily associated with the collection and processing of information, then data flow may be all that needs to be modeled. For dynamic systems, any time delays associated with data acquisition and processing, or the ability of the mechanical elements to act, must be modeled. Fundamentally,

the more complicated the system, the more DRMs are needed to scope the minimum acceptable operations. However, almost no one can make decisions regarding more than three DRMs as that requires just too many potentially contradictory situations to be simultaneously comprehended. The key is to make a single DRM's features more difficult to achieve, rather than create more dissimilar DRMs.

3. Establish the fewest, simplest equations that represent the system doing the scenario along with the interfaces to the necessary external entities that can be used to establish requirements to implement the system. This step requires extreme care. Obviously, whatever condition that is left out of the model will not manifest a corresponding translation of inputs to outputs. This fact is often forgotten, much to the chagrin of those who design structures that fail due to the differences between dynamic and static loads, or lose fortunes because they failed to model the lack of liquidity that will occur if many people pursue simultaneously the same investment strategy.

Attempting to model a system is an extraordinary powerful proxy for assessing if the system should be made real. For example, if the system is to go very fast in the atmosphere, both the drag and heat buildup associated with its speed are critical to model, or the representation is inherently flawed. But items going very fast in the atmosphere also create a sonic boom, as well as create exhaust gases high in the atmosphere; these consequences may irritate people so much that many may lobby to prevent the systems from ever becoming a reality. The modeler is probably the first person who will discern if the system is achievable and practical. The modeler has the responsibility to make it known if the proposed system is not possible or necessary.

The systems engineer's job is to identify the functions and performance requirements for the systems. The designer and manufacturers are tasked to figure out how to best implement a solution to these requirements. The standard systems engineering heuristic is to stipulate requirements that are as neutral as possible in their implementation approach so the designers and manufacturers have the largest possible trade space to search for the implementing solution. The logic is that by doing so, the implementers are free to explore many implementation options to find the best one. Clearly, this is a noble goal, but to make mathematical models, with appropriate fidelity, some level of physical manifestation must be presumed. One means to avoid implementation prejudice is to model system performance using utility functions. Utility functions enable one to numerically score a systems option on a scale from 0 to 1. The utility functions may be artfully designed to be linear, exponential, or S shaped to try to mimic the approximate increase in utility (goodness or badness) in proportion to some feature. Multiple utility functions may be scored to assess a system, with numerical weights used to combine the scores. This approach allows many options to be assessed quickly with respect to goodness criteria without ever modeling the physical behavior of the system. My experience is that customers and users rarely find utility functions useful. The different system option scores obtained using utility functions differ on the order of hundredths or even thousandths of a unit-less number. Typically customers and users find this difficult to relate to. Since the total utility score is unit-less, and a combination of many diverse parameters, it is very difficult for customers and users to comprehend what improvements they are getting from one option versus another for different levels of cost. Perhaps most damningly, it is extraordinarily difficult to obtain useful utility function scores. Often only personal opinions are used to determine both the independent inputs and the shape of the utility curve. Though expert opinion is certainly worth something, it is not a substitute for physical reality. Indeed, studies have shown that the term *expert* is often a misnomer, and that for many decisions, the mean opinion of a large number of people is often closer to the truth than the opinion of a few experts.[12] In plain English, the designated expert often is not. Customers and users typically want to know how their options vary in terms of understandable physical parameters. So, when modeling a system mathematically, if the various options for the system implementation have different physical features, create different models true to the features, common only in that the outputs are the key criteria associated with the system.

2.2.2 MATHEMATICALLY MODEL ARCHITECTURES

The complexity of architectures increases the temptation to use utility functions to model effectiveness. Architectures, being combinations of systems, have profoundly large numbers of inputs and outputs, often of very different natures, so the effort to achieve physics-based models first appears overwhelming. But ironically, this very complexity extenuates the need to present candidate architecture goodness data in realistic dimensional terms for customers and uses; otherwise, they must find themselves having to blindly trust their analysts. Now, I have tremendous respect for analysts, but given the reality that all models are wrong and some models are useful, no decision maker should ever blindly trust analysts. I recommend you avoid making one humongous "be all do all" model. Rather, utilize a series of simpler models at three to four levels as illustrated in Figure 2.7. At the lowest level, use physical behavior to reasonably model relative performance and size (determine the basic features) of the candidate systems that make up the architecture. For the second analysis level, model how well a single system contributes to achieving a quantifiable mission goal. For the third analysis level, model how finite numbers of multiple systems can achieve the mission goal. Often, the third level is sufficient to make architecture decisions. Sometimes, the measure of goodness at the third level is still too abstract for customers or users. For example, suppose the measure of goodness at the third level is warning time, but what the user wants to know is which family of systems minimizes their losses in a hypothetical conflict? Warning time alone might contribute to minimizing losses, but does not by itself minimize loss. So at a minimum, the modeler must show the decision makers how by making a decision to maximize warning time they are also almost certainly minimizing their losses. To do so, a fourth level of analysis is needed, in which the quantifiable mission goal (in this example, warning time) is one of several inputs to an analysis that predicts what the decision makers most care about, in this case, losses. This fourth level of analysis is called the *campaign level*, since it is at this level that military utility analyses are conducted. Campaign-level analyses need to address behavior well beyond physics. So, usually campaign analyses combine equations based on physics with canned formulas and utility curves. Some customers may doubt the credibility of such approximations; if so, then it is useless to conduct campaign analyses.

The nine things to do to mathematically model an architecture are as follows:

1. Establish what at least one architecture reference mission (ARM) is to accomplish, for which it is possible to grade the goodness of candidate family of systems with respect to performance and cost. This ARM illustrates the key desired accomplishment of the architecture and constraints under which the architecture must operate. Please note I used a different term for the mission used to determine a system from that used to determine an architecture. A sample DRM for a system is to deliver at least 65,000 lbs. to a specified orbit from a specified launch site subject to specified constraints. A sample ARM for an architecture is at any given instant to maximize the knowledge of the position and characterization of all objects in space utilizing a to-be-determined set of existing or new systems with multiple owners and operators as well as a large defined set of existing systems. A system DRM, though perhaps extraordinary difficult to accomplish, is substantially narrower in scope than an architecture ARM.
2. Identify all the existing systems currently used to achieve the mission, and analyze the current extent to which the mission is being performed and at what total ownership cost.
3. Identify all the performance, capability, mission, and if necessary campaign-level shortfalls.
4. For each shortfall, identify optional ways of using existing systems or new systems to potentially close the shortfall. For each shortfall, identify which options are inherently superior to other options, if any. That is, if to close a shortfall, one can think of six possible ways, and two of those ways are inferior with respect to all selection criteria to the other four, then it is only necessary to document why the two inferior options are not considered.
5. Fully define the candidate systems using physics to predict ability to achieve the mission and estimate the total cost of ownership.
6. Identify all the possible combinations of systems that could conceivably achieve this mission. This step calls for being able to model the performance and cost of existing systems as well as conceptualized systems. But this creates a paradox: the cost of existing systems and their performance are usually well known, while the real performance and cost for conceptual systems are but a guess. All the blemishes of existing systems are known. The blemishes of conceptual systems are likely not yet known. So to compare existing and yet-to-exist systems, one must take care to credit existing systems for improvements that might negate current weaknesses, along with the associated cost to implement as well as modify the claimed cost and performance for proposed new systems in proportion to the technological maturity of the proposed systems. A methodology to do so is presented

in this chapter. Frankly, the modeler will find it difficult to downgrade the new system, as advocates for new systems are inherently motivated to do the opposite. Historical data are your best defense. Claims to achieve something substantially better or cheaper than something similar done in the past may be shown to lack credibility based on physics. But quite often the new performance or cost claim is based on some new technology that is at least physically possible, but not fully developed today. If the performance claim is based on something that doesn't exist (say, a very low-mass tile that can withstand 2500°C for nearly 30 minutes for which there is no obvious means to attach to the titanium primary structure), then any cost estimate is a guess. Now, for the particular mission to be accomplished, with any hope of cost-effectiveness, just such a phenomenal tile may be needed. So the gamble to produce it must be taken or there is no system. But at this point, the decision is clear. Unless the magic tile is real, there is no option worth pursuing, so one bets on the magic tile and proceeds to plan a way to make the magic tile real, fully understanding that both more time and more money may be needed than originally anticipated, but also knowing that, if necessary, the effort can be abandoned if the difficulty of achieving the magic shows no sign of alleviating.

7. Establish the fewest and simplest equations that enable determining how well the candidate systems achieve the architecture ARM. If time is critical to the ARM, include in this model any and all delays associated with obtaining the information necessary for the candidate systems to be able to actually act, due to the need to obtain material or information from internal or external systems. Also include a representation of how the constituent systems will be tasked to engage in the mission. To the extent the candidate architecture may affect the environment in which its mission is to be conducted in such a manner as to affect the performance of the architecture, then this feedback must be incorporated into the equations. For the reasons we have stated for system modeling, avoid using utility functions graded by experts. A very good illustration of how to do this is provided by reference 13. In this paper, to avoid examining a large number of options, the authors utilized the Taguchi method, which will be examined in Chapter 3. The Taguchi method is very valuable if the cost to evaluate the options is very high. If the cost to evaluate the options is low, which is the case for the problem addressed in reference 13, then go ahead and evaluate all the options. Indeed, if the authors had evaluated all the options, they would have noticed in their Table 4 that trial 11 is better per all their criteria than the solution they pick using the Taguchi method, which is trial 24.

8. Establish the cost to obtain and operate the candidate systems while achieving the architecture design reference mission. All new systems require the expenditure of effort to design, produce, and test. When comparing existing and conceptual systems, the development costs for the new systems must be included in the cost estimate. The family of system costs could be as simple as adding up the individual system costs. But quite

often for architectures, some systems considered to include are provided by agencies distinct from the agencies looking to enhance the mission. How are the costs of these vital shared systems to be included? One must compare costs at the level truly borne. For example, suppose architecture option A consists of systems 1, 2, and 3, where 2 and 3 are provided by another agency, and our agency will pay to acquire system 1. Let architecture option B consists of system 4, which our agency must pay for. Suppose options A and B achieve the same mission value and have equal secondary considerations such as manpower required and technological maturity. If system 4 costs us more than system 3, we will prefer A, because with respect to our expenditures, we will get the same result at less expense. If the taxpayer is the true source for funding for all the systems, then we should make the decision including all the costs to the taxpayer. To do so, we must include the costs for systems 2 and 3 in the cost comparison, and if the sum of costs for systems 1, 2, and 3 exceeds the costs for system 4, now B is preferred. But, as systems 2 and 3 are also achieving other missions for their sponsors, in addition to contributing to accomplishing our mission, from the taxpayers' view systems 2 and 3 should be perceived as having more value than just contributing to our mission. So, if considering the perspective from the taxpayers' point of view, the architectural analysis must increase in scope to consider which systems are best to achieve both our mission and the mission that systems 2 and 3 are already contributing to. The recommended approaches for representing the costs of systems shared across mission areas are as follows:

Option 1. Determine and include in the cost estimate for a single mission what the external agencies will charge for using their systems. In our example, find out what the cost to use systems 2 and 3 would be and add this cost to the estimate for Option A.

Option 2. Increase the scope of the architecting analysis to include all the missions all the candidate systems might perform. To do this, recommend against creating some ponderously complicated reference mission, but rather continue to assess candidates against separate mission values.

9. Plot each candidate family of systems with respect to a vertical axis denoting mission goodness versus a horizontal axis of total cost of ownership. What will result is an efficient frontier of families of systems at a given cost for which no other family of systems can be found that provides superior mission goodness. Which of all the families of systems on or near the efficient frontier is preferred is partly a function of what can be afforded, but also the consideration of factors in addition to the single index of performance used to grade each candidate family of systems. If one is assessing candidate systems for multiple missions, include all the costs for all systems in one mission plot. If a system in that plot is also included as part of a family of systems on another mission plot, do not include the cost for that system on any other plots, except as may be needed to add more quantities

in the second plot than may be utilized in the first plot. This way the full cost of the system is accounted for in one mission, at it is essentially free for other missions. If the system shows up in a candidate family of systems near the efficient frontier for several missions, then it is clearly worth having. Otherwise, though the system is making multimission contributions, the contributions are insufficient in proportion to the cost relative to other family-of-system options.

2.3 MATHEMATICALLY MODEL UNCERTAINTY

Uncertainty can manifest itself in several ways. First, the underlying behavior may be random. That is, the system may initiate behavior from various starting points and reach an end point in a manner that is knowable probabilistically, as it is knowable that a die will land on one of its six sides. Second, the underlying behavior may be uncertain but bounded. For example, when predicting the cost to complete an item that has not previously been made, it may be possible to estimate the lowest cost likely and the highest cost likely, and guess as to the most likely cost. These costs have degrees of belief from 0 for the lowest and highest, to 1 for the most likely. This is not random behavior, as the costs are just uncertain with a range of possible values, or "fuzzy." Third, the uncertainty may be due to inherent complexity, or our ignorance regarding how to approximate the behavior in some aggregate means. Examples are prices for items in a large market, the behavior of a population in response to legislation, and global weather. In such cases, the uncertainty is in how to model the aggregate behavior of many individual elements for which behavior may be well understood. The following sections illustrate using *Mathematica*® to model uncertainty using regression, Monte Carlo analysis, fuzzy logic, agent-based programming, and fractals.

2.3.1 USING REGRESSION

Regression is the process of deriving an equation of independent variables to predict the values of dependent variables. The equation may be linear or nonlinear, with one or multiple independent and dependent variables. Typically, the equation is derived using many more data points than independent variables. A wonderful feature is that in addition to the equation, a confidence level associated with the likelihood the equation is correct is also produced. Regression results are best used for independent variable values within the range of those used to derive the result. Extrapolating beyond the range of the independent variables used to formulate the equations is a fool's bet. There are numerous books available to describe the process, and *Mathematica* provides routines to implement the methodology along with copious documentation and examples. A simple example follows, which will be revisited to show how fuzzy logic is used to achieve the same end. Reference 14 includes the following data pairs, relating an independent variable (the first of each pair) to a dependent variable (the second of each pair):

```
In[1]:= trainingdata = {{.09, .22},
                       {.28, .57},
                       {.34, .76},
                       {.65, 1.47},
                       {.67, 1.64},
                       {1.53, 0.14},
                       {1.66, -0.31},
                       {1.90, -1.40},
                       {2.35, -2.75},
                       {2.95, -1.24},
                       {3.30, 1.17},
                       {3.61, 3.23},
                       {3.64, 3.49},
                       {3.70, 4.30},
                       {3.91, 4.94},
                       {4.25, 3.72},
                       {4.37, 4.25},
                       {4.43, 3.16},
                       {5.12, -6.24},
                       {5.21, -8.01},
                       {5.42, -10.15},
                       {5.48, -6.48},
                       {5.72, -7.23},
                       {5.97, -5.79},
                       {7.50, 1.00}};
```

The *Mathematica* routine *ListPlot* enables visualizing the data:

```
In[2]:= ListPlot[trainingdata, PlotStyle → PointSize[0.02]]
```

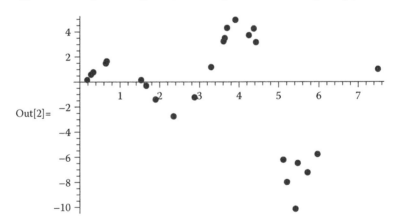

Out[2]=

Since there are 25 data points, up to 25 coefficients can be used to combine functions to fit the data. As there are at least five changes of slope for a curve through these data, at least a sixth-order polynomial is needed, which needs seven coefficients to be determined. *Mathematica* provides *NonlinearModelFit* to determine the fit function coefficients, which are saved in an output given the name **sixthorderfit** as follows:

```
In[3]:= sixthorderfit = NonlinearModelFit[trainingdata,
         c0 + c1 * x + c2 * x^2 + c3 * x^3 + c4 * x^4 + c5 * x^5 + c6 * x^6, {c0, c1, c2, c3, c4, c5, c6}, x]
```

Out[3]= FittedModel[$-3.76444 + \ll 6\gg + \ll 1\gg - 0.0297249\,x^6$]

The estimate, standard error, t statistic, and P-value of **sixthorderfit** are easily obtained:

In[4]:= **sixthorderfit["ParameterTable"]**

Out[4]=

	Estimate	Standard Error	t-Statistic	P-Value
c0	-3.76444	2.64303	-1.42429	0.171469
c1	28.5011	11.4696	2.48492	0.0230178
c2	-42.5329	14.1366	-3.00871	0.00754112
c3	23.8561	7.48463	3.18735	0.00510236
c4	-6.02547	1.92617	-3.12822	0.00580915
c5	0.694241	0.236332	2.93757	0.00880057
c6	-0.0297249	0.0110132	-2.69902	0.0146821

as is other information about the goodness of the fit using other *Mathematica* parameters. We can plot **sixthorderfit** against **trainingdata** using the *Mathematica* functions *Show* and *ListPlot*:

In[5]:= **Show[ListPlot[trainingdata, PlotStyle → PointSize[0.02]], Plot[sixthorderfit[x], {x, 0, 8}]]**

Out[2]=

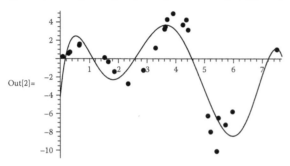

Not a bad fit. Would a seventh-order polynomial provide a better fit? To find out, here is the result called **seventhorderfit**:

In[6]:= **seventhorderfit = NonlinearModelFit[trainingdata,**
c0 + c1 * x + c2 * x^2 + c3 * x^3 + c4 * x^4 + c5 * x^5 + c6 * x^6 + c7 * x^7,
{c0, c1, c2, c3, c4, c5, c6, c7}, x]

Out[6]= FittedModel$\left[\boxed{1.89376 - 18.0367 x + \ll6\gg + \ll1\gg - 0.0295861 x^7}\right]$

Plotting **seventhorderfit** against the **trainingdata**:

In[7]:= **Show[**
ListPlot[trainingdata, PlotStyle → PointSize[0.02], PlotRange -> {-10, 10}],
Plot[seventhorderfit[x], {x, 0, 8}, PlotRange -> {-10, 10}]]

Out[7]=

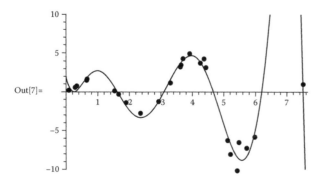

The seventh-order polynomial fits the data very well, except for the spike in the range of 6 to 8, as there are no data in that range to justify such a spike or absolutely negate the spike. The data themselves would seem to justify only a near-linear transition from the group of points in the range from 5 to 6 to the one point at about 7.5. This is a common issue regarding regression analysis. The most prudent course of action would be to get some data between 6 and 7. Alternatively, **seventhorderfit** can be used for 0 to 6, but a substitute regression should be prepared from the range of 5 to 8.

2.3.2 Using Monte Carlo Analysis

The Monte Carlo analysis method is now well known and routinely used. This section illustrates Monte Carlo analyses to enable comparison with the fuzzy logic method presented subsequently. Monte Carlo analysis starts with the deterministic equations used to model the system or architecture. The input parameters are modeled as random variables, that is, each input is presumed to have a known probability density function. Then a draw of each input variable is determined consistent with its stipulated probability density function. Then the equations are solved with the draw values for each of the inputs, and the results noted. Draws are repeated **N** times. Using the **N** results, the mean and variance for each dependent variable are calculated. Thus, for each dependent variable, the probability of being less than or greater than a specified value can be determined by adding or subtracting the calculated number of standard deviations to the calculated mean value. How many draws need to be conducted? A heuristic is at least 30 for each independent variable. So, if a problem has **M** independent random variables, at least 30 * **M** draws should be obtained before relying on the summary statistics for the dependent variables. To be prudent, one should experiment with the number of draws. The mean and variance of the dependent variables should be obtained using different numbers of draws differing by orders of magnitude. Then use the fewest number of draws for which the mean and variance have the desired precision for which no larger number of draws alters their values.

2.3.2.1 Multiple-Input Single-Output Monte Carlo Example

Our task is to develop a high-energy laser to be used to make holes into a material within a specified time with the key performance features of the system uncertain. *LaserTimeMargin* determines the margin in the time for a high-energy laser to achieve a half tear length of a specified size in a material requiring a specified flux to achieve the tear.

To illustrate using the Monte Carlo method to model this situation, all the independent variables are assumed to follow triangular distributions, except for the laser wavelength, which is assumed to be certain. Since there are 11 input variables, 30 * 11 = 330 draws will be defined. The potential variability for each of the 11 input variables is simulated using *Mathematica*'s *RandomReal*, with type set to *TriangularDistribution* with associated {{lowest, highest}, most likely values} as inputs. The twelfth input, laser wavelength, is presumed certain at $2.8*10^{-6}$ meters.

```
In[8]:= laserpower = 10^6 * RandomReal[TriangularDistribution[{.8, 1.2}, 1], 330];
        beamdiameter = RandomReal[TriangularDistribution[{2.36, 2.40}, 2.38], 330];
        Transmission = RandomReal[TriangularDistribution[{0.8, 0.9}, 0.85], 330];
        obscuration = RandomReal[TriangularDistribution[{0.6, 0.70}, 0.65], 330];
        sigmawfe = .14 * RandomReal[TriangularDistribution[{0.7, 1.4}, 1.0], 330];
        Rangetotarget = 3 * 10^5 + RandomReal[TriangularDistribution[{-20000, 20000}, 0], 330];
        sigmaj = 3 * 10^-7 * RandomReal[TriangularDistribution[{0.8, 3.16}, 1.8], 330];
        angle = RandomReal[TriangularDistribution[{-15 Degree, 15 Degree}, 0], 330];
        requiredlength = RandomReal[TriangularDistribution[{.087, .093}, .09], 330];
        Fluencerequired =
          5 * 10^6 + RandomReal[TriangularDistribution[{-129 400, 129 400}, 0], 330];
        timerequired = 10 + RandomReal[TriangularDistribution[{-.05, .05}, 0], 330];
        laserwavelength = 2.8 × 10^-6;
```

Then *LaserTimeMargin* is evaluated for each of the 330 draws with the result held in an array called **margin**:

```
In[19]:= margin = Table[LaserTimeMargin[{laserpower[[i]], beamdiameter[[i]], Transmission[[i]],
            obscuration[[i]], sigmawfe[[i]], Rangetotarget[[i]], sigmaj[[i]], angle[[i]],
            requiredlength[[i]], Fluencerequired[[i]], timerequired[[i]], laserwavelength}],
          {i, 1, 330}];
```

A histogram is used to plot the 330 margin values:

```
In[168]:= Histogram[margin]
```

Out[168]=

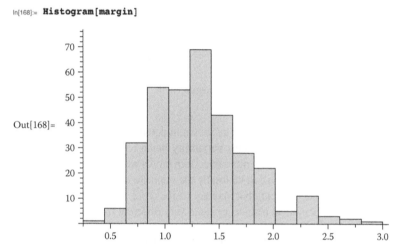

What is important is the probability to achieve a margin of at least 1, that is, the laser will succeed to produce the fluence for a sufficient time to produce the desired tear length. This can be approximately determined by dividing the number of draws that resulted in margins less than 1 to the total number of draws:

```
In[169]:= N[Count[Table[margin[[i]] >= 1, {i, 1, 330}], True] / Dimensions[margin]]
```

Out[169]= {0.718182}

So there is better than about a 70% chance the needed margin will be obtained.

The value of the method is plain. For relatively little work, the possible range of outputs is determined. The weaknesses of the method are also plain. This analysis concludes there is about a 30% chance of failure. Is that acceptable? Deciding will be difficult for people uncomfortable with probability. There are a few more subtle problems that can result in the perfectly good analysis being a perfectly inappropriate model of reality. First, clearly, if we underestimated the input uncertainty, we will underestimate the

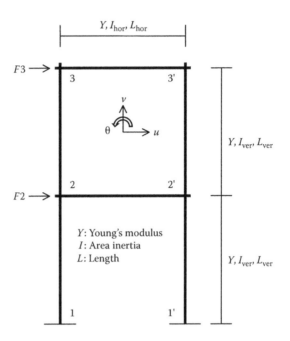

FIGURE 2.8   Two-story planar frame structure with two horizontal forces.

output uncertainties. Worse, as will be shown in detail in Section 2.3.3.1, the mathematics results in the dependent variable variability being reduced by the quantity of input variables. So, paradoxically, the more uncertain inputs we include, the less uncertainty in the result that we get, which can be totally the opposite of reality. Finally, we presumed the inputs are all independent; that may not be the case, and that we failed to model this can mean the mathematical results are again very different from what reality will be.

2.3.2.2   System of Linear Equations: Monte Carlo Example

Another very common mathematical model for a system or architecture is a system of linear equations. To illustrate this application, the displacement for a planar, rigid, two-story space frame is modeled for horizontal loads (see Figure 2.8). The structure actually has eight degrees of freedom, but symmetry allows the analysis to be done using only four degrees of freedom, two horizontal displacements, and two rotations. *TwoStoryFrameDisplacements* obtains the four displacements.

Again, for illustration purposes, each input variable is presumed to be modeled by a triangular distribution with {{lowest, highest}, most likely} values, and $7 \times 30 = 210$ draws are obtained.

```
In[170]:= F2 = RandomReal[TriangularDistribution[{.8, 1.2}, 1.0], 210];
          F3 = RandomReal[TriangularDistribution[{1.2, 1.8}, 1.5], 210];
          Y = RandomReal[TriangularDistribution[{14.9, 15.1}, 15.], 210];
          Ihor = RandomReal[TriangularDistribution[{.99, 1.01}, 1.0], 210];
          Lhor = RandomReal[TriangularDistribution[{14.99, 15.01}, 15.0], 210];
          Iver = RandomReal[TriangularDistribution[{49, 51}, 50], 210];
          Lver = RandomReal[TriangularDistribution[{24.9, 25.1}, 25], 210];
```

Then using *TwoStoryFrameDisplacements*, the output vector, **v,** is determined for each of the input cases.

In[177]:= **v = Table[TwoStoryFrameDisplacements[F2[[i]], F3[[i]],**
    Y[[i]], Ihor[[i]], Lhor[[i]], Iver[[i]], Lver[[i]]], {i, 1, 210}];

The *Mathematica* routines *Mean* and *StandardDeviation* are used to calculate the mean and standard deviation for each of the displacement vector components:

In[178]:= **Mean[v]**

Out[178]= {24.1626, 0.0829024, 39.1047, 0.0315959}

In[179]:= **StandardDeviation[v]**

Out[179]= {1.45188, 0.00543664, 2.55797, 0.00258879}

The first and third numbers in each array are for the first- and second-floor ceiling translations due to the loads applied to the structure. The second and fourth numbers in each array are for the two rotations, which will occur at the intersection of the vertical and horizontal beams. A histogram enables visualizing the distribution for the 210 trials for the two translations:

In[180]:= **Histogram[Table[v[[i, 1]], {i, 1, 210}]]**

Out[180]=

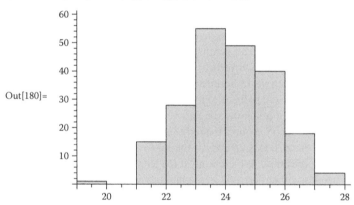

In[181]:= **Histogram[Table[v[[i, 3]], {i, 1, 210}]]**

Out[181]=

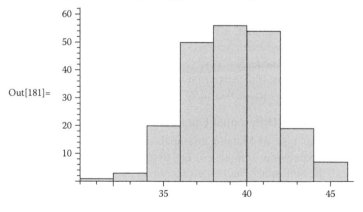

Since, by nature of the model, each of the output vector components is the result of linear relationships, there is a high degree of correlation between any two of them. The routine *Correlation* returns the correlation coefficient for two vectors. In this case, we'll determine it for the horizontal displacements of the second floor and ceiling, as follows:

In[182]:= **Correlation[Table[v[[i, 1]], {i, 1, 210}], Table[v[[i, 3]], {i, 1, 210}]]**

Out[182]= 0.966377

As expected, the results are almost perfectly correlated. This is true for all Monte Carlo analyses based on simultaneous linear equations. This strong correlation between the outputs usually allows conclusions to be made by looking at as few as one variable, rather than every output, since all the variables are so highly correlated, one is a good proxy for them all.

How appropriate is Monte Carlo analysis for this type of problem? Inputs such as loads certainly could be random. They could also just be unknown, but certain to be in a range. Presuming a random load for an unknown load with a range of value is a bad model. What is most important may be the behavior under the worst-case loads, which can be modeled deterministically. Structural member material properties and dimensions will very likely vary randomly in a manner that can be determinable by sampling. But the statistics will vary with vendor and production lot, so care is needed to use the statistics that match the actual items. Perhaps the most serious potential flaw in the model is the presumed symmetry. An actual two-story frame fixed at its base has 12 degrees of freedom, a vertical and horizontal displacement, and one rotation at each the four intersections of horizontal and vertical members. For our illustration, the degrees of freedom were reduced to four by limiting the model to two horizontal forces, which in turn allowed assuming the four vertical displacements were all zero, and noting structural symmetry ensures the right-hand side of the structure will move the same way as the left-hand side. Certainly if the load magnitude can vary, then so could the angle of load application. And certainly if the lengths of the members can vary, then it is likely no beam is either perfectly horizontal or vertical. These differences can be modeled by introducing direction cosines for each element of the structure and are important to include if one wants to accurately predict system behavior. The model assumes all of the beams were put in place unloaded, that is, none was pulled or pushed prior to being connected. If any of the beams were indeed forced into position, then this model provides incorrect predictions. If the amount any beam was preloaded or deformed is known, that could be modeled. And so it goes, if the loads either are impulsive or change with time, a different model is needed that includes the mass as well as the stiffness of the structural members. If the structural members may be heated dissimilarly, a different model is needed to include thermally inducted displacements and loads.

2.3.2.3   System of Differential Equations: Monte Carlo Example

To illustrate this type of Monte Carlo analysis, the flight of a two-stage rocket to achieve a stipulated orbit is modeled. The first stage flies an open-loop steering profile through the atmosphere and is subject to two environmental dispersions, both of which vary with altitude (atmospheric density and wind) and eight system dispersions (initial mass, first-stage vacuum thrust, first-stage specific impulse, axial and normal aerodynamic coefficients, initial propellant mass, first-stage inert mass, and first-stage thrust vector misalignment). The second stage flies an open-loop linear tangent steering law and is subject to three system dispersions (second-stage vacuum thrust, second-stage specific impulse, and second-stage thrust vector misalignment), so the number of Monte Carlo draws (**numbermcr**) will be $2 + 8 + 3 = 13 * 30 = 390$:

In[2]:= **numbermcr = 390;**

Motion is constrained to be in a plane. The Earth is assumed to be a spherical body with a uniform gravitational force dependent on the distance from the center of the Earth. The constants needed are the gravity constant mu in m^3/sec^2, the spherical radius of the earth, **rearth** in meters, and **g = mu / rearth^2**:

In[3]:= **mu = 3.986005 × 10^14;**
rearth = 6 378 140.;
g = mu / rearth^2;

To approximate the effect of atmospheric density and winds on the first-stage trajectory, randomly generated density and wind profiles are needed. These can be obtained by direct measurement, or synthetically derived if the mean and variance are known. The basic process to generate a synthetic profile is to find the mean and the one sigma possible deviation. The synthetic profile is then the mean plus a normally distributed random number times the standard deviation.

First, we model a synthetic density dispersion. Generally, cold gases are denser than warm gases. So, for a column of air, if the air near the earth is denser for a cold day, then the air at altitude must generally be less dense; otherwise, the denser higher altitude air will sink down to the lower altitudes. This means the difference between a 3 sigma cold density and a 3 sigma hot density won't be a physically realistic atmospheric density at altitudes. So rather than using the difference of two extremes, only one extreme is used. Reference 15 provides a graph of the 2 sigma cold density, for which there is only a 2% chance for the air to be denser. The variable **density2sigmacoldvalues** holds data read from the plot in {altitude in meters, density impact in kg/meter^30} as follows:

In[6]:= **density2sigmacoldvalues =**
{{0., .05},
{5000., 0.02},
{10 000., -0.03},
{15 000., -0.05},
{20 000., -0.04},
{25 000., -0.03},
{30 000., -0.03},
{40 000., -0.06},
{50 000., -0.10},
{60 000., -0.12},
{70 000., -0.11},
{80 000., -0.10},
{90 000., -0.0},
{100 000., .10}};

The *Mathematica* routine *Interpolation*, with **InterpolationOrder** set to 3 so cubic splines are fit between the data, is used to define **density2sigmacold**, which is then plotted versus altitude as follows:

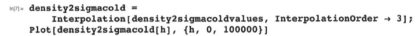

```
In[7]:= density2sigmacold =
            Interpolation[density2sigmacoldvalues, InterpolationOrder → 3];
        Plot[density2sigmacold[h], {h, 0, 100000}]
```

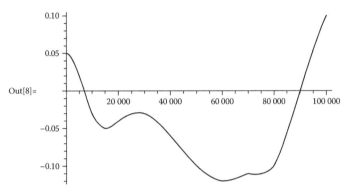

To create the synthetic random density, random multiples of 0.5***density2sigma-cold** (an approximation of one standard deviation) are added to the mean profile. The mean profile is assumed to be

$$\text{IF } \mathbf{r} \geq \mathbf{rearth} + 100000 \text{ OR } \mathbf{r} < \mathbf{rearth} \text{ THEN } \mathbf{density} = 0$$

$$\text{IF } \mathbf{r} \geq \mathbf{rearth} \text{ THEN } \mathbf{density} = 1.225 * Exp[-h[\mathbf{r}] / 8600]$$

The random multiples are determined from a normal distribution with mean 0 and 1 as the standard deviation. The random draws are recorded in **densityfactorset**:

```
In[9]:= densityfactorset = RandomReal[NormalDistribution[0, 1], numbermcr];
```

Assuming the rocket launches due east, a wind with an azimuth of 270° is a head wind (that is, winds blowing toward the west are flowing against the motion of the rocket), so such a wind has a negative sign relative to a position axis that is positive due east from the launch site. A wind with an azimuth of 90° is a tail wind (that is, winds blowing due east are flowing in the same direction as the rocket, so such a wind has a positive sign relative to a position axis that is positive due east from the launch site). To construct a synthetic wind versus altitude in the plane of the rocket's motions, we need the mean wind, a representation of the standard deviation for the wind, and a way to model gusts. The following illustrate how to construct random synthetic winds versus altitudes. To start, the 95% head wind data points are read off the reference graph in reference 15, which is in {feet, feet/second}, converted to {meters, meter/second} by multiplying by 0.348 meters/foot, and stored in the variable **windspeed95percentat 270az**:

In[10]:= **windspeed95percentat270az = 0.3048 ***
```
{{0,  20},
{10 000,  -110},
{20 000,  -170} ,
{30 000,  -230},
{40 000,  -270},
{50 000,  -210},
{60 000,  -120},
{70 000,  -65},
{80 000,  -65},
{90 000,  -90},
{100 000,  -120},
{110 000,  -170},
{120 000,  -190},
{130 000,  -220},
{140 000,  -240},
{150 000,  -270},
{160 000,  -280},
{170 000,  -300},
{180 000,  -315},
{190 000,  -340},
{200 000,  -395},
{210 000,  -360},
{220 000,  -330},
{230 000,  -295},
{240 000,  -120},
{250 000,  0}};
```

The *Mathematica* routine *Interpolation*, with **InterpolationOrder** set to 3, is used to obtain a cubic spline fit of the data to define the **windspeed95at270** profile versus altitude of the 95% wind at 270° azimuth. The result is plotted for altitudes from 0 to 250,000 feet:

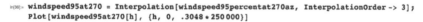

In[36]:= **windspeed95at270 = Interpolation[windspeed95percentat270az, InterpolationOrder -> 3];**
Plot[windspeed95at270[h], {h, 0, .3048 * 250000}]

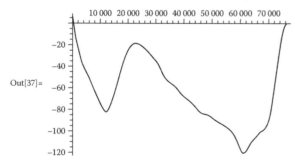

The same is done to define an interpolation for the 95% wind at 90° azimuth:

```
In[38]:= windspeed95percentat90az = 0.3048 *
            {{0, 20},
            {10 000, 30},
            {20 000, 20} ,
            {30 000, 20},
            {40 000, 20},
            {50 000, 25},
            {60 000, 10},
            {70 000, 40},
            {80 000, 30},
            {90 000, 50},
            {100 000, 35},
            {110 000, 40},
            {120 000, 45},
            {130 000, 55},
            {140 000, 45},
            {150 000, 45},
            {160 000, 30},
            {170 000, 25},
            {180 000, 40},
            {190 000, 50},
            {200 000, 40},
            {210 000, 35},
            {220 000, 35},
            {230 000, 30},
            {240 000, 10},
            {250 000, 0}};
        windspeed95at90 = Interpolation[windspeed95percentat90az, InterpolationOrder -> 3];
        Plot[windspeed95at90[h], {h, 0, 250 000 * 0.3048}]
```

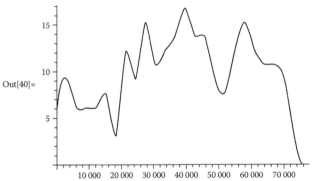

Out[40]=

The mean wind data are read off the reference graph (in {feet, feet/second}, converted to {meters, meters/second}), and for the third time *Interpolation* is used to create **windspeedmean** to fit the mean wind data.

```
In[13]:= windspeedmeanhv = 0.3048 *
            {{0, 5},
            {10 000, -40},
            {20 000, -75},
            {30 000, -110},
            {40 000, -145},
            {50 000, -110},
            {60 000, -50},
            {70 000, -10},
            {80 000, -5},
            {90 000, -10},
            {100 000, -45},
            {110 000, -65},
            {120 000, -70},
            {130 000, -80},
            {140 000, -95},
            {150 000, -115},
            {160 000, -130},
            {170 000, -145},
            {180 000, -170},
            {190 000, -190},
            {200 000, -205},
            {210 000, -180},
            {220 000, -160},
            {230 000, -145},
            {240 000, -60},
            {250 000, 0}}};
windspeedmean = Interpolation[windspeedmeanhv, InterpolationOrder -> 3];
Plot[windspeedmean[h], {h, 0, 250 000 * 0.3048}]
```

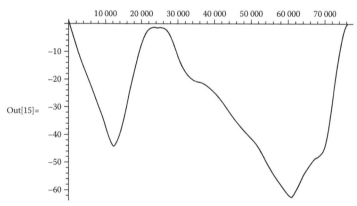

The variable **windleveset** holds the Monte Carlo draws for formulating the synthetic wind. Each is generated from a normal distribution with mean 0 and a standard deviation of 1:

```
In[16]:= windlevelset = RandomReal[NormalDistribution[0, 1], numbermcr];
```

In this case, the reference provided the extreme values as percentiles. To determine the corresponding number of standard deviations, *Mathematica* provides the routine *InverseCDF*. For a normal distribution, 95% is found to be about 1.64 standard deviations.

In[26]:= **InverseCDF[NormalDistribution[0, 1], .95]**

Out[26]= **1.64485**

The wind **speedataltitude** is the mean wind plus the chosen **windleveset** value times the standard deviation of the wind speed, which is approximated by taking the difference of the 95% wind with an azimuth of 270°, minus the 95% wind with an azimuth of 90°, divided by 2*1.64. Here is an example with **mcr** set to 1 so we use the first set of the random inputs:

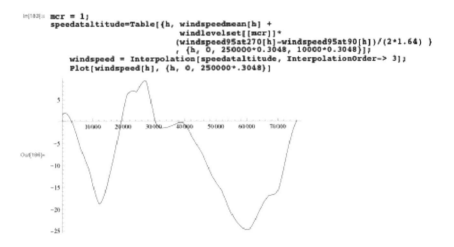

```
In[182]:= mcr = 1;
         speedataltitude=Table[{h, windspeedmean[h] +
                        windlevelset[[mcr]]*
                        (windspeed95at270[h]-windspeed95at90[h])/(2*1.64) }
                      , {h, 0, 250000*0.3048, 10000*0.3048}];
         windspeed = Interpolation[speedataltitude, InterpolationOrder-> 3];
         Plot[windspeed[h], {h, 0, 250000*.3048}]
```

Real winds also have gusts, locally occurring increases and decreases in velocity. To model the gusts, at 5000-feet intervals, a normally distributed random variable is added with a zero mean and a standard deviation of 3 meters/second[16]. First, the random variables needed to produce the gusts are set for each of 51 data points from 0 to 250,000 feet and are stored in **gustamountset**:

The array **speedwithgusts** are the wind speed values with the **gustamontset** values added at 5000 foot intervals. An illustration for the first draw (**mcr** = 1) follows, this time with **InterpolationOrder** set to 2, so a quadratic is fit to the data to make the result less smooth than if used cubic splines:

In[187]:= **gustamountset = Table[Random[NormalDistribution[0, 3]], {numbermcr}, {51}];**

The array **speedwithgusts** are the wind speed values with the **gustamountset** values added at 5000 foot intervals. An illustration for the first of these draws (**mcr** = 1) follows. This time with *InterpolationOrder* set to 1, so a quadratic is fit to the data to make the results less smooth than if used cubic splines:

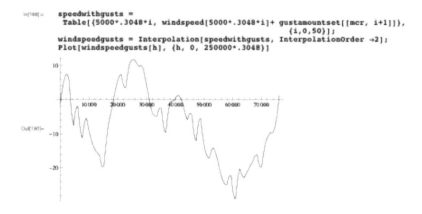

```
In[188]:= speedwithgusts =
          Table[{5000*.3048*i, windspeed[5000*.3048*i]+ gustamountset[[mcr, i+1]]},
                                                              {i,0,50}];
          windspeedgusts = Interpolation[speedwithgusts, InterpolationOrder →2];
          Plot[windspeedgusts[h], {h, 0, 250000*.3048}]
```

The simulation starts after the vehicle has cleared a hypothetical tower and has obtained some initial height (**hinitial**, in meters) and horizontal and vertical inertial speeds (**uinitial**, **vinitial**, in meters per second):

```
In[48]:= hinitial = 60.;
         uinitial = 25 * Cos[1.5];
         vinitial = 25 * Sin[1.5];
```

Though certainly these values could vary for each launch, there is little value in simulating them as random variables, since whatever small variation may exist at this point will be overwhelmed by the environmental and system dispersions that will manifest themselves subsequently. However, the movement of the vehicle in the first few seconds is extremely important, particularly if there is a tower nearby that might be hit. If that is the issue of concern, then the simulation should model just these first few moments with the fidelity needed to determine all locations on the vehicle relative to the tower.

For our example, the mean values for the system dispersions are as used in reference 16. The standard deviations are the same percentage of mean as exhibited by the Space Shuttle.[17] The following physical attributes of the first stage are the assumed constant engine exit area (**exitareastg1**, in meters squared) and the reference area for calculating aerodynamics forces on the vehicle (**s** in meters squared):

```
In[51]:= exitareastg1 = 19.115;
         s = 55.18;
```

Both the exit area and the reference area of the rocket could be uncertain. But, since both **exitareastg1** and **s** multiply other inputs, we might as well presume all the sample data are purely the result of the other values, since both these variables are essentially impossible to measure directly during the test firing of a rocket.

The nine first-stage system dispersions are initial mass (**massinitialset**, in kilograms), vacuum thrust (**thrustfvacstg1set**, in Newtons), specific impulse (**ispstg1set**, in seconds), axial aerodynamic coefficient (**caset**, unit-less), normal aerodynamic coefficient (**cnset**, unit-less), mass of the propellant loaded (**masspropfsinitial**, in kilograms), time of staging (**tstage**, in seconds; since this is when the first-stage propellant mass is all consumed, it is not an independent random variable, but must be calculated using the equation **mass / thrust / (g* ispstg1set)**, the inert mass of the first stage (**stagingmassloss**, in kilograms), and the first-stage thrust vector misalignment (**fstvmaset**, in degrees). Each of the source dispersions is chosen to be modeled as a normal distribution. *Mathematica* provides the routine *NormalDistribution*[**mean**, **1sigma**] to randomly generate numbers consistent with a normal distribution of mean m and standard deviation **1sigma**.

```
In[53]:=  massinitialset = RandomReal[NormalDistribution[890 150., .001 * 890 150.], numbermcr];
          thrustvacstg1set =
              RandomReal[NormalDistribution[1.2975 × 10^7, .00425 * 1.2975 × 10^7], numbermcr];
          ispstg1set = RandomReal[NormalDistribution[430.55, 2.3 / 3], numbermcr];
          caset = RandomReal[NormalDistribution[0.3, 0.0015], numbermcr];
          cnset = RandomReal[NormalDistribution[3.1, 0.15], numbermcr];
          masspropfsinitial = RandomReal[NormalDistribution[599 748., .001 * 599 748.], numbermcr];
          tstage1 = masspropfsinitial / (thrustvacstg1set / (g * ispstg1set));
          stagingmassloss = RandomReal[NormalDistribution[29 920., .001 * 29 920], numbermcr];
          fstvmaset = RandomReal[NormalDistribution[0, 0.25 Degree], numbermcr];
```

As second-stage flight is above the atmosphere, the equations of motion simplify and we are concerned only with the second-stage system dispersions. These are vacuum thrust for the second stage (**thrustvacstg2set**, in Newtons), specific impulse for the second stage (**ispstg2set**, in seconds), and second-stage thrust vector misalignment (**sstvmaset**, in degrees). As with the first stage, the nozzle exit area (**exitarestg2**) is presumed constant, since in the equations of motion it is multiplied by other variables, which are easier to model as random variables.

```
In[62]:=  thrustvacstg2set =
              RandomReal[NormalDistribution[2.595 × 10^6, .00425 * 2.595 × 10^6], numbermcr];
          ispstg2set = RandomReal[NormalDistribution[430.55, 2.3 / 3], numbermcr];
          sstvmaset = RandomReal[NormalDistribution[0, 0.25 Degree], numbermcr];
          exitareastg2 = 3.823;
```

Now we need control laws to steer each rocket stage. For the first stage, there is no closed-form solution for the optimal profile. The stage flies open loop; the thrust angle is stipulated using a fourth-order Hermite polynomial in time. For the second stage, there is a well-known approximate optimal control law that the tangent of the thrust angle varies linearly with time, that is:

tan[**thrust_angle**] = **constant_1 + constant_2 * time**

The first stage thrusts until all propellant is consumed. For the second stage, the objective is to achieve a specified orbit, which is uniquely determined by the final horizontal and vertical velocities as well as the altitude. To achieve these three conditions, three controls are needed. Two are the coefficients associated with the linear tangent steering law. The third is the time to burn propellant during the second-stage flight. So the control variables are a seven-element vector: four constants to multiply times Hermite polynomials to steer the first stage, two constants for the linear tangent function to steer the second stage, and the burn time for the second stage. Here is a possible set where the first four numbers are for the constant, linear, quadratic, and cube Hermite polynomials; the fifth is **constant_1**; the sixth is **constant_2**; and the last is the burn time for the second stage:

In[66]:= **coef** = {0.770768,-0.343217,-0.0480433,0.0166494,0.47862,-0.00162042,288.136};

FirstStage simulates the first-stage behavior. The final-stage vector (mass, altitude, horizontal, and vertical velocities) comprises the initial conditions for the second-stage flight. *SecondStage* simulates the behavior of the second stage. *RocketTraj* executes the first- and then the second-stage calculations. Two utility formal requirements will be useful. *FSOut* plots the time history of first-stage trajectory parameters. *SSOut* plots the second-stage trajectory parameters.

Before performing the Monte Carlo analysis, the trajectory for the first set of system and environmental dispersion values is determined. The controlled variables, **coef**, is printed to record what the inputs are, then each of the system dispersions is assigned the first value from the random set previously defined:

```
In[67]:= coef
        thrustvacstg1 = thrustvacstg1set[[1]]
        ispstg1 = ispstg1set[[1]]
        ca = caset[[1]]
        cn = cnset[[1]]
        fstvma = fstvmaset[[1]]
        tstage1[[1]]
        stagingmassloss[[1]]
        massinitial = massinitialset[[1]]
        thrustvacstg2 = thrustvacstg2set[[1]]
        ispstg2 = ispstg2set[[1]]
        sstvma = sstvmaset[[1]]
```

Out[67]= {0.770768, -0.343217, -0.0480433, 0.0166494, 0.47862, -0.00162042, 288.136}

Out[68]= 1.29141×10^7

Out[69]= 429.935

Out[70]= 0.300158

Out[71]= 2.77579

Out[72]= 0.00331536

Out[73]= 195.786

Out[74]= 29 906.

Out[75]= 889 248.

Out[76]= 2.59685×10^6

Out[77]= 431.218

Out[78]= 0.00347663

RocketTraj is run to integrate the first- and second-stage equations of motion. The output is an array of the form

{first_stage_output, second_stage_output}

The **first _stage _output** =

{maximum_dynamic_pressure, final_mass, altitude, horizontal_inertial _velocity, vertical_inertial_velocity, flight_path_angle, maximum_angle_of_ attack, minimum_angle_of_attack}

The **second_stage_output** =

{mass, altitude, inertial_horizontal _velocity, inertial_vertical_velocity, flight_path_angle}

```
In[226]:=  RocketTraj[coef, tstage1[[1]], stagingmassloss[[1]]]
Out[226]=  {{28 303.8, 289 052., 91 327.4, 3217.59, 575.969, 10.1488, 11.166, -14.5209},
           {82 054.5, 151 178., 7905.77, 33.3819, 0.241928}}
```

FSOut plots the time histories of first-stage trajectory parameters. The first stage ends at **tstage1[[1]]**:

```
In[658]:=  FSOut[solstage1, tstage1[[1]]]
```

First Stage

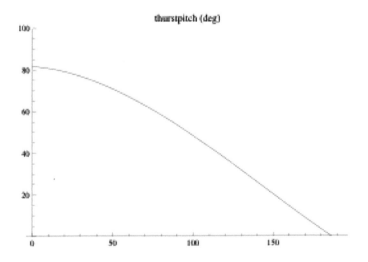

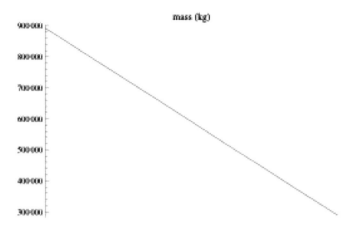

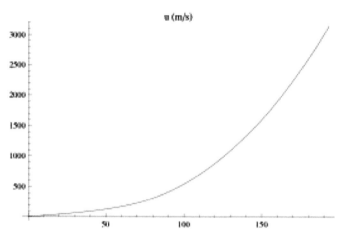

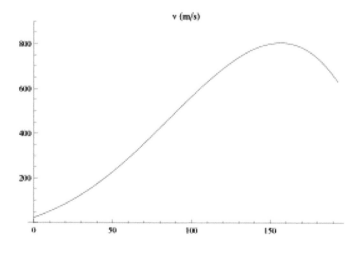

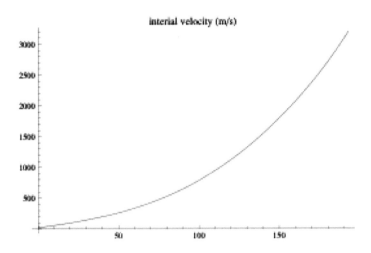

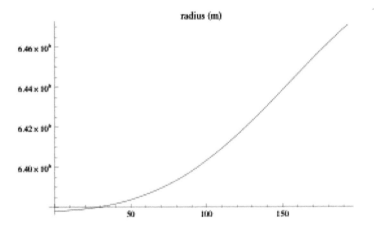

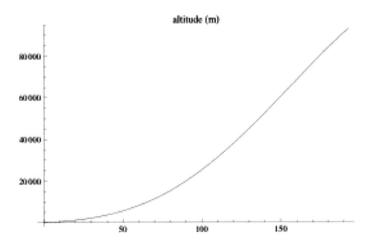

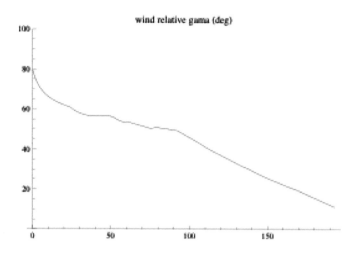

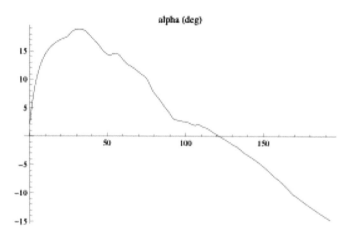

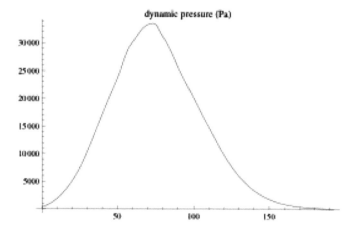

SSOut plots the time histories of second-stage trajectory parameters. The second-stage duration is **coef[[7]]**:

Second Stage

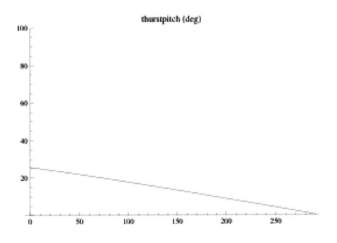

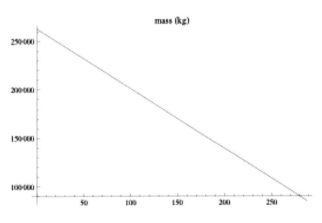

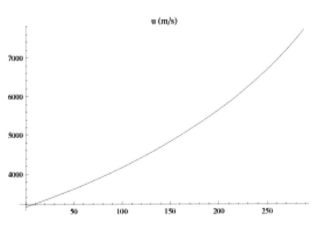

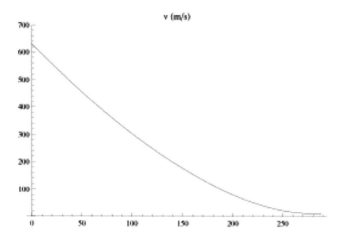

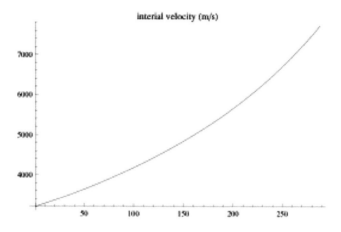

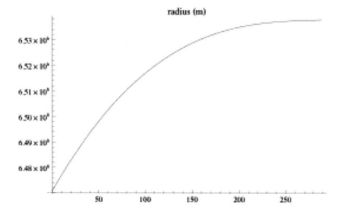

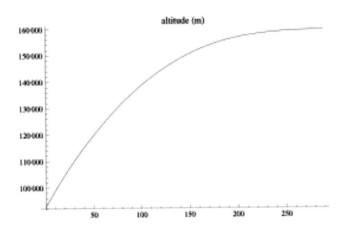

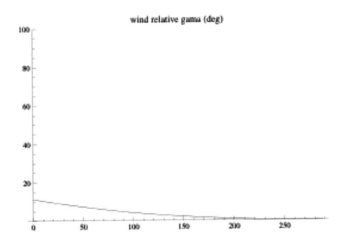

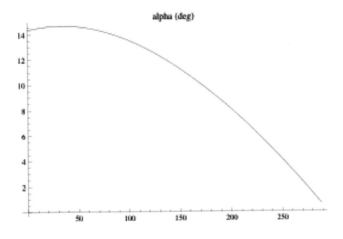

To perform the Monte Carlo analysis, the formal requirements *OpenLoopEnvandSysDispersionCase* evaluate the **mcr** random draws and report the resulting mean and standard deviation obtained for the resulting trajectories.

Finally, by executing the *OpenLoopEnvandSysDispersionCase*, the mean and standard deviation of the impact of the environmental and system dispersions are estimated:

In[229]:= **OpenLoopEnvandSysDispersionCase[coef, numbermcr]**

```
Staging Results
Mean[Max q]32 653.7
SD[Max q]1707.96
Mean[h_at_staging]92 163.7
SD[h_at_staging]1963.61
Mean[u_at_staging]3187.43
SD[u_at_staging]24.6391
Mean[v_at_staging]584.693
SD[v_at_staging)]39.0289
Mean[speed_at_staging]3240.88
SD[speed_at_staging)]19.9944
Mean[gama_at_staging]10.3971
SD[gama_at_staging]0.735548
Mean[Max_alpha]18.2478
SD[Max_alpah]3.1306
Mean[Min_alpha]-15.0218
SD[Min_alpha]1.11224


State Vector at Nominal t_final
Mean[mass_at_t_tfinal]83 333.2
SD[mass_at_t_tfinal]1291.6
Mean[h_at_t_final]149 442.
SD[h_at_t_final]13 153.7
Mean[u_at_t_final]7832.62
SD[u_at_t_final]61.2789
Mean[v_at_t_final]-2.18773
SD[v_at_t_final]44.8946
Mean[speed_at_t_final]7832.75
SD[speed_at_t_final]61.2637
Mean[gama_at_t_final]-0.016269
SD[gama_at_t_final]0.328702
```

Here are a few examples of how these results can be used. Since the maximum dynamic pressure standard deviation is about 1700 Pa, if one wanted 3 sigma certainty that the vehicle never exceeded a maximum dynamic pressure of 35,500 Pascals, one needs to design the mean trajectory to be 35,500 – 3 * 1700 Pa. Also note that standard deviation of the mass at **t_final** is about 1300 kg. If one wants to be certain one has at least 3 sigma protection that they will not run out of propellant prior to achieving the orbit condition, then 3*1370 kg of propellant must be stored in reserve, and the useful payload would be the mean mass at **t_final** – 3*1370, the second-stage inert mass. Finally, since the final orbit is a function of the altitude and velocities (or speed and flight path angle) at **t_final**, the mean and standard deviations of these terms enable us to predict the mean and standard deviations of the resulting orbits that we will achieve. Clearly, the Monte Carlo analysis method can be a very valuable means to make decisions in the presence of random uncertainties.

2.3.3   USING FUZZY LOGIC

Fuzzy numbers are an alternate way to model uncertainty.[14,18,19] If the inputs or model parameters are uncertain but not random, then fuzzy logic modeling is preferred to using Monte Carlo analysis since the representation of the uncertainty is truer. Fuzzy numbers have a range: the wider the range, the more uncertainty. Fuzzy numbers have a degree of belief between zero and one associated with each point in the range. The degree of belief is not a probability of occurrence. The degree of belief curve does not need to integrate to 1, and integrating it does not produce a cumulative density function. There is no probability associated with the extremes of the range, or any point within the range.

For the triangular fuzzy number

$$<\mathbf{l, m, h}>$$

the degree of belief is zero from minus infinity to **l** and from **h** to positive infinity. The degree of belief at **m** is 1, from which the degree of belief varies linearly to zero at both **l** and **h**. Using triangular fuzzy numbers, the number "5" might be represented as <3, 5, 9>, <4, 5, 6>, <5, 5, 5.5>, or <5, 5, 5>, depending on the potential values the number might have other than exactly 5. A trapezoidal fuzzy number is denoted as

$$<\mathbf{l, m1, m2, h}>$$

where again the degree of belief is 0 from minus infinity to **l** and from **h** to positive infinity and is 1 between **m1** and **m2**, varying linearly from 0 to 1 between **l** and **m1** and linearly from 1 to 0 from **m2** to **h**. In general, for fuzzy numbers, the degree of belief can vary in any manner appropriate from 0 to 1, but it will be numerically convenient if the degree of belief is unimodal over the uncertain range.

Triangular fuzzy numbers are popular because the resulting linear fuzzy arithmetic is so simple for linear operations. For triangular fuzzy numbers of the form **<l, m, h>**:

$$a * <l, m, h> + b = <al + b, am + b, ah + b>$$

$$<l1, m1, h1> + <l2, m2, h2> = <l1 + l2, m1 + m2, h1 + h2>$$

$$<l1, m1, h1> * <l2, m2, h2> = <l1 * l2, m1 * m2, h1 * h2>$$

If a deterministic number is needed to implement the result of an analysis, then the fuzzy output must be defuzzified. There are many possible ways to formally defuzzify a number. A convenient geometric argument is for the centroid of the fuzzy range to be an appropriate deterministic representation. This is particularly convenient for triangular fuzzy numbers of the form **<l, m, h>**, for which the centroid and thus the defuzzified values are $(l + m + h)/3$.

Evaluating nonlinear functions of fuzzy numbers requires an iterative evaluation of the function at intermediate degrees of belief values, not just at degrees of belief of 0 and 1. The formal requirement *fuzFoffuzX* determines the fuzzy result of an arbitrary function of fuzzy numbers, and is based on the algorithm presented by Hanss,[20] here modified to work for a multivalue function.

First let's use *fuzFoffuzX* to evaluate a simple linear sum of three fuzzy numbers:

```
In[231]:= F[p_] := p[[1]] + p[[2]] + p[[3]];
          ps = {{1, 2, 3}, {4, 4.5, 5}, {0, 1, 2}};
```

For the interim evaluation points set to 1:

```
In[233]:= fuzFoffuzX[ps, 1, F]
```

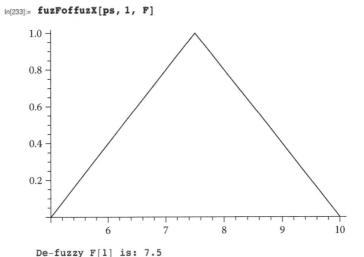

```
De-fuzzy F[1] is: 7.5

The minimum value of F[1] is: 5.

The maximum value of F[1] is: 10.
```

which is the expected result, the sum of the low, degree of belief of 1, and high values. Notice, since this $F[\mathbf{p}]$ is a linear combination of fuzzy numbers, we get the same result if we increase the number of interim evaluations to 5 or 10:

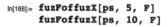

In[168]:= **fuzFoffuzX[ps, 5, F]**
         fuzFoffuzX[ps, 10, F]

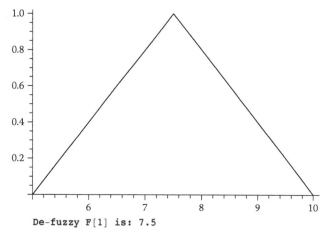

De-fuzzy F[1] is: 7.5

The minimum value of F[1] is: 5.

The maximum value of F[1] is: 10.

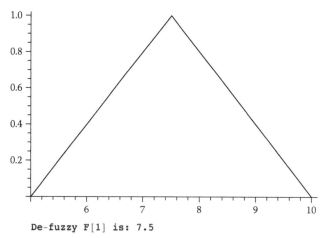

De-fuzzy F[1] is: 7.5

The minimum value of F[1] is: 5.

The maximum value of F[1] is: 10.

Now we'll use *fuzFoffuzX* to evaluate the simple nonlinear function:

$$F[\mathbf{p}] = (\mathbf{p1} + \mathbf{p2})/\mathbf{p1},$$

with $\mathbf{p1} = <1, 2, 3>$ and $\mathbf{p2} = <4, 4.5, 5>$, for the number of intermediate values of 1, 5, and 10:

```
In[170]:= F[p_] := (p[[1]] + p[[2]]) / p[[1]];
         ps = {{1, 2, 3}, {4, 4.5, 5}};
         fuzFoffuzX[ps, 1, F]
         fuzFoffuzX[ps, 5, F]
         fuzFoffuzX[ps, 10, F]
```

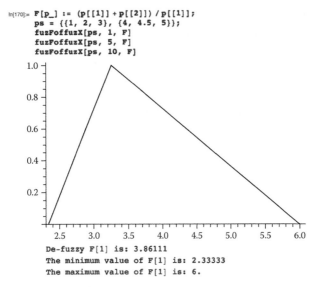

De-fuzzy F[1] is: 3.86111
The minimum value of F[1] is: 2.33333
The maximum value of F[1] is: 6.

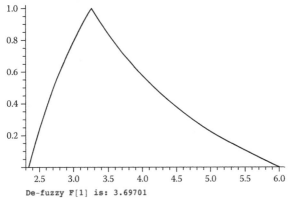

De-fuzzy F[1] is: 3.69701
The minimum value of F[1] is: 2.33333
The maximum value of F[1] is: 6.

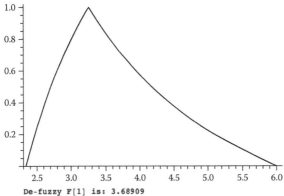

De-fuzzy F[1] is: 3.68909
The minimum value of F[1] is: 2.33333
The maximum value of F[1] is: 6.

We see that the minimum, degree of belief of 1, and maximum values are the same regardless of the number of intermediate values. The maximum value is the largest value that **p2** can have (5) divided by the smallest value p1 can have (1), while the smallest value is the smallest value **p2** can have (4) divided by the largest value **p1** can have (3). But the degrees of belief of the intermediate values change based on the number of incremental evaluations, which in turn changes the shape of the fuzzy result, which in turn changes to the defuzzified value, since we are interpreting the centroid of the shape to be this value.

It is possible that a nonlinear function of fuzzy numbers will result in a non-unimodal fuzzy number. Here is an example:

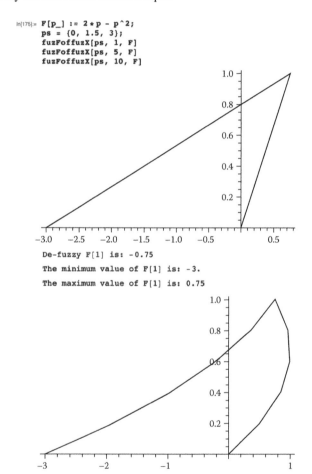

```
In[175]:= F[p_] := 2*p - p^2;
          ps = {0, 1.5, 3};
          fuzFoffuzX[ps, 1, F]
          fuzFoffuzX[ps, 5, F]
          fuzFoffuzX[ps, 10, F]
```

De-fuzzy F[1] is: -0.75

The minimum value of F[1] is: -3.

The maximum value of F[1] is: 0.75

De-fuzzy F[1] is: -0.39
The minimum value of F[1] is: -3.
The maximum value of F[1] is: 0.99

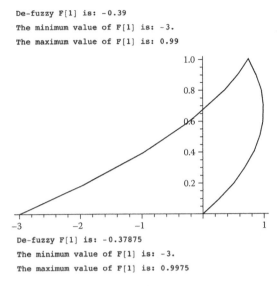

De-fuzzy F[1] is: -0.37875
The minimum value of F[1] is: -3.
The maximum value of F[1] is: 0.9975

What is happening is that the fuzzy result is consistent with the truth that there are two solutions to the quadratic that can result in a value of 0, namely, **p** equal to 0 and 2. For values greater than 0, the higher degree of belief curve is associated with the input near 2, since the degree of belief of the input at 2 is 0.5. The lower value degree of belief curve is associated with the solution of input near 0, since the degree of belief of the input at 0 is 0. Notice that the largest value F can have is indeed 1, which occurs when **p** has the value 1, which has a degree of belief of about 2/3.

A fuzzy rule is an inference that a set of fuzzy numbers implies another set of fuzzy numbers. For two deterministic inputs **input_1** and **input_2**, which infer two fuzzy outputs, **output_1**, and **output_2**, the inference rule is

IF **input_1** = **<l1, m1, h1>** AND **input_2** = **<l2, m2, h2>** THEN **output_1** = **<l3, m3, h3>** AND **output_2** = **<l4, m4, h4>**.

The shorthand depiction of the same is

IF **<l1, m1, h1>** AND **<l2, m2, h2>** THEN **<l3, m3, h3>** AND **<l4, m4, h4>**.

The ultimate shorthand depiction is

If **f1** AND **f2** THEN **f3** and **f4**.

where **f1, f2, f3**, and **f4** are the fuzzy numbers associated with input and output.

Since many applications require the conclusion to be implemented as a nonfuzzy number, a more convenient fuzzy rule representation is

IF **f1** AND **f2** THEN **o1** AND **o2**.

Here the outputs, **o1** and **o2**, are regular deterministic numbers.

To evaluate a fuzzy rule, first we need to define how to determine the degree of fulfillment of each input (that is, how much the input equals $<l, m, h>$), then we need to define how to determine the net truth of the combined inputs (that is, how to combine the degree of fulfillment of multiple IF clauses), and last we must define how to establish the THEN clause(s) consistent with the determined fuzzy truth of all the IF clauses.

The degree of fulfillment of an input x with the fuzzy inference $<l, m, h>$ is denoted by $d(x, <l, m, h>)$, and for linear triangular fuzzy numbers it is calculated by

$$\text{IF } x > l \text{ AND } x <= m \text{ THEN } d = (x - l) / (m - l)$$

$$\text{IF } x > m \text{ AND } x < h \text{ THEN } d = (x - m) / (m - h)$$

So the degree of fulfillment is a number between 0 and 1, where the number is 0 if the IF proposition is completely false, 1 if the IF proposition is completely true, and an intermediate value otherwise. For triangular fuzzy inferences, geometrically, this is simply the degree of belief of the deterministic input variable for the fuzzy number that it is being compared to.

The next step is to interpret the combination of multiple input clauses to also yield a value between 0 and 1, for completely false to completely true, respectively. There are at least two equally valid optional ways to interpret the combination of multiple IF clauses. One is to multiply all input degrees of fulfillment for each IF clause (also known as *product inference*), and the other is to take the minimum of all the degrees of fulfillment of each for each of the IF clauses (also known as *minimum inference*). Herein, minimum inference is used to combine IF clauses, unless otherwise stated.

Now that the IF clauses can be evaluated to yield a number between 0 and 1, the resulting inference is used to determine how much of the THEN clauses to impose. If the THEN clauses are fuzzy numbers, the inference value is taken as the limiting degree of belief for each THEN clause. Geometrically, the IF inference value is used to establish a cap on each THEN fuzzy number. If the IF inference is 0, then the cap is 0, so the fuzzy output is 0. If the If inference is 1, then the cap is 1, so the complete fuzzy number is output (which to be used may be de-fuzzified to its centroid). If the IF inference is between 0 and 1, and the fuzzy output is a triangular fuzzy number, the THEN clause is the trapezoid formed by the triangular fuzzy number truncated at the degree of belief that corresponds to the IF clause inference value. If there are multiple rules for the same output, and a deterministic number is needed to implement the output, the output value is the centroid of all the output value areas. If the THEN clauses are regular numbers (**o1** and **o2** in the example), then the inference value is taken as a weight to apply to the THEN clause outputs. If there are multiple rules addressing the same output, the net output is the weighted sum of all the rules that may affect the output. The precise arithmetic to implement these narratives will be specified in Section 2.3.3.2.

2.3.3.1 Using Fuzzy Logic to Predict Results Based on Uncertain Inputs

Fuzzy mathematics is a powerful means to quantify things that are uncertain but not random. When creating a new system, both the developer and customer want to estimate the cost and schedule with some margin to compensate for the risks associated with making the system operational. Particularly if some aspect of the endeavor has never been done before, both the cost and the time to complete are uncertain, but not random. Yet, often today probability is used to estimate the risk-adjusted cost and schedule.[21,22]

To estimate system cost using probability, the system is first divided into pieces, typically in the form of a contract work breakdown structure (CWBS). To present the customer a bid, the developer estimates the cost of each CWBS item. If the customer is to pay for the development, then the cost estimate includes the activities to design, manufacture, verify, and deliver the item. If the customer is to pay for the life cycle costs, then estimates are also needed for operations, support, training, and disposal costs. To obtain these estimates, each item's developer may use prior actual costs (adjusted for scope) and/or mathematical cost-estimating relationships (usually based on regression analysis of actual costs with respect to some physical parameter, such as mass), extrapolate from experience, or simply guess. Then, one or more experts identify risk sources that might prevent that item from being delivered for the specified cost. To aid this effort, a risk source versus level matrix (also known as a Maxwell scale) is provided. The rows of the matrix enumerate the potential sources of risk, for example the required technical advancement, technology status, complexity, interactions or dependencies, process controls, manufacturing precision, reliability, produceability, critical status to the mission, and cost. The columns of the matrix state the risk level, perhaps as a numerical scale, or perhaps simply low, medium, or high. The risk level attempts to gauge the probability of the risk occurring combined with the severity of the consequence. With these risks in mind, one or more experts are asked to identify the potential risk level for each source and then to produce a probability density function for the item's cost. The experts may actually draw a probability distribution, or a distribution is created for them, based on how they respond to questions about the minimum, most likely, and maximum cost. Once probability density functions are obtained for all the system's items, a Monte Carlo simulation is run to obtain a cumulative distribution for the total cost and schedule. Note, then, if the system was made up of n items each with normally distributed cost estimates of mean u and standard deviation s, the sum has mean nu and standard deviation $Sqrt[n]* s$. The ratio of total cost standard deviation to the total cost mean is $(s / u) / Sqrt[n]$, which is smaller than the ratio (s / u) for each item's ratio of standard deviation to item mean. In plain English, if a probabilistic cost estimate is constructed as described, the standard deviation relative to the total cost estimate reduces as the number of itemizes increases, which is clearly nonsensical. This observation causes most cost analysts to require the item costs to be correlated, either directly through use of equations in the Monte Carlo simulation or approximately by using for the total cost standard deviation the average of the perfectly correlated standard deviation and the not correlated standard deviation, which is $(1/2) (n + 1 / Sqrt[n]) s$. Clearly, this is very worrisome, because by definition, the items are defined to be independent, so why must their costs now be correlated?

The same basic approach is used to estimate the duration of the development effort. The tasks to design, manufacture, and verify the item are defined with appropriate dependencies. The times to complete each task are estimated using probability density functions, and Monte Carlo simulations are run to determine the total duration of the effort recognizing the task dependencies that exist. Note that such an approach will also return the probability that a given task is on the critical path. Perhaps, more logically, cost and schedule risks can be determined simultaneously by using probability density functions for duration, and the fixed and labor costs associated with each task.

Though rigorous, Monte Carlo–based probability and schedule estimates should be avoided. I recommend this to avoid the phenomenon of an ever-shrinking net uncertainty as the number of tasks increases. Practically speaking, the Monte Carlo methodology is actually very difficult to implement. While it is good practice to ask domain experts to identify potential causes for cost increases or schedule delays, combining these overlapping sources into a single risk impact is a very subjective decision. Many domain experts don't know how to assess cost as well as they can predict technical performance. For a totally new component, any cost is a guess. Fitting a probability curve to a guess is reasonable, but there is no way to know for sure if the domain expert is conservative or an optimist, so it is not knowable if the ends of the distribution he or she provides are at the 1 sigma or 10 sigma level. Finally, if the program manager and/or the customer are not comfortable with probability, they won't know what to do with the result. Even if both the program manager and the customer can comfortably interpret a cumulative distribution, they may doubt its accuracy. All these issues motivate finding an alternative approach to model cost and schedule uncertainty. Fuzzy logic provides such an alternative.

First, let's define some terms to clarify what we will discuss. We postulate two factors are predominantly responsible for the cost uncertainty. The first factor is the item's technical maturity—is it something done before or just an idea? We define a scale to allow us to quantify an item's technical maturity at least in a fuzzy sense. The second factor is literally our certainty with our own estimate. We call this *proposal risk*, and we define a scale to enable it to be quantified. The cost margin that is needed will be calculated based on fuzzy rules concerning the technical maturity and the proposal risk of the items.

Table 2.1 broadly defines levels of technical maturity on a scale from 0 to 8. For any given end item, it is devilishly difficult to declare its technical maturity. For example,

TABLE 2.1
Technical Maturity Level Definitions

Technical Maturity Definition	Technical Maturity Level
Successfully operating	8
Qualified but not yet used operationally	7
Prototype demonstrated in operational environment	6
Prototype demonstrated in relevant environment	5
Portions validated by analysis or in laboratory	4
Analysis or experiments show concept could work	3
Concept defined	2
Solution scientifically feasible	1
No known approach to meet requirements	0

TABLE 2.2

Fuzzy Technical Maturity Levels

Fuzzy Maturity	Technical Maturity Levels
Very mature	<6, 8, 8>
Mature	<3, 5, 7>
Immature	<1, 2, 4>
Very immature	<0, 0, 2>

suppose the system needs a new propellant tank with twice the volume as any previously made tank, but to be built exactly the same way as the smaller tanks that have successfully operated. The item certainly isn't successfully operating, so maturity is less than 8. But the to-be-made tank certainly has a known approach to meet requirements, so its technical maturity level is more than 0. The extent that the to-be-made tank has intermediate levels of technical maturity is best approached as a degree of belief, so fuzzy modeling is appropriate. The to-be-made tank's technical maturity could be represented as the fuzzy number <3, 5, 7>, if we have done an analysis that convinces us the tank can be made and meets its requirements (the 3), and we give ourselves some credit for the fact that we have made, qualified, and successfully operated similar, albeit smaller, tanks (the 7), and our previously made tanks are effectively the prototype for the new larger tank (the 5). Please notice how much easier it is to potentially agree on the range of technical maturity than it is to agree on the explicit level given the inevitably ambiguous definitions. This is precisely the beauty of the fuzzy logic method—unobtainable precision is avoided. Fuzzy logic methods enable the explicit articulation of the inherent uncertainty. Table 2.2 defines four technical maturity states. Table 2.3 shows a scale to broadly define levels of cost estimate certainty.

As with technical maturity, fuzzy numbers allow us to be ambiguous about the cost estimate certainty, too. If we are very certain our estimate has little certainty, we can categorize our proposal risk as <1, 1, 1>, while if we have some fear our estimate

TABLE 2.3

Cost Estimate Certainty Definitions

Cost Estimate Certainty Definitions	Certainty Level
No prior similar efforts and for many reasons actual cost may be different than predicted.	0
Mostly never done before with more cost uncertainties than certainties.	1
Similar to prior efforts with reasonable certainty that costs can be predicted accurately.	2
Nearly identical to prior efforts and costs are highly predictable.	3

TABLE 2.4

Fuzzy Proposal Risk Levels

Fuzzy Proposal Risk	Cost Uncertainly Levels
Low	<1, 2, 3>
Medium	<0.5, 1.5, 2.5>
High	<0, 0, 2>

---

is off, using the scale in Table 2.3 we might categorize our proposal certainty as <1, 2, 3>. Table 2.4 defines three proposal risk states.

Technical maturity level and proposal risk level are used to define fuzzy rules of the form:

> IF **Technical_Maturity_Level** AND **Proposal_Risk_Level**, THEN **Margin_recommended**.

Reference 23 is used to define the **Margin_recommended** fuzzy numbers. These values are used to create a set of 12 fuzzy rules, summarized in Table 2.5.

For example, an item with high proposal risk that is very immature has the **Margin_recommended** set to <0.25, 0.75, 1.50> to indicate the final cost could be 25% to 150% more than the estimate, with 75% the degree of belief of 1 value. Similarly, an item with low proposal risks that is very mature has the **Margin_recommended** set to <−0.10, 0.00, 0.15> to indicate the final cost could be from 10% less than the estimate to 15% more than the estimate, with a degree of belief of 1 value of 0, to indicate the final cost will be the estimate.

The 12 rules are captured as follows; notice by substituting the defuzzified value for the 12 THEN clauses:

```
In[146]:= rulesTMPRM = {{{0, 0, 2}, {1, 3, 3}, (.1 + .25 + .75) / 3},
                        {{0, 0, 2}, {.5, 1.5, 2.5}, (.25 + .5 + .75) / 3},
                        {{0, 0, 2}, {0, 0, 2}, (.25 + .75 + 1.5) / 3},
                        {{1, 2, 4}, {1, 3, 3}, (0 + .15 + .25) / 3},
                        {{1, 2, 4}, {.5, 1.5, 2.5}, (.1 + .25 + .75) / 3},
                        {{1, 2, 4}, {0, 0, 2}, (.25 + .5 + 1) / 3},
                        {{3, 5, 7}, {1, 3, 3}, (-.1 + 0.05 + .25) / 3},
                        {{3, 5, 7}, {.5, 1.5, 2.5}, (0 + .15 + .5) / 3},
                        {{3, 5, 7}, {0, 0, 2}, (.1 + .25 + .75) / 3},
                        {{6, 8, 8}, {1, 3, 3}, (-.1 + 0 + .15) / 3},
                        {{6, 8, 8}, {.5, 1.5, 2.5}, (-.05 + .05 + .25) / 3},
                        {{6, 8, 8}, {0, 0, 2}, (0 + .15 + .25) / 3}
                        };
```

TABLE 2.5

Fuzzy Rules for Determining Fuzzy Margin Required

Proposal risk level	High	< 0.25, 0.75, 1.50 >	< 0.25, 0.50, 1.00 >	< 0.10, 0.25, 0.75 >	< 0.00, 0.15, 0.25 >
	Medium	< 0.25, 0.50, 0.75 >	< 0.10, 0.25, 0.75 >	< 0.00, 0.15< 0.50 >	< −0.05, 0.05, 0.25 >
	Low	< 0.10, 0.25, 0.75 >	< 0.00, 0.15, 0.25 >	< −0.10, 0.05, 0.25 >	< −0.10, 0.00, 0.15 >
Fuzzy margin		Very mature	Immature	Mature	Very Mature
		Technical maturity level			

The *Mathematica* routine *Plot3D* is used to show the recommended margin as a function of all possible inputs for the *rulesTMPRM*:

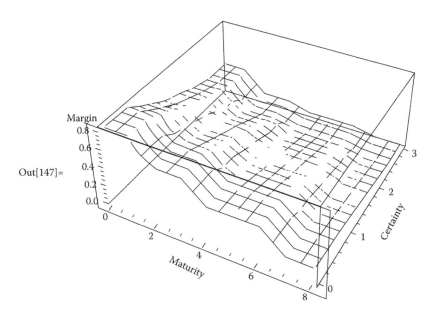

Out[147]=

Table 2.6 shows the result of applying the process to a four-element system. First, itemize the work content into independent but complete tasks using a contract work breakdown structure (CWBS) (column 1). Then, for each CWBS item, determine its technical maturity as a fuzzy variable using the previously stipulated technical

TABLE 2.6
Example Fuzzy Cost Estimate

1: CWBS	2: Technical Maturity	3: Cost Estimate ($K)	4: Cost Certainty	5: Cost Estimate with Margin ($K)
1	<6, 7, 8>	1500	<1, 3, 3>	1550
2	<3, 5, 8>	200	<1, 2, 3>	228
3	<1, 3, 5>	2000	<0, 1, 2>	2950
4	<7, 7, 7>	500	<2, 3, 3>	508
Total	—	4200	—	5237

maturity scale (column 2). Then, for each CWBS item, determine a cost estimate (column 3), and denote your certainty with the cost estimate using the previously stipulated cost estimate certainty scale (column 4). To estimate the likely final cost of each item, input to *EvalFuzRulesIFA1andA2THENBconst* the defuzzified values for technical maturity and cost estimation certainty, and multiply each item cost estimate times 1 plus the recommended margin, to estimate the likely final cost of each item (column 5).

The values in column 5 are obtained as follows:

```
In[148]:=  Cost1withmargin =
           1500 * (1 + EvalFuzRulesIFA1ANDA2THENBconst[ (6 + 7 + 8) / 3,  (1 + 3 + 3) / 3,  rulesTMPRM])
Out[148]=  1550.

In[149]:=  Cost2withmargin =
           200 * (1 + EvalFuzRulesIFA1ANDA2THENBconst[ (3 + 5 + 8) / 3,  (1 + 2 + 3) / 3,  rulesTMPRM])
Out[149]=  228.333

In[150]:=  Cost3withmargin =
           2000 * (1 + EvalFuzRulesIFA1ANDA2THENBconst[ (1 + 3 + 5) / 3,  (0 + 1 + 2) / 3,  rulesTMPRM])
Out[150]=  2950.

In[151]:=  Cost4withmargin =
           500 * (1 + EvalFuzRulesIFA1ANDA2THENBconst[ (7 + 7 + 7) / 3,  (2 + 3 + 3) / 3,  rulesTMPRM])
Out[151]=  508.333
```

Fuzzy logic models can also be used to estimate the time to accomplish a project. If the input uncertainty is being guessed, rather than based on observed statistics, then it is more appropriate to model each input as a fuzzy number with a stated minimum and maximum possible values for which the degrees of belief are both 0, and a best-guess value, for which the degree of belief is 1. Suppose there are eight tasks to be accomplished, with the dependencies as shown in Figure 2.9. What is of interest is the possible total time to assemble the item. Based on dependencies shown in Figure 2.9, this time is

$$Max[Max[\textbf{T1}, \textbf{T2}], Max[\textbf{T3}, \textbf{T4}]] + Max[\textbf{T5}, \textbf{T6}, \textbf{T7}] + \textbf{T8}$$

where $\textbf{Ti}$ is the time it takes to complete task $\textbf{i}$.

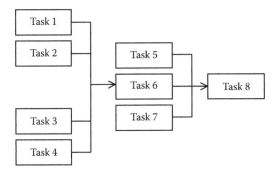

FIGURE 2.9 Example activities with dependencies.

Each task time **Ti** can be estimated directly, or as a consequence of the number of required operations, divided by the sum of the products of number of performers times the performer's rate of accomplishment. For this illustration, let

T1 through **T7** be estimated as fuzzy times (**E1** through **E7**)

and

T8 = 100 operations / (10 machines x **MachineOps8**)

where **MachineOps8** is the machine operations per time.

Each task has some chance of having to be redone due to the need to fix or rework the effort. The task **i** rework rate is denoted by **RWRi**, which increases the time to complete each task i by 1 + **RWRi**, so:

$$\text{Ti} = (1 + \text{RWRi}) * \text{Ei}, \text{ i} = 1 \text{ to } 7$$

$$\text{T8} = (1 + \text{RWR8}) * (100/(10* \text{MachineOps8})$$

A representation of the total time to do all tasks is then provided by

```
In[152]:= Criticalpath[iv_] := Module[{times, i},
              times = Append[Table[(1 + iv[[i + 8]]) * iv[[i]], {i, 1, 7}],
              (1 + iv[[16]]) * (100 / (10 * iv[[8]]))];
              Return[times[[8]] + Max[times[[7]], times[[6]], times[[5]]] +
              Max[Max[times[[3]], times[[4]]], Max[times[[1]], times[[2]]]]]]]
```

Here are illustrative example inputs:

```
In[153]:= E1 = {6, 8, 10};
          E2 = {7, 8, 11};
          E3 = {4, 6, 9};
          E4 = {5, 8, 10};
          E5 = {5, 8, 12};
          E6 = {4, 6, 8};
          E7 = {5, 8, 12};

          MachineOps8 = {10, 12, 14};

          RWR1 = {.05, .1, .2};
          RWR2 = {.05, .1, .2};
          RWR3 = {.025, .05, .15};
          RWR4 = {.025, .05, .15};
          RWR5 = {.1, .2, 3.};
          RWR6 = {.15, .25, .35};
          RWR7 = {.1, .2, .3};
          RWR8 = {.1, .2, .3};

          iv = {E1, E2, E3, E4, E5, E6, E7,
              MachineOps8, RWR1, RWR2, RWR3, RWR4, RWR5, RWR6, RWR7, RWR8};
```

The fuzzy total time to accomplish all eight tasks is then found by

```
In[170]:= fuzFoffuzX[iv, 10, Criticalpath]
```

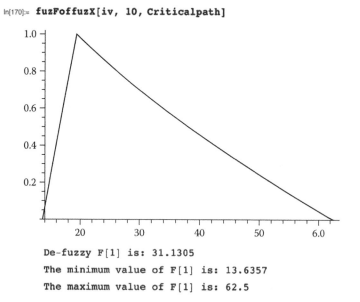

```
De-fuzzy F[1] is: 31.1305
The minimum value of F[1] is: 13.6357
The maximum value of F[1] is: 62.5
```

The time to budget to accomplish all eight tasks is the defuzzified value of about 31 days. The degree of belief is 0 that the eight tasks can be accomplished in less than 13.6 days or take longer than 62.5 days. Please note, however, that the 60-day outcome could happen. So while the current plan for the project is to complete in about

31 days, contingency plans should be prepared in case the project requires as many as 62.5 days.

Virtually any uncertain outcome can be assessed by one or the other of these fuzzy logic methods: either produce fuzzy rules for what to do or what will happen based on fuzzy inputs, or model the situation deterministically and assess the consequence of the fuzzy inputs.

2.3.3.2   Using Fuzzy Logic to Predict Results Based on Observed Data

Fuzzy logic rules can be constructed to fit observed data as an alternative to using regression. To compare the fuzzy logic approach to regression, we'll use the same training data as used previously. Rather than a polynomial, fuzzy logic rules of the form:

IF **Ai** THEN **Bi**,

where **i** = 1 to the number of rules, are used to fit the training data. The fuzzy **Ai**'s are at the discretion of the user. A way to pick the **Ai**'s is to divide the independent variable range into overlapping regions of fuzzy numbers. As it was noted in the regression example, the training data had at least six changes in slope, and so at least a six-order polynomial was needed; similarly, the fuzzy **Ai**'s can be grouped to overlap more in ranges of high data variation and spread out for ranges of little variation or near-linear variation.

Two optional methods to determine corresponding **Bi** values are counting and least-squares fit. For the counting method, let $b[\mathbf{Ai}]$ denote the set of training data dependent variable values that have independent variables in the range **Ai**. The counting method sets **Bi** to be the defuzzified value of the triangular fuzzy number of the form:

$$<Min[b[\mathbf{Ai}]], Mean[b[\mathbf{Ai}]]], Max[b[\mathbf{Ai}]] >$$

which is

$$(Min[b[\mathbf{Ai}]] + Mean[b[\mathbf{Ai}]] + Max[b[\mathbf{Ai}]])/3$$

For example, for **trainingdata** shown for the regression example, suppose our first **Ai** is the fuzzy number <0, 0, 1>; the counting method determines that the following data points fall within that range:

$$\{\{.09, .22\},$$

$$\{.28, .57\},$$

$$\{.34, .76\},$$

$$\{.65, 1.47\},$$

$$\{.67, 1.64\}\}$$

because these are the only data points for which independent variables are between 0 and 1. So, $b[<0,0,1>]$ are the corresponding dependent variables, $\{0.22, 0.57, 0.76, 1.47, 1.64\}$, and

$$Min[b[<0,0,1>] = 0.22$$

$$Max[b[<0,0,1>]] = 1.64$$

$$Mean[b[<0,0,1>]] = 0.93$$

and the rule becomes

IF <0, 0, 1> THEN (0.22+0.93+1.64)/3.

Additional rules are formulated for additional **Ai**'s distributed over the range of independent variables for which the predicted results are needed and there is data.

The least-squares method solves a set of linear equations to determine **Bi** to minimize the squared difference between the observed dependent variable values and the fuzzy logic rules assessed at the observed independent variables.[14]

To begin, let's try eight rules, the left-hand sides of which each span a portion of the data range from 0 to 8, as stipulated below in the variable called **A8rules**:

```
In[171]:= A8rules = {{0, 0, 1},
              {0, 1, 2},
              {1, 2, 3},
              {2, 3, 4},
              {3, 4, 5},
              {4, 5, 6},
              {5, 6, 7},
              {6, 7, 8}};
```

The formal requirements *FuzRulestoFitDatabyCounting* determine **Count8rules**, which holds the eight **Ai** and **Bi** values obtained using the counting method with **trainingdata** and **A8rules** as inputs:

```
In[173]:= Count8rules = FuzRulestoFitDatabyCounting[trainingdata, A8rules]

Out[173]= {{{0, 0, 1}, 0.930667}, {{0, 1, 2}, 0.20875}, {{1, 2, 3}, -1.24067}, {{2, 3, 4}, 1.35571},
           {{3, 4, 5}, 3.21417}, {{4, 5, 6}, -3.18037}, {{5, 6, 7}, -7.75222}, {{6, 7, 8}, 1.}}
```

The formal requirements *EvalIFAiTHENBconsti* evaluate rules of the form IF **Ai** THEN **Bi**, where **Bi** is a constant.

The following plots the **trainingdata** points, the **sixthorderfit** obtained by using regression, and the fuzzy logic fit using the counting method:

```
In[176]:= Show[ListPlot[trainingdata, PlotStyle → PointSize[0.02]],
    Plot[{sixthorderfit[x], EvalIFAiTHENBconsti[x, Count8rules]},
    {x, 0, 8}, PlotStyle → {Directive[Dashed, Black], Black}]]
```

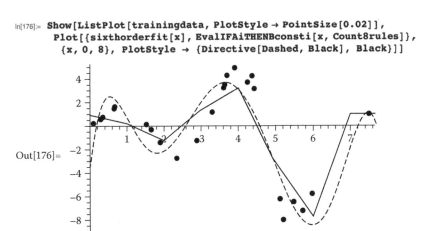

Out[176]=

The **sixthorderfit** is the smooth dashed curve, and the fuzzy rules provide the piecewise linear fits (solid). Let's see if more rules improve the fit to the **training-data**. We'll define 13 left-hand sides for 13 rules as **A13rules**:

```
In[177]:= A13rules = {{0, 0, .5},
    {0.5, 1., 1.5},
    {1, 1.5, 2},
    {1.5, 2., 2.5},
    {2, 2.5, 3},
    {2.5, 3, 3.5},
    {3, 3.5, 4},
    {3.5, 4, 4.5},
    {4, 4.5, 5},
    {4.5, 5, 5.5},
    {5., 5.5, 6},
    {5.5, 6, 6.5},
    {6, 8, 8}};
```

For this **A13rules**, the rules for **trainingdata** by the counting method become

```
In[178]:= Count13rules = FuzRulestoFitDatabyCounting[trainingdata, A13rules]

Out[178]= {{{0, 0, 0.5}, 0.498889}, {{0.5, 1., 1.5}, 1.555}, {{1, 1.5, 2}, -0.594444},
    {{1.5, 2., 2.5}, -1.23}, {{2, 2.5, 3}, -1.995}, {{2.5, 3, 3.5}, -0.035},
    {{3, 3.5, 4}, 3.17867}, {{3.5, 4, 4.5}, 3.99}, {{4, 4.5, 5}, 3.70667},
    {{4.5, 5, 5.5}, -8.03667}, {{5., 5.5, 6}, -7.75222}, {{5.5, 6, 6.5}, -6.51}, {{6, 8, 8}, 1.}}
```

Again, let's plot the **trainingdata**, the **sixthorderfit**, and the result fuzzy fit using **Count13rules**:

In[179]:= **Show[ListPlot[trainingdata, PlotStyle → PointSize[0.02]],**
Plot[{sixthorderfit[x], EvalIFAiTHENBconsti[x, Count13rules]},
{x, 0, 8}, PlotStyle → {Directive[Dashed, Black], Black}]]

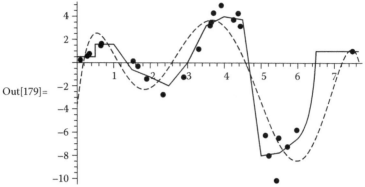

Out[179]=

The fuzzy rules obtained are arguably a better fit to the data than that obtained by regression for the range of 0 to 6. There is no evidence to substantiate the claim for the predicted values in the range from 6 to 8, but the fuzzy rules give a reasonable guess, perhaps more so than that made by the seventh-order polynomial and almost certainly better than that obtained by the seventh-order polynomial.

FuzRulestoFitDatabyLeastSquares determines rules using the least squares method. This method solves a set of linear equations for the **Bi** values so that the application of the rules minimizes the error between the training value and those obtained by the rules.

Here are the rules using the least-squares method on **trainingdata** for the left-hand sides denoted by **A8rules**:

In[160]:= **LS8rules = FuzRulestoFitDatabyLeastSquares[trainingdata, A8rules]**

Out[180]= {{{0, 0, 1}, -0.143461}, {{0, 1, 2}, 2.56849}, {{1, 2, 3}, -2.12517}, {{2, 3, 4}, -1.96354},
{{3, 4, 5}, 7.45634}, {{4, 5, 6}, -7.01524}, {{5, 6, 7}, -7.08886}, {{6, 7, 8}, 1.}}

Which yields the following result:

In[181]:= **Show[ListPlot[trainingdata, PlotStyle → PointSize[0.02], PlotRange → {-10, 10}],**
Plot[{sixthorderfit[x], EvalIFAiTHENBconsti[x, LS8rules]}, {x, 0, 8},
PlotRange → {-10, 10}, PlotStyle → {Directive[Dashed, Black], Black}]]

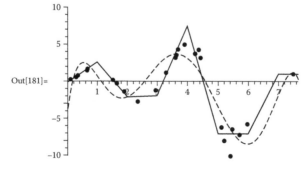

Out[181]=

This is perhaps the best fit so far, except in the neighborhood of 4. Let's try the least-squares method for the larger number of rules associated with **A13rules**. Determine the rules using *FuzRulesFitDatabyLeastSquares*:

```
In[182]:= LS13rules = FuzRulestoFitDatabyLeastSquares[trainingdata, A13rules]

Out[182]= {{{0, 0, 0.5}, 0.516667}, {{0.5, 1., 1.5}, 1.555},
        {{1, 1.5, 2}, 0.305639}, {{1.5, 2., 2.5}, -1.80372},
        {{2, 2.5, 3}, -3.14541}, {{2.5, 3, 3.5}, -1.08349}, {{3, 3.5, 4}, 2.85864},
        {{3.5, 4, 4.5}, 5.37787}, {{4, 4.5, 5}, 3.05486}, {{4.5, 5, 5.5}, -6.62168},
        {{5., 5.5, 6}, -8.4153}, {{5.5, 6, 6.5}, -5.64022}, {{6, 8, 8}, 1.}}
```

This provides the following prediction, which appears to be very accurate, with the understood uncertainty in the range of 6 to 8:

```
In[183]:= Show[ListPlot[trainingdata, PlotStyle → PointSize[0.02`], PlotRange → {-10, 10}],
        Plot[{sixthorderfit[x], EvalIFAiTHENBconsti[x, LS13rules]}, {x, 0, 8},
        PlotRange → {-10, 10}, PlotStyle → {Directive[Dashed, Black], Black}]]
```

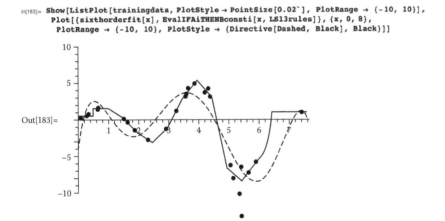

Out[183]=

The fuzzy logic rules using a least-squares fit and **A13rules** fit the data extremely well. (Again, as with the regression result, we really should get more data in the range from 6 to 8.) Fuzzy logic can therefore be used for predictions based on observed data at least as well as regression. Though the above examples are limited to predicting an outcome based on a single independent variable, the same techniques apply for multiple independent and dependent variables. My experience is that the counting algorithm is preferred as more variables become involved, but one should always experiment with both methods to find the best for the situation.

2.3.3.3 Multiple-Input Single-Output Fuzzy Logic Example

For this example, we'll redo the "time for a laser to make a hole" margin analyses we did using the Monte Carlo method. This time the inputs are triangular fuzzy numbers with ranges matching the triangular probability distributions used in the Monte Carlo analyses. For convenience, scale factors are defined as **PO** for power, **WE** for wavefront error, **AE** for jitter, **RE** for range, lambda for wavelength, **TLF** for half tear length, **TLFdelta** for high and low uncertainty, **FF** for fluence required, **FFdelta** for high and low offset, **TR** for time required, and **TRdelta** for high and low offset.

```
In[184]:= PF = 10^6;
          WE = 0.14;
          AE = 3 10^-7;
          RE = 300000.;
          TLF = 0.09;
          TLFdelta = .0003;
          FF = 5 10^6;
          FFdelta = 129400;
          TR = 10.;
          TRdelta = .05;
          lambda = 2.8 10^-6;
```

Define fuzzy power (**p**) as

```
In[195]:= p = PF*{0.8, 1.0, 1.2};
```

Define fuzzy beam diameter (**d**) as

```
In[196]:= d = {2.36, 2.38, 2.40};
```

Define fuzzy transmission factor (**T**) as

```
In[197]:= T = {0.80, 0.85, 0.90};
```

Define fuzzy obscuration factor (**o**) as

```
In[735]:= o = {0.60, 0.65, 0.70};
```

Define fuzzy sigma wavefront error (**sigmawfe**) as

```
In[199]:= sigmawfe = WE*{0.7, 1.0, 1.4}
```

```
Out[199]= {0.098, 0.14, 0.196}
```

Define fuzzy range (**R**) as

```
In[200]:= R = {RE-20000, RE, RE+20000};
```

Define fuzzy sigma jitter (**sigmaj**) as

```
In[201]:= sigmaj = AE*{0.8, 1.8, 3.16};
```

Define fuzzy angle of incidence (**theta**) as

```
In[202]:= theta = {-15. Degree, 0., 15. Degree};
```

Define fuzzy half tear length (**htl**) as

```
In[203]:= htl = {TLF - TLFdelta, TLF, TLF + TLFdelta};
```

Define fuzzy fluence required (**Frequired**) as

```
In[204]:= Frequired = {FF-FFdelta, FF, FF+FFdelta};
```

Define fuzzy time required (**trequired**) as

In[205]:= **trequired = {TR-TRdelta, TR, TR+TRdelta};**

The input variables are collected in a single array named **iv**:

In[206]:= **iv = {p, d, T, o, sigmawfe, R, sigmaj, theta, htl, Frequired,
           trequired, {laserwavelength, laserwavelength, laserwavelength}};**

fuzLaserMargin is defined to provide the margin obtained for the independent fuzzy numbers representing each of the inputs.

Evaluating *fuzLaserMargin* for 10, 50 and 100 incremental degrees of belief:

In[244]:= **iv**

Out[244]= $\{\{800\,000., 1. \times 10^6, 1.2 \times 10^6\}, \{2.36, 2.38, 2.4\}, \{0.8, 0.85, 0.9\},$
$\{0.6, 0.65, 0.7\}, \{0.098, 0.14, 0.196\}, \{280\,000., 300\,000., 320\,000.\},$
$\{2.4 \times 10^{-7}, 5.4 \times 10^{-7}, 9.48 \times 10^{-7}\}, \{-0.261799, 0., 0.261799\}, \{0.0897, 0.09, 0.0903\},$
$\{4\,870\,600, 5\,000\,000, 5\,129\,400\}, \{9.95, 10., 10.05\}, \{2.8 \times 10^{-6}, 2.8 \times 10^{-6}, 2.8 \times 10^{-6}\}\}$

In[245]:= **fuzFoffuzX[iv,10, fuzLaserMargin]**

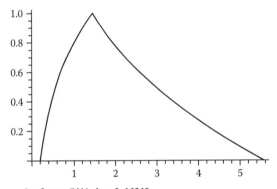

De-fuzzy F[1] is: 2.16245
The minimum value of F[1] is: 0.207388
The maximum value of F[1] is: 5.5722

In[246]:= **fuzFoffuzX[iv, 50, fuzLaserMargin]**

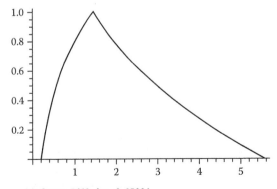

De-fuzzy F[1] is: 2.15994
The minimum value of F[1] is: 0.207388
The maximum value of F[1] is: 5.5722

fuzFoffuzX[iv,100, fuzLaserMargin]

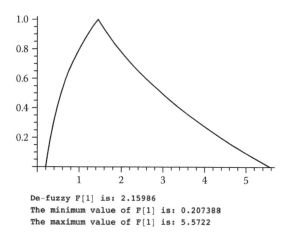

```
De-fuzzy F[1] is: 2.15986
The minimum value of F[1] is: 0.207388
The maximum value of F[1] is: 5.5722
```

Notice that, unlike a Monte Carlo analysis in which the number of trials needs to be determined experimentally, for a fuzzy model, as few as 10 incremental degrees of belief between 0 and 1 provide enough precision for answer. Second, notice that the margin of about 1.5 has degree of belief of 1. This means the individual margin that is most expected has about 50% contingency. Third, the defuzzified margin is about 2.16, which means that given all the presumed uncertainty, the likely deterministic margin is 2.16, so there is more than 100% contingency. Finally, notice the degree of belief that the margin could be less than 1 begins at about 0.7 and declines to 0 for a margin about 0.2.

How to interpret these results? There are three useful outputs: the degree of belief of 1 value, the defuzzified value, and the extreme value at degree of belief of 0, be it the maximum or the minimum. In general, a situation where the defuzzified value is adequate can be taken as strong evidence the modeled uncertainties will result in an acceptable result. Recall that the defuzzified value was chosen to be the centroid of the output distribution degree of belief, but this is not the 50% certain point as it is for a probabilistic assessment. If both the defuzzified value and the degree of belief of 1 values are adequate, that is even stronger indication the real outcome will be acceptable. The ultimate certainty is that the extreme values also meet the criteria. If the model is reasonably accurate, such a result indicates complete and total likelihood the real result will always be acceptable. That is an extraordinary conservative position.

Still, some caution is urged. Studies have shown people often greatly underestimate the actual variability that can occur with respect to something they know little about.[12] When using fuzzy logic models, be careful to input the true likely variation, and be conservative, that is, err by assuming more variation, particularly in the direction that can lead to a bad result. Care is especially needed when the uncertainty is asymmetric and the real level of bad things that can happen is larger than that of good things that can happen. If the model fails to represent the true possible range of bad things that can happen, then the fuzzy logic model will overstate the goodness of the result relative to reality, and decisions based on the model will be overly optimistic and therefore failure is more likely to occur than perceived.

Recognizing the possibility of errors in the model relative to the reality of the real item, many disciplines utilize the concept of a factor of safety, **FS**. For example, if something must be greater than **X** to be acceptable, a solution is sought which an analysis shows is **FS** * **X**, with **FS** > 1. Appropriate factors of safety to use are a compromise between safety and cost, generally established based on lessons learned, larger **FS** are gradually reduced until failures start occurring, while smaller **FS** gradually increased until failures stop occurring. In addition to providing protection from modeling errors, factors of safety provide design robustness, as the real entity can now operate beyond its explicit design criteria. For situations where the analysis is highly questionable or the consequence of error is loss of life, **FS** values typically around 10 are appropriate. For situations where the model is believed to be sufficiently validated against reality, and either the consequences are minor or it is simply not tolerable to be so conservative, lower **FS** values are utilized. Indeed, in the aerospace industry, where additional mass is so detrimental to system performance, despite the high potential consequences of failure, an **FS** value as high as 3 is typically used for only pressure vessels designs based on analysis only, reducing to 1.5 for pressure vessel designs based on actual demonstration, and 1.1 to 1.2 for all other structural considerations. Fuzzy logic model results can be utilized in conjunction with factors of safety. The defuzzified value should be more than a factor of safety multiple of the required value.

2.3.3.4 Fuzzy Solutions to Simultaneous Linear Equations Example

The model equations are of the form:

$$A * x = b,$$

with **A** and **n** × **n** matrix, and **b** and **n** × 1 array, and any of the elements of matrix **A** and array **b** represented by fuzzy numbers. The same two-story frame structure analyzed by the Monte Carlo method will serve as the example for the fuzzy logic analysis alternative. *fuzTwoStoryFrame* provides the two horizontal displacements and two rotations as a function of fuzzy inputs.

A matrix is defined that contains each of the needed fuzzy numbers, namely, the two horizontal forces (**fF2**, **fF3**), the Young's modulus (**fE**), the horizontal beam area moment of inertia (**fIhor**) and length (**fLhor**), as well as the vertical column area moment of inertia (**fIver**) and length (**fLver**). For this example, each fuzzy number input is given the same numerical values used for the triangular probability density functions used for the Monte Carlo simulation.

```
In[146]:=  fF2 = {0.8, 1.0, 1.2};
           fF3 = {1.2, 1.5, 1.8};
           fE = {14.9, 15.0, 15.1};
           fIhor = {0.99, 1, 1.01};
           fLhor = {14.99, 15.0, 15.01};
           fIver = {49, 50, 51};
           fLver = {24.9, 25, 25.1};
           pin = {fF2, fF3, fE, fIhor, fLhor, fIver, fLver};
```

Then, again use *fuzFoffuzX* to solve for the four outputs from the model, two horizontal displacements and two rotations. Here is the result:

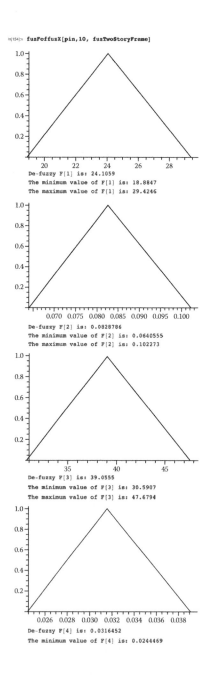

The results are almost linear, but not quite. The output with the degree of belief of 1 is clearly determined. The range of possible output is clearly identified with lesser degrees of belief. The defuzzified value for each output can be taken as the presumed output. The defuzzified values can be compared directly to requirement values or the results and be compared to the requirements multiplied by a factor of safety.

2.3.3.5 Model Uncertainty Using Fuzzy Ordinary Differential Equations

The last example addresses a system of ordinary differential equations. Since the solution to all such equations can be approximated by numerical methods, and since these methods basically generate a time history $F[x]$, where x are the fuzzy factors, be they terms in the equations or uncertainty associated with the initial conditions, the same method we just used on simultaneous linear equations can be used for differential equations, where

$$dx/dt = F[x[t], t], \text{ and } x[0] = xo$$

To begin, here is a very simple single-variable differential equation in which two constants, **k1** and **k2**, are fuzzy:

$$dx/dt = (k1^\wedge 2) * x + k2,$$

with $x[0] = 0$, and **k1** = <0.5, 1.0. 1.5> and **k2** = <0.75, 1.0, 1.25>.

The routine to determine the derivative value is

In[226]:= **derivativeofx[x_, t_, fuzp_] := fuzp[[1]]^2*x + fuzp[[2]]**

Although *Mathematica* provides routines to symbolically or numerically integrate ordinary differential equations, let's define our own formal requirements *MidPointIntegrate* to do the numerical integration.

To use *fuxFoffuxX* to integrate a system of ordinary differential equations, we first need to define an *F* to return the integrated value at **tfinal** based on specific values for the fuzzy numbers in the equation:

```
In[156]:= F[p_] := Module[{fpin},
           fpin = p;
           Return[MidPointIntegrate[0, 0, 1, .01, fpin, derivativeofx]]
           ]
```

Here is the solution obtained for **x[tfinal]** for

$$pin[[1]] = <0.5, 1, 1.5>$$

$$pin[[2]] = <0.75, 1, 1.25>$$

evaluated at the low, high, and degree of belief 1 value of the inputs:

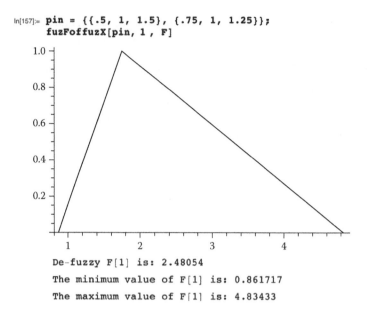

In[157]:= **pin = {{.5, 1, 1.5}, {.75, 1, 1.25}};**
fuzFoffuzX[pin, 1 , F]

```
De-fuzzy F[1] is: 2.48054
The minimum value of F[1] is: 0.861717
The maximum value of F[1] is: 4.83433
```

Here is the result when evaluating using intermediate degrees of belief, so the nonlinear nature of the equations manifests itself into the value of **x** at **tfinal**:

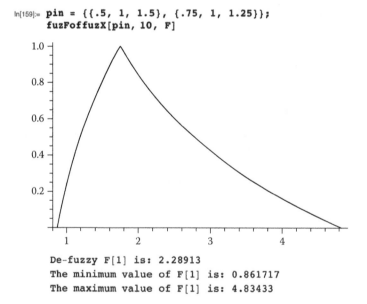

In[159]:= **pin = {{.5, 1, 1.5}, {.75, 1, 1.25}};**
fuzFoffuzX[pin, 10, F]

```
De-fuzzy F[1] is: 2.28913
The minimum value of F[1] is: 0.861717
The maximum value of F[1] is: 4.83433
```

Examine the result; the extremes and degree of belief of 1 value can be determined by the extremes of the inputs, but the fuzzy nature of **x[tfinal]** is not linear and the defuzzified value needs to be determined using intermediate degrees of belief.

The Monte Carlo analysis of the two-stage rocket trajectory subject to wind and system dispersions is not a good candidate for fuzzy logic representation. The atmospheric density and wind velocity variations are random, not fuzzy, and the possibility of nominal system dispersions can also appropriately be modeled as random behavior since it is possible to run enough tests to determine valid statistics.

But suppose we have not built the rocket yet and we intend to fly it on a planet we have not visited before. Rather than assuming probability density functions for the key modeling parameters, for which we have no basis but a guess, we simply model them as fuzzy numbers. A much simpler model, proposed in reference 24, is used to illustrate the fuzzy solution of a system of ordinary differential equations:

$$dx[[1]](t)/dt = x[[3]]$$

$$dx[[2]](t)/dt = x[[4]]$$

$$dx[[3]](t)/dt = A * Cos[theta]$$

$$dx[[4]](t)/dt = A * Sin[theta] - g$$

with

$$x[[1]](0) = x[[2]](0) = x[[3]](0) = x[[4]](0) = 0$$

and

$$x[[2]](tfinal) = H$$

$$x[[3]](tfinal) = U$$

$$x[[4]](tfinal) = 0$$

where

x[[1]] for horizontal position;
x[[2]] for vertical position (which is to be **H** at time equal **tfinal**);
x[[3]] for horizontal velocity (which is to be **U** at time equal **tfinal**); and
x[[4]] for vertical velocity (which is to be 0 at time equal **tfinal**).

Reference 24 uses these equations to show how a calculus of variations derives the open-loop optimal steering law, that is, the **theta** time history so that the terminal conditions at **tfinal** are achieved in the least amount of time, which, since this rocket is assumed to be accelerating at a constant rate **A**, also is the minimal fuel consumptions profile. Even though the obtained steering law is clearly based on very crude assumptions such as no atmosphere, constant gravity and thrust, and a flat planet, it turns out to be an extraordinarily good control law for real rockets flying over real planets provided one updates the solution as fly.

The calculus of variations derived steering law is

$$Tan[\textbf{theta[t]}] = \textbf{constant\_1} * \textbf{t} + \textbf{constant\_2}$$

Reference 24 shows for the case:

A = 20.82 meters/second;
g = 5.32 meters/second^2;
H = 50,000 meters; and
U = 5444 meters/second;

the constants for the optimal theta time history are

constant_1 = -1.598*10^-3 radians;
constant_2 = 0.4887 radians/second; and
tfinal = 272.4 second.

To illustrate an application of integrating a fuzzy system of differential equations, we'll assume both the system performance (as represented by **A**) and the environment (as represented by **g**) are fuzzy, and *derivativeofx* returns the derivative values for these fuzzy inputs:

```
In[161]:= derivativeofx[x_, t_, pfuzzy_] := Module[{theta},
        theta = ArcTan[-1.598 * 10^-3 * t + 0.4877];
        Return[
          {x[[3]], x[[4]], pfuzzy[[1]] * Cos[theta], pfuzzy[[1]] * Sin[theta] - pfuzzy[[2]]}]
        ]
```

Fuzzy **A** (**fA**) and fuzzy **g** (**fg**) are taken to be

```
In[167]:= fA = {19, 20.82, 21.5};
        fg = {5, 5.32, 6};
```

Next we formulate *F*[**p**] to return the **x[tfinal]**, with *F*[[1]] = **x**[[1]] (range), *F*[[2]] = **x**[[2]] (altitude), *F*[[3]] = **x**[[3]] (horizontal velocity), and *F*[[4]] = **x**[[4]] (vertical velocity), with

$$\textbf{tinital} = 0;$$

$$\textbf{finitial} = \{0,0,0,0\};$$

$$\textbf{tfinal} = 272.4; \text{ and}$$

$$\textbf{h} = .01$$

as follows:

```
In[169]:= F[p_] := Module[{A, g},
        Return[MidPointIntegrate[0, {0, 0, 0, 0}, 272.4, .01, p, derivativeofx]]
        ]
```

The fuzzy terminal state vectors are then determined using 10 incremental degrees of belief:

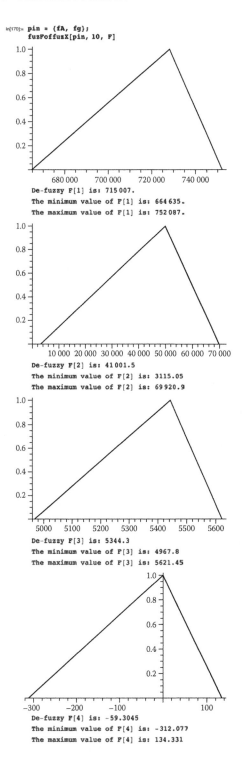

In[170]:= pin = {fA, fg};
fuzFoffuzX[pin, 10, F]

De-fuzzy F[1] is: 715 007.
The minimum value of F[1] is: 664 635.
The maximum value of F[1] is: 752 087.

De-fuzzy F[2] is: 41 001.5
The minimum value of F[2] is: 3115.05
The maximum value of F[2] is: 69 920.9

De-fuzzy F[3] is: 5344.3
The minimum value of F[3] is: 4967.8
The maximum value of F[3] is: 5621.45

De-fuzzy F[4] is: -59.3045
The minimum value of F[4] is: -312.077
The maximum value of F[4] is: 134.331

The results indicate that given uncertainty in **A** and **g**, we get uncertainty in our final position ($F[1]$, $F[2]$) as well as velocity ($F[3]$, $F[4]$). Notice our final orbit is low, slow, and heading down since the defuzzified final altitude ($F[2]$) is 41,001.5 feet, rather than the desired 50,000 feet; the defuzzified final horizontal velocity ($F[3]$) is 5344.4 feet/second, rather than the desired 5444 feet/second; and the final defuzzified vertical velocity ($F[4]$) is −59.3 feet/second, rather than 0. So we immediately know we need better decisions for steering. I will show how fuzzy logic can be used to obtain superior steering decisions in Chapter 3.

2.3.4 WITH AGENTS

The components of an agent based model are as follows:[25]

- *Agents*: Coded rules that take inputs from the environment and other agents to produce outputs. The rules enable perception of the environment or other agents, performance instantiated by motion, communication or action, memory, and policy.
- *Environments*: Codes that create a situation external to the agents.

Several agents of the same, or dissimilar, definition interact within the environment, and their impact on each other or the environment is noted. The motivation is that by causing many relatively simple events occurring simultaneously, the grander scheme will be observable. These models seek to simulate behavior when an aggregating principle remains unknown. Modeling uncertainty using agent-based models is a developing procedure. The ability to utilize large numbers of parallel processors for massive amounts of calculations enables such simulations. As is common with new things, the method has extreme skeptics as well as extreme advocates. I know of one person who is adamant that the only appropriate means to model a military conflict is by agent-based models, his claim being that to do so by any other method was inherently dishonest. His argument, in a nutshell, is that the results of military conflicts are so uncertain, with so much that can depend on the instances of a few key interactions, that no aggregation method would ever be truthful. Yet, ample evidence exists that the aggregate models based on Lanchester equations accurately predict many military conflicts.[26]

In systems engineering and architecting, the most valuable models are predictive. They enable us to anticipate something before it happens, thereby using that information to take action either to alter the events or to better prepare for the inevitable. Predictive models are only as good as the accuracy of their prediction and the opportunity time they provide between the receipt of the prediction and the occurrence of the predicted event. There is much comfort in using a validated aggregation rule because one is reasonably certain the result will be correct and useful. Validation of the agent-based model remains the biggest challenge to acceptance. For example, one of the earliest agent-based models was of two sets of agents, one red and one green, initially randomly placed on a large two-dimensional grid. At each step in the simulation, each agent checked its eight possible neighbors, determining the fraction of those of the same color. The control variable was a tolerance level. If the fraction was above it, the agent stayed put; if the fraction was below it, the agent moved one spot away if any open spots were available (no two agents could occupy the same spot). It was found that for tolerance levels of 0.3 or higher, after numerous simulations, the environment became a checkerboard of agents of the same color as the agent. The inference was that a group of people with any intolerance to live near-dissimilar people will naturally form neighborhoods of similar people. But is this conclusion valid? Can it be used for prediction? I honestly do not know the answers to these questions, and because I don't, I am uncertain what to do with such a model. This issue is common to many agent-based models.

Clearly the technology will evolve and its value become more established. Also clearly, our job as system engineers and architects is to provide models to help make decisions. For those situations in which an aggregation rule is unknown, then agent-based models enable analysis to proceed. Here is an illustration. Suppose we are considering improving the intelligence, surveillance and reconnaissance, and communications (ISRC) abilities of our military forces. Intuitively, each should be beneficial. And people accept this. But there are a lot of good ideas out there. How much benefit would these improvements provide, say, relative to other measures? An agent-based model gives us the ability to approximately answer these questions.

The environment consists of two 50 × 50 unit squares, one considered Blue territory, and the other Red. A location on either is denoted by an **i** and **j** value. Initially, there are 25 Blue agents. Each will be assigned to a fixed random location in their territory. Initially, there are 50 Red agents; each will be assigned a fixed random (**i, j**) location in their territory as well. Each Blue and Red agent has a device that, up to a range of 75 units, can destroy another agent with probability of kill (**Pkill**, presumed to be 50%). A unit is the distance between a spot (**i1, j1**) and a spot (**i2, j2**) measured by

$$Sqrt[(i1 - i2)\wedge 2 + (j1 - j2)\wedge 2]$$

In the first simulation, none of the agents know where the enemy agents are, nor can they communicate, or maneuver. Of interest is how long does it take the Red agents to destroy the Blue agents, and what is the relative loss rate? The form requirements *BluevsRedAreaFire* create the environment, and randomly locate the Blue and Red agents. Then time is advanced, and at the start of each time interval, all the remaining agents fire randomly into enemy territory, and the number of surviving agents is noted.

After 50 simulations, the results are

```
In[172]:= BluevsRedAreaFire[50, 25, .5, 1, .5, 1, 50]

       Mean Blue to Red loss ratio 2.14942

       Mean Blue shots 472.

       Mean Red shots 1012.
```

As expected, since Blue is outnumbered two to one, it has a fatality rate more than twice as fast as Red's.

To investigate the ISRC benefits, a new formal requirement, *BluevsRedwith BlueDirectedFire*, is written to simulate that Blue takes advantage of an ISRC capability that lets Blue know where all the Red targets are and each Blue agent then targets a random Red agent at its known location. In plain English, in this simulation, Blue agents now fire directly at Red agents, without coordination, while the Red agents still must randomly fire into Blue territory.

Here is the result of 50 simulations with Blue having perfect knowledge of Red's location, but no coordination of fire:

```
In[173]:= BluevsRedwithBlueDirectedFire[50, 25, .5, 1, .5, 1, 50]

         Mean Blue to Red loss ratio 0.0947644

         Mean Blue shots 50.

         Mean Red shots 60.
```

Perfect knowledge of Red's location, with Red oblivious to Blue agent locations, has reduced the Blue-to-Red kill ratio from over 2 to 1 to about 0.1 to one. What is still happening is each of the Blue elements is randomly selecting a Red target, so as the targets reduce, there is a lot of duplicated targeting. This is beneficial for Blue, as even two engagements of the same target increase the effective kill probability to 75%. Blue would be more efficient if they coordinate their attacks, which requires communications.

In this final modification, the formal requirements *BluevsRedwithBlueDirected FireandComm*, two Blue entities are chosen at random to engage the same identified target.

Here is the result of simulations of Blue having perfect knowledge of Red agent locations; Red agents are oblivious to Blue agent locations, and Blue agents can communicate to coordinate their attacks:

```
In[174]:= BluevsRedwithBlueDirectedFireandComm[50, 25, .5, 1, .5, 1, 50]

         Mean Blue to Red loss ratio 0.0186881

         Mean Blue shots 100.

         Mean Red shots 8.
```

Directed fire with communications enables the Blue forces to concentrate fire on Red agents, thereby essentially guaranteeing kill as a result of each coordinated attack. The relative loss ratio is now hugely in Blue's favor even though Blue is initially outnumbered by 2 to 1. All 50 Red agents are killed with essentially two Blue shots each. The Red agents are killed so swiftly that they barely get any shots off at all.

The agent-based models suggest ISRC information reduces relative loss rates on the order of 200 times, from worse than 2 to 1 to almost 0.01 to 1. So, clearly agent-based models enable quantification of behavior that at first appears difficult or impossible to discern using aggregation principles. Yet, difficult does not mean impossible. So, if an aggregation principle is not known, certainly utilize agent-based models to enable some level of prediction. But do not confuse your own inability to produce a set of equations based on aggregation principles with the presumption that no aggregation principle exists.

2.3.5 With Fractals

The nature and use of fractals to model real-world physical phenomenon are well documented.[27] Fractals are relatively simple rules recursively exercised to produce a result. Find the right rules, and one can simulate many different things, from what mountain ranges look like, to the nooks and crannies of a coastline, to the location, diameter, and length of branches of capillary systems such as plant appendages, rivers, arteries, and nerves. Fractal shapes can have desirable properties. Antennas with fractal shapes cover a very small area but are sensitive to a wide range of frequencies, making them very practical for small devices such as cell phones. Research continues on how to use the fractal as a predictive tool. For example, if one thing is measured (say, the branching of tree limbs or capillary networks), then other things are predicted (say, the total number of leaves in a forest given a sampling of trunk circumferences, or the presence or absence of a tumor). For readers unfamiliar with the method, I urge they undertake further education in this area. What follows is an illustration that I hope will motivate such investigation. Suppose one needs to produce many representations of a fern. We do not want all the ferns to be exactly alike, nor do we want them to be randomly altered in some manner that would result in them not looking like real ferns. *FernBuilder*, based on an algorithm presented in reference 28, produces a possible fern, but never produces the same fern twice. (Note that the color was set to black to make producing this book more convenient; however, it is easy to modify the routine to introduce appropriate colors.)

Here are ferns using 100, 1000, and 10,000 iterations:

In[175]:= **FernBuilder[100]**

Out[175]=

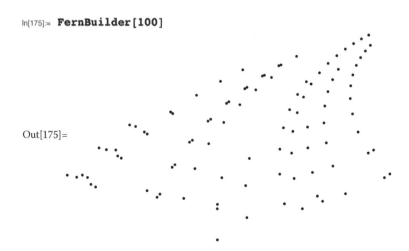

In[176]:= **FernBuilder[1000]**

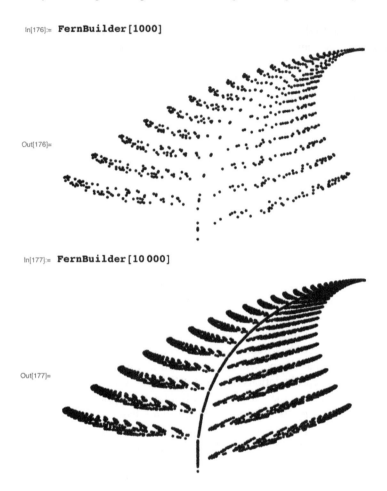

Out[176]=

In[177]:= **FernBuilder[10 000]**

Out[177]=

2.4 MONITORING SYSTEM OR ARCHITECTURE TECHNICAL PERFORMANCE MEASURES

While a system or architecture is being produced, modeling may be used to periodically predict the expected performance of the few key technical performance measures (TPMs).

There are two types of TPMs: those associated with a continuing process, and those representative of the physical features of the system or architecture. Examples of process metrics are defect rates, throughput quantities, or delivery times. Process TPMs are monitored using run charts.[29,30] Run charts need to be made showing the corresponding control limits. This is critical, as only processes that exhibit statistical control can be improved. If the process is out of control, the first priority must be to establish control.

The following formal requirements provide the different types of control charts. Each example uses data found in reference 30.

The first type is the *c chart*, which is for the number of nonconforming items from multiple samples of identical quantities.

The variable **ctest** holds data representing the number of blemishes on samples each with a fixed area. Then *cControlChart* is used to produce the control chart.

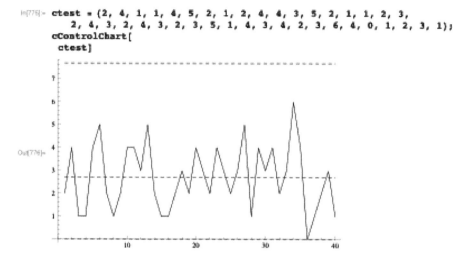

```
In[775]:= ctest = {2, 4, 1, 1, 4, 5, 2, 1, 2, 4, 4, 3, 5, 2, 1, 1, 2, 3,
          2, 4, 3, 2, 4, 3, 2, 3, 5, 1, 4, 3, 4, 2, 3, 6, 4, 0, 1, 2, 3, 1};
        cControlChart[
          ctest]
```

The production process is out of control as one of the non-conformance counts touches the lower control limit of 0. Process changes may be attempted to reduce the non-conformance variability, to reduce the mean non-conformance rate.

The second type is the *u chart*, which is for the number of non-conforming items from multiple samples with different quantities.

The variable **utest** is an array of two elements, the first element is the number of defective items, the second number is the number produced that day. Then *uControlChart* is used to produce the control chart.

```
In[178]:= utest = {{14, 39}, {4, 45}, {5, 46}, {13, 48},
          {6, 40}, {2, 58}, {4, 50}, {11, 50}, {8, 50}, {10, 50}, {3, 32},
          {11, 50}, {1, 33}, {3, 50}, {6, 50}, {8, 50}, {5, 50}, {2, 50}};
        uControlChart[
          utest]
```

The production process may now be in control, as just on the first day of production the fraction of errors was greater than the upper control limit.

The third type of control chart is the *np chart*, which is the count of items possessing an attribute out of a fixed number of items.

The variable **nptest** holds data representing the number of rejected parts per basket of 60 items, for 21 baskets. Then *npControlChart* is used to produce the control chart.

In[180]:= **nptest =**
        {11, 20, 19, 24, 19, 18, 16, 42, 18, 24, 15, 17, 19, 26, 19, 22, 21, 32, 22, 33, 30};
        npControlChart[nptest, 60]

Out[181]=

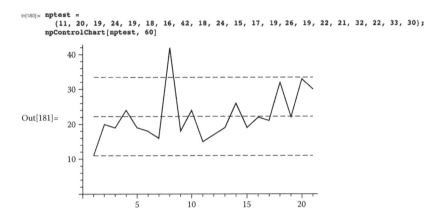

The production process is out of control. Immediate effort should be limited to get the process of rejecting parts under control.

The fourth type of control chart is the *p chart*, which does the same thing as the np chart, but allows for varying sample sizes.

The variable **ptest** holds data representing a daily number of items with errors out of a total number of items. Then *pControlChart* is used to produce the control chart.

In[182]:= **ptest = {{20, 98}, {18, 104}, {14, 97}, {16, 99}, {13, 97}, {29, 102},**
        {21, 104}, {14, 101}, {6, 55}, {6, 48}, {7, 50}, {7, 53}, {9, 56},
        {5, 49}, {8, 56}, {9, 53}, {9, 52}, {10, 50}, {9, 52}, {10, 47}};
        pControlChart[
        ptest]

Out[183]=

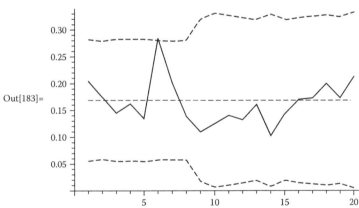

The production process is out of control since on day 6 the upper control limit was exceeded.

The fifth and final type is the average and range chart, also called an *XR control chart*; these are the most versatile.

The **xrtest** holds the average and range for two sampled items for 30 samples. Then *xrControlChart* is used to produce the control chart.

```
In[184]:=  xrtest = {{255, 10}, {330, 20}, {280, 100}, {235, 10}, {230, 40},
            {240, 0}, {280, 20}, {235, 50}, {240, 20}, {315, 30}, {325, 10},
            {280, 60}, {260, 20}, {275, 50}, {330, 100}, {250, 40}, {320, 80},
            {260, 60}, {275, 30}, {295, 30}, {225, 30}, {300, 40}, {330, 0}, {275, 10},
            {290, 20}, {295, 10}, {265, 10}, {280, 20}, {285, 10}, {295, 10}};
          xrControlChart[
            2,
            xrtest]
```

Xbar Control Chart

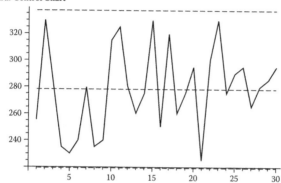

Out[185]=

Range Control Chart

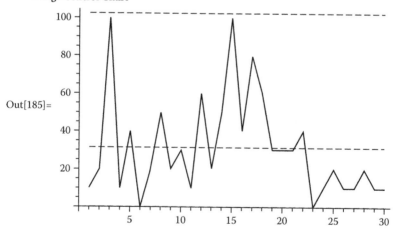

Though none of the average or range value exceeds the control limits, the process is still out of control, because there are many consecutive range samples which fall below the mean range.

For a physical feature TPM, any apparent trend in the plotted values versus date of evaluation is usually an illusion. Physical feature TPMs change because the assumptions used to calculate the prediction changed, or how the metric is calculated changed. For example, mass margins almost always shrink as presumed mass estimates are replaced with estimates that take into account parts not anticipated, as well as the reality of the actual part mass versus prior assumptions. This phenomenon is so expected that it is actually possible to assign contingency percentages to add to a current mass estimate based on the maturity of the design.[31] The date of the evaluation has nothing to do with the mass estimate, so the mass margin history versus time cannot be used to predict mass in the future. For physical feature TPMs, what is most useful to plot are three values: *was*, *is now*, and *will be*. *Was* denotes the prior estimate as a reference to explain why an action may now be required that was not then, or vice versa. *Is now* denotes what the value will be if nothing is changed. *Will be* denotes the predicted value if the recommended actions are taken.

As discussed throughout this chapter, the prediction must include allowance for bad things that can yet happen—with the badness estimated using a Monte Carlo or fuzzy logic analysis. If using fuzzy logic, plot the defuzzified value, unless there is a need to be very conservative, in which case multiply the defuzzified value by a factor of safety. If using a Monte Carlo analysis, then plot the value at the probability level for which you can tolerate failing to achieve the prediction. Note that this means for a constraint, plot how close it could get to or exceed the constraint, while for a desired value, plot how distant it might be. What is usually most useful to show is the margin, or predicted difference between the target or limit and the prediction. To help comprehension, always define *margin* so that positive is good, and negative is bad, with the legend used to remind everyone the limit or desired value. Provided the limit or desire is not 0, it is also useful to plot the margin percent, which is the difference divided by the target or limit. Finally, since the purpose of the plots is to notice a situation developing so that decisions can be made to do something about them, quite often program managers or customers will insist that a trigger line be established, so it becomes immediately apparent that action is needed once the margin falls below the trigger amount, but is still well away from the limit.

The formal requirements *MarginMonitor* track margins. An example follows. Suppose the mass limit is 30,000 kg, the trigger is 5% of the limit, and the predicted history is as shown in Table 2.7.

TABLE 2.7
Example Margin Monitor

When	Mass Estimate (kg)	Contingency Mass Increase (kg)
Was	22,500	5000
Is	27,500	3900
Will be	28,000	1000

Then, the inputs to the formal requirements and the results are

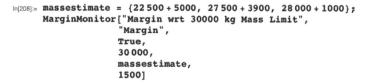

```
In[208]:= massestimate = {22 500 + 5000, 27 500 + 3900, 28 000 + 1000};
        MarginMonitor["Margin wrt 30000 kg Mass Limit",
                      "Margin",
                      True,
                      30 000,
                      massestimate,
                      1500]
```

Out[209]=

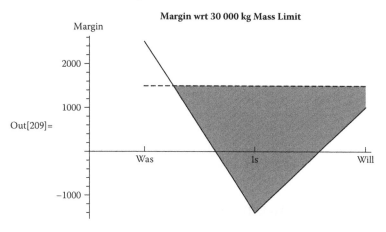

Similarly, for a key parameter that should not go below a 60,000 kg lift mass limit, the predicted history is shown in Table 2.8.

TABLE 2.8
Example "More Is Better" Margin Monitor

When	Mass Estimate (kg)	Contingency Mass Increase (kg)
Was	70,000	5000
Is	65,000	2500
Will be	61,000	500

Providing *Margin Monitor* the data in Table 2.8 results in

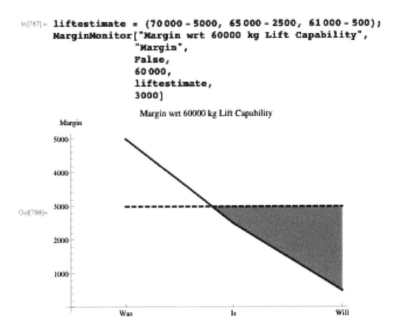

```
In[787]:= liftestimate = {70000 - 5000, 65000 - 2500, 61000 - 500};
        MarginMonitor["Margin wrt 60000 kg Lift Capability",
                     "Margin",
                     False,
                     60000,
                     liftestimate,
                     3000]
```

Margin wrt 60000 kg Lift Capability

2.5 MODELING HEURISTICS

- Models are valuable in proportion to how quickly and inexpensively they provide trusted results to make decisions.
- The modeler's adage is, "All models are wrong; some models are useful." Determining the appropriate level of model fidelity is an art, not a definable process.
- Build on models used to produce successfully operating systems and architectures.
- Every machine performs a little differently every time it is turned on, so when human life is at stake or a high probability of mission success is to be achieved, it is necessary to model the impact of these system dispersions.
- When human life is at stake or a high probability of mission success is to be achieved, the impacts of environmental variability must be modeled.
- Determining what to model is as important as determining how to make the model.
- That a model is correct must be proven with evidence.
- If real-world test data are not available, the best method to verify a mathematical simulation is to compare results with at least two other independently developed and operated mathematical simulations.
- Do only the analysis needed to make the decision needed.
- The model maker is responsible for constructing a model that can be used by the decision makers.
- The value of the model is determined by those who use the results, not by the model makers.

- Make lots of small models, which are easier to verify and thus easier to be trusted, rather than one big model.
- Architecture-level analysis needs to be a notch higher in abstraction than system-level analysis, just as system-level analysis is a notch higher in abstraction than subsystem-level analysis.
- The following types of diagrams help comprehend a system or architecture:
  1. Reference mission
  2. Key performance parameters
  3. Customer, user, and operator hierarchies
  4. Constraints
  5. System boundary
  6. State transition diagram
  7. Functional hierarchy
  8. Functional flow
  9. Functional timeline
  10. Information structure
  11. Trade tree
  12. Configured end item hierarchy
  13. Functional allocation
  14. Interface diagram
- At any time, different methods to make such diagrams will be in vogue. Research the methods, and use the one that allows the most reuse from existing systems similar to the one being developed.
- Useful diagramming prescriptions are
  1. IDEF
  2. Model framework
  3. Department of Defense Architecture Framework
  4. SySML
  5. UML
  6. Colored Petri nets
  7. Higraphs
- Four very useful sources for reusable mathematical models are
  1. Wolfram Research website
  2. MATLAB® website
  3. National standards website
  4. Cambridge numerical recipes
- Establishing models for systems is more art than execution of a well-defined process.
- The three things to do to mathematically model a system are as follows:
  1. Define functions that constitute the systems.
  2. Establish at least one worst-case representative design reference mission for the system to accomplish, such that if it is achieved, all other potential uses of the system are acceptable.
  3. Establish the fewest and simplest equations that represent the system doing the scenario along with the interfaces to the necessary external entities that can be used to establish requirements to implement the system.

- Three design reference missions (DRMs) are usually sufficient for any system. For very complicated systems, make fewer reference missions that are more difficult to achieve, rather than many easy-to-do DRMs.
- Attempting to model a system is an extraordinarily powerful proxy for assessing if the system should be made.
- The modeler has the responsibility to make it known if the proposed system is impossible or not necessary.
- Avoid using utility curves to model systems because customers and users find it difficult to make decisions based on such models. Make models based on physics instead.
- "Designated experts" usually aren't.
- The nine things to do to mathematically model an architecture are as follows:
  1. Establish at least one representative architecture reference mission for the system to accomplish, such that if it is achieved, all other potential uses of the system are acceptable.
  2. Identify all the existing systems currently used to achieve the mission and analyze the current extent to which the mission is being performed and at what total ownership cost.
  3. Identify all the performance, capability, mission, and if necessary, campaign-level shortfalls.
  4. For each shortfall, identify optional ways of using existing systems or a new system to potentially close the shortfall.
  5. Fully define the candidate systems using physics to predict the ability to achieve the mission and estimate the total cost of ownership.
  6. Identify all the possible combinations of systems that could conceivably achieve this mission.
  7. Establish the fewest and simplest equations that enable determining how well the candidate systems achieve the reference mission.
  8. Establish the cost to obtain and operate the candidate systems while achieving the architecture design reference mission.
  9. Plot each candidate family of system with respect to a vertical axis denoting mission goodness versus a horizontal axis denoting total cost of ownership.
- The weaknesses of existing systems are known, while the weaknesses of conceptual systems are yet to be known.
- When comparing systems with multiple owners and users, calculate their costs to all those who truly bear their burden.
- To be useful, models of uncertainty must be true to the nature of the uncertainty—random, fuzzy, evolving from underlying simple behavior, or as a consequence of noninteger dimensions.
- Extrapolating beyond the range of independent variables used to derive a regression is a fool's bet.
- Monte Carlo analysis predicts the consequences of random uncertainty.
- Fuzzy numbers and logic predict the consequence of value uncertainty.
- Agent-based models predict the consequences of aggregation uncertainty.

- If an aggregation principle is not known, utilize agent-based models to enable some level of prediction. But do not confuse your own inability to produce a set of equations based on aggregation principles with the presumption that no aggregation principle exists.
- Fractal models predict the consequences of dimensional uncertainty.
- There are two types of technical performance measures, those associated with a continuing process, and those representative of physical features of the system or architecture.
- Process metrics should be monitored using run charts.
- Physical feature metrics should have their *was, is now,* and *will be* values monitored with respect to margin trigger levels, with values discounted for the bad things that can still happen, using either a Monte Carlo or fuzzy logic analysis.

REFERENCES

1 http://www.idef.com.
2. Hotly, Derek, Hruschka, Peter, and Pirbhai, Imtiaz, *Process for System Architecture and Requirements Engineering*, New York: Dorset House, 2000.
3. http://www.architectureframework.com/dodaf/.
4. http://www.sysml.org.
5. http://www.uml.org.
6. Jensen, Kurt, *Coloured Petri Nets*, Berlin: Springer-Verlag, 1997.
7. Fogarty, Kevin, and Austin, Mark, "System Modeling and Traceability Applications of the Higraph Formalism," *Systems Engineering* no. 2 (Summer 2009): 117–40.
8. http://www.wolfram.com.
9. http://www.mathworks.com.
10. http://www.nist.gov.
11. http://www.cambridge.org/us/numericalreceipes.
12. Taleb, Nicholas Nassium, *The Black Swan: The Impact of the Highly Improbable*, New York: Random House, 2007.
13. Huynh, Thomas V., Kessler, Andrew, Oravec, Joseph, Wark, Shaunnah, and Davis, Jennifer, "Orthogonal Array Experiment for Architecting a System of System Responding to Small Boat Attacks," *Systems Engineering* 10, no. 3 (2007): 241–59.
14. Bardossy, Andras, and Duckstein, Lucien, *Fuzzy Rule-Based Modeling with Applications to Geophysical, Biological, and Engineering Systems*, New York: CRC Press, 1995.
15. Johnson, D. L. (editor), *Terrestrial Environment (Climatic) Criteria Guidelines for Use in Aerospace Vehicle Development*, Huntsville, AL: NASA, 1993.
16. Bless, Robert R., Hodges, Dewey H., and Seywald, Hans, "Finite Element Method for the Solution of State-Constrained Optimal Control Problems," *Journal of Guidance, Control and Dynamics* 18, no. 5 (September–October 1995): 592–9.
17. *Space Shuttle Design Criteria: System and Environmental Dispersions*, NSTS 08209, vol. 33, Huntsville, AL: NASA, 1992.
18. Jamshidi, Mohammad, Vadiee, Nader, and Ross, Timothy J. (editors), *Fuzzy Logic and Control: Software and Hardware Applications*, Englewood Cliffs, NJ: Prentice Hall, 1993.
19. Kosko, Bart, *Neural Networks and Fuzzy Systems: A Dynamical Systems Approach to Machine Intelligence*, Englewood Cliffs, NJ: Prentice Hall, 1992.
20. Hanss, Michael, *A Nearly Strict Fuzzy Arithmetic for Solving Problems with Uncertainties*, University of Stuttgart, Germany: Institute A of Mechanics.

21. Dienemann, Paul F., *Estimating Cost Uncertainty Using Monte Carlo Techniques*, Santa Monica, CA: Rand Corporation, 1966.
22. Kwak, Young Hoon, and Ingall, Lisa, "Exploring Monte Carlo Simulation Applications for Project Management," *Risk Management* 9 (2007): 44–57.
23. Larson, Wiley J., and Wertz, James R., *Space Mission Analysis and Design*, 2nd ed., Norwell, MA: Microcosm, Inc., and Kluwer Academic Publishers, 1992.
24. Citron, S. J., *Elements of Optimal Control*, New York: Holt, Rinehart and Winston, 1969.
25. Gilbert, Nigel, *Agent-Based Models*, Thousand Oaks, CA: Sage Publications, 2008.
26. Przemieniecki, J. S., *Mathematical Methods in Defense Analysis*, 3rd ed., AIAA Educational Series, Reston, VA: AIAA, 2000.
27. Mandelbrot, Benoit B., *Fractals: Form, Chance, and Dimension*, New York: W. H. Freeman and Company, 1977.
28. See http://en.wikipedia.org/wiki/Fractal.
29. Deming, W. Edwards, *Out of Crisis*, Cambridge: Massachusetts Institute of Technology, 1989.
30. Wheeler, Donald J., and Cambers, David S., *Understanding Statistical Process Control*, Knoxville, TN: Statistical Process Control Institute, 1986.
31. *AIAA Standard: Mass Properties Control for Space Systems*, S-120-2006e, Reston, VA: AIAA, 2006.

3 Make Decisions

Decisions must be made to establish system or architecture requirements, and to determine the implementation approach. Rational decisions follow conscious effort, as our brains are wired to make intuitive decisions.[1] Table 3.1 summarizes some of the errors intuitive decisions are prone to. What constitutes a good decision? To answer that simple question, we need to address seven aspects of the decision:

1. What needs to be decided?
2. When is the decision needed?
3. Who needs to make the decision?
4. What options are there to choose from?
5. What is it about each option that distinguishes it from the other options with respect to what needs to be decided?
6. How can appropriate and accurate information be obtained to assess each option with respect to what needs to be decided?
7. How robust is the decision should the goodness criteria assessment be flawed?

Making a good decision means getting all seven factors "right," which is rarely an easy thing to do for complicated situations. Formal methods usually improve only our ability to assess the robustness of the decision. Activities to identify the "correct" decisions to make, identify the most appropriate decision maker, fully explore the option space, and distinguish relevant and useful data from bad data are all more an art than a science. Only after a type of decision was made many times, so all the seven factors have knowable bounds, can formal methods find the best decision. For example, deciding what car to buy with what features or what medical treatment to perform for a well-known disease with a cure can both be formulated in such a way that all seven questions can be addressed in a formal way. But deciding the best car to make, or the details of a successful medical treatment, is not fully determinable by formal methods alone.

The following case studies illustrate real-world examples of difficult decisions to make. The first case concerns whether to adopt a system configuration that appears to provide better performance, but at higher cost.

Case Study 3.1: Deciding System Features

BACKGROUND

The government's request for proposal for the National Polar–orbiting Operational Environmental Satellite System (NPOESS) included a reference design. In that

TABLE 3.1

Potential Causes for Faulty Intuitive Decisions

Potential Cause for Intuitive Decision Error	Example from Case Studies
Jump to conclusions, that is, generalize on insufficient evidence.	Do not need day of launch I-load update because Shuttle launches have never been scrubbed due to winds.
Too much weight on own experience.	The next proposal must be lowest cost because prior proposals were lost because the cost was too high. AND Dinged tiles won't cause the loss of a Shuttle.
Confirmation bias—when first number or option mentioned limits the options considered.	Providing crew with an escape method is more important than other Category 1 failures.
Hubris.	Can solve a vexing technical issue others have repeatedly failed to solve. AND Can win in one year more proposals than won in prior decade.
Disaster myopia.	Cold weather cannot cause Shuttle loss.
Averse to ambiguity—selecting the surer option when the uncertainty option is better.	Rather use an I-load created for a season than a to-be-determined I-load for the day.
Procrastination and laziness.	Continue to use utility curves for option assessment when decision makers express dissatisfaction with the method.

design, two existing, government-owned, high-latitude ground stations were to collect the instrument data from the satellite, and then forward those data to computational facilities in the United States, where the instrument data were transformed into many weather products. Since the satellite orbits sometimes failed to fly within range of either of these ground stations, on those occasions, up to two orbits would transpire before the raw data stored on the satellite were finally sent to the ground for processing. This was significant because the government had very stringent data delivery timeliness requirements. Whenever the orbit failed to over fly one of the ground stations, all the timeliness requirements were not met for a large fraction of two orbits of data. Engineers on the proposal team developed an alternative architecture, which they called SAFETYNET™ and subsequently patented. SAFETYNET had numerous antennas spread around the world, each near a major fiber hub. With SAFETYNET, each satellite was assured to come into range of at least one of these receivers every orbit. Indeed, most orbits overflew more than one of the ground stations. So the data timeliness requirement was actually exceeded. But this came at the cost of buying, installing, and maintaining many more ground receivers, rather than relying on established government facilities already in place. The SAFETYNET option was repeatedly briefed to the customer during the contract activity that paralleled the proposal period. But prior to submitting the proposal to obtain the contract, the customer gave no explicit indication they preferred it to their own reference. The decision the capture team had to make was whether to propose the government reference solution, or the SAFETYNET solution. The decision made was to propose the SAFETYNET architecture.

Many years later, after the NPOESS effort had experienced massive time delays and cost growth, the same customer issued a request for proposal for a replacement for their Geostationary Operational Environmental Satellites, called GOES-R. Satellite missions are accomplished by their payloads, which in this case were weather instruments. The government had previously awarded parallel, competitive payload development contracts first, before giving serious attention to the spacecraft, or the ground-based data-processing systems necessary to turn the instrument data into weather products. They sought to get all the instruments mature before beginning detailed spacecraft and ground system engineering and production. This was a well-thought-out attempt to minimize program risk. Space systems often turn out to be much more difficult to accomplish than originally envisioned. Considerable delays and cost growth occur if a spacecraft is built to host a satellite with a mass of 100 kg, needing 100 W of power that must be pointed to within 0.1 of a degree, then must be modified to accommodate more instrument mass or power or precise pointing.

The government wanted to fly on GOES-R a new large instrument called a *sounder*, since it would provide a profile of a column of the atmosphere. Producing the new sounder quickly became more difficult than anticipated. Evidence mounted that the sounder's costs were unpredictable and it would not be available in time for the first GOES-R. So the government canceled the development of the sounder. Up to that point, the three companies competing to be the spacecraft provider were studying how to best provide a new spacecraft to host all the instruments, including the sounder, which was by far the instrument needing the most spacecraft resources.

Once the largest and most demanding instrument was removed from the manifest, a decision had to be made about what to require of the spacecraft. One option would be to lower spacecraft cost by "shrinking" it to provide just enough volume, power, thermal control, pointing stability and accuracy, and mass allowance for the reduced instrument manifest. Or, the spacecraft could be left the same with the available resources providing margin for all the other instruments in development, or as a reserve capacity for a new instrument.

The proposal team studied the options, concluding that the larger spacecraft increased life cycle cost about 3%, but offered considerable margin for the remaining instruments, essentially mitigating their cost and schedule growth risks. However, should the sounder ever be remanifested, the larger spacecraft would probably exceed the lift capability of the desired launch vehicle, forcing the government to use a larger and more expensive launch vehicle. Prior to receiving the proposal to award the contract, the proposal team briefed the customer once, recommending that the GOES-R spacecraft be required to provide the resources needed for the now-vanished sounder, because for only a few percent larger life cycle cost, the total cost risk would be significantly reduced, since the spacecraft would need few modifications to host a sounder, should the development continue. The customer listened politely, but gave no indication of agreement.

We favored going with the larger spacecraft for competitive reasons as well. Each of our competitors had spacecraft assembly lines so they could offer a standard spacecraft less expensively than we could. But these standard spacecraft were primarily for communications satellites, not weather satellites, so they were all on the "small" side. The more demanding the GOES-R requirements, the more the competitors needed to grow their assembly line spacecraft, negating their apparent cost advantages. Our spacecraft was based on a successfully operating design built a decade ago, but we had not built one in the interim. Our NPOESS spacecraft was also based on the prior design. Though on a briefing chart our prior, GOES-R, and NPOESS spacecrafts looked very similar, in reality their sizes and internal components were very different. This meant we had to offer a new spacecraft for GOES-R, with all the inherent risks, but this seemed preferable to claiming a modification of the NPOESS spacecraft that still existed only on paper and had associated with it a lot of bad press. The customer was a sophisticated buyer. They had been buying and operating GOES spacecraft from different providers for decades. They understood they were asking for a new spacecraft, not one off an assembly line. They recognized the differences between the spacecraft we were offering and that under development for NPOESS.

The proposal team needed to decide what spacecraft to offer. The proposal team that had worked with the customer for over two years was convinced the customer saw GOES-R as a risky project. The instruments were mostly new and the spacecraft was new with pointing requirements significantly more difficult to achieve than the current GOES. We decided the customer wanted the lowest risk approach. So we elected to baseline the larger satellite based on our opinion that although the customer didn't literally say it, the larger spacecraft was the lower risk approach.

The year 2007 shaped up to be a competitive turning point for our company. Six major proposals needed to be responded to during the year. The GOES-R proposal was slated to be the last one received. Indeed, the company's senior leadership thought it would be delayed at least a year, since the existing GOES spacecraft appeared to have plenty of operational life remaining, and their replacements were yet to be launched. Also, since we already had the NPOESS contract, which was not going particularly well, the senior leadership judged it unlikely that we would also be given the geosynchronous weather satellite work. The logical conclusion was that GOES-R was the lowest priority capture effort. Those of us who worked with the customer were certain the proposal would come out late in the year because the customer had expressed profound concern that the spacecraft being launched now may not achieve their design lives and they wanted GOES-R ready to go soon, and because the leader of the customer team had always met his publicly announced dates. We judged that if he said the request for proposal would come out on a specified date, it would. As it turned out, we lost all but two of the major proposals, spending well over $40M in discretionary money on one futile bid to win the Tracking and Data Relay Satellite (TDRS) replacement contract away from an incumbent. Then, the GOES-R proposal came out in December, as advertised. Suddenly, GOES-R went from the lowest priority effort to a must win. The proposal asked that the spacecraft be able to accommodate the larger, eliminated instrument for the second and third satellites in the series.

New management was brought in to run the capture, and they concluded that the customer wanted lowest cost rather than the low-risk solution. This put us at a major disadvantage to our competitors, who each could offer modifications of spacecraft coming off their respective communications satellite assembly lines. The strategy to offer the lowest cost resulted in three fateful decisions: (1) team with Loral, who had a spacecraft assembly line, and had built earlier GOES spacecraft, but had not been engaged in the GOES-R effort; (2) propose that the GOES-R spacecraft was "the same" as the NPOESS spacecraft, thereby arguing that the NPOESS effort would pay for the bulk of the nonrecurring costs and that the GOES spacecraft would be fourth, fifth, and sixth off the (not yet currently functioning) assembly line; and (3) redesign our baseline offering so that it was smaller and could be launched by a less costly launch vehicle. All these decisions were made after the draft proposal was received, so there was no means to discuss them privately with the customer to gauge their opinion.

It was obvious to anyone with a spacecraft background that Loral's existing spacecraft could not achieve the GOES-R requirements. A good fraction of the proposal response time was consumed convincing senior management. Why? Because they had communications payload backgrounds, rather than spacecraft backgrounds. The new teammate could credibly offer two subsystems: the communication subsystem needed to get the instrument data delivered to the ground, which was very similar to systems they had produced, and we had not; and their lithium ion battery–based electrical power subsystem. The customer required this type of power system, and we lagged in the industry since we were still using nickel hydride batteries in our satellites. Loral was given a cost target for the communication subsystem, which they then declined to meet showing that the parts alone would cost more than the target, despite any negotiations that might be achieved with the providing suppliers. Loral did design a suitable lithium ion battery–based power system, but there was one catch. Loral built communications satellites using 100 volts to be efficient. For GOES-R, the customer required all the instruments to use 28 volt power and the rest of our spacecraft components also utilize this voltage. So a large number of proven voltage converters had to be added to Loral's heritage power subsystem. The extra mass of all these converters made the spacecraft arguably too heavy to fit on the desired launch vehicle. The mad last-minute design effort to come up with a smaller first spacecraft came to naught when it was shown we had little hope of reducing the mass enough to be able to be launched by the less costly launch vehicles and it was pointed that out our total costs for two similar but different spacecraft would be more than our one cost for the same spacecraft.

What Happened

The NPOESS contract was won and SAFETYNET was cited as one of the reasons why, but there is uncertainty regarding just how significant this feature was to the win. Prior to awarding large government contracts, the competitors often work with the government for many years. A significant factor with respect to winning such awards is to achieve good working relationships during these preliminary stages, and the NPOESS team did so, most likely better than the competition, who were long-time incumbents in the weather satellite business. It's possible the customer was tired of working with the current contractor and was simply ready for a new relationship. Also, our teammate Raytheon offered a superior data-processing system, which was critical. In addition, the instruments proposed to fly on NPOESS were, at time of award, so immature and so different from anything that flew before, that no existing spacecraft could be used without significant modification, so our proposed modifications to a successfully operating spacecraft were judged very credible and low risk.

Of the three GOES-R bidders, we were ranked last, though we were the lowest cost. We were given a significant plus grade for offering the larger spacecraft. However, our offering was graded negatively for being too heavy and for many other features of the spacecraft that were not liked. Indeed, the complaints echoed all the known complaints of the NPOESS spacecraft, which was still in development. It was clear the competition was to provide the lowest risk spacecraft, not the lowest cost. Both the higher rated proposals were perceived superior in this regard, and the winner much more so. Though the words were not explicitly used, it was obvious the customer was not about to award a contract for their geosynchronous spacecraft to the same organization providing them the troubled low–earth orbit spacecraft, NPOESS, so efforts to explicitly tie the GOES bus to NPOESS completely backfired.

Lessons Learned

When making proxy decisions, test the candidate decision with the proxy. A proxy decision is one where you have to make a decision in anticipation of someone else. When you submit a proposal, inevitably, given the ambiguity in the customer's desires, you have to make decisions as though you were the customer, but you are not. You will find it virtually impossible to avoid your own prejudices in determining the answers to the seven aspects of the decision-making process, despite all your attempts to the contrary. So, it is vitally important to test your decision with the real decision maker before executing it. The NPOESS team was able to do this with SAFETYNET; the GOES-R proposal team made almost every decision wrong from the customer's perspective, but they didn't know this because they couldn't communicate with the customer privately during the proposal phase.

Complicated situations require a hierarchy of decisions to be made, so make sure the top decision is made correctly. The decision to offer SAFETYNET was an important but not overriding decision. If decisions regarding which people to tell the customer would constitute the NPOESS team, ground segment design, or spacecraft design had been faulty, that we offered SAFETYNET by itself would not have been enough to win. For GOES-R, the key decision was whether to offer the lowest risk or lowest cost spacecraft. Once that decision was made incorrectly, all the subordinate decisions, though perhaps appropriate for achieving lowest cost, were wrong by default.

The wrong people will make wrong decisions. The decisions on how to approach the GOES-R capture were made by the most senior management of the organization after the proposal was received. Virtually none of these people had spent any time with the customer and therefore could only guess what the customer might really want. During that year they had received feedback that they had lost four major bids in a row because our costs were too high. They were determined not to make that mistake again. And they didn't. Based on the customer's feedback, our GOES-R cost was the lowest and judged credible. Unfortunately, the customer wasn't looking for the lowest advertised cost; they were looking for the lowest risk spacecraft. The customer found little to admire about our spacecraft relative to the competitors'. Since all the key decision makers knew communication payloads much better than they did spacecraft, none instinctively recognized a good spacecraft from a bad one. Three months after the loss, the organization was taken over by another part of the corporation and most of the principals were reassigned or replaced.

The second case study concerns deciding whether to keep a problematic system as a part of a system of systems.

Case Study 3.2: Kill or Fix a Key System?

Background

The space superiority family of systems included a system that was years late and way over budget. That the system was not meeting requirements was clear, but it was uncertain what to do to correct the problems. Even figuring out what the problems were was difficult. Some customers were certain a working system would add a very important capability. Some customers perceived that stopping now, after a huge outlay of resources, would be a waste of all those resources. The customer's key question was whether to continue to build the system or stop. The contractor's key decision was whether to take responsibility to complete the system from the incumbent. Two parallel efforts were kicked off simultaneously to determine what to do. The architecture team was asked to determine the mission impact if the system was terminated, and what alternative systems might be best pursued. Meanwhile, a tiger team was assembled to review the system in depth to see if and how it could be salvaged. At the time, the architecting process described in the previous chapter was still being developed and all the needed processes and modeling tools were not yet in place, so the architecture assessment was done in an ad hoc manner. The architecture team concluded the system was not critical to the mission, given that some newer systems were about to be fielded successfully that were not even envisioned when the troubled system was initiated. In addition, the architecture team concluded that even if the troubled system was successfully developed, it would be extremely difficult to use operationally and maintain. Finally, the architecture team concluded parts of the troubled system did work well, and could be salvaged and utilized to make an effective alternative system that would have value and be easier to maintain. Basically, the ideal was to subsume the troubled system into an enhancement of the successful system that was procured recently. As the architecture team was assembled in a classified conference room formulating the briefing to present these findings, the leader of the tiger team came into the conference room, announced a decision was made that we would be taking over responsibility to complete the troubled system, so the architecture team recommendations were no longer needed, and walked out.

WHAT HAPPENED

The effort to produce the system proved to be tremendously more difficult than anticipated. Many issues were fixed. Perhaps symbolically, the system was literally hit by lightning during a test when it had been left unattended! The prime issue, that a non-real-time operating system was failing to operate the system in real time, was never resolved. The problem manifested itself most acutely in system tests. The people who designed and built the system were able to operate it, while the government's operators could not. Over a year after the salvage effort had commenced, the same government leader who had asked us to attempt to fix the system asked again whether the system was still worth salvaging. By this time the architecture process had matured tremendously. Within one day, convincing evidence was obtained and presented that the system had negligible value and its operating and sustenance difficulties would continue the pain. The tiger team lead, who had become the program manager in charge of fixing the system, was given a few weeks to report what to do to fix the system. He didn't have much to say. The government program manager canceled the system. The government leader was so unhappy that the system had to be terminated that he reduced our award fee for that period, and for every period in the future for which he was our award fee determination authority.

LESSONS LEARNED

Hubris comes before the fall. Don't get talked into fixing other people's problems. You may think you're better than the other guy, and you might be, but that does not mean he messed things up so badly that you can figure it out and make the necessary fixes.

Make decisions when you have to, not before. Before spending a lot of time and effort to fix something that is not working, make sure it's worth fixing. One more day was all that was needed to wait for the architecture team's recommendation. The evidence may or may not have swayed the day, but the evidence would have recommended against proceeding. As a result, we would have lost the sales that came with the attempt to fix the system, but we may have earned somewhat less but still good sales providing the less risky alternatives, and we certainly would have avoided the zero award fees that ultimately led to the termination of the entire support contract, not just the salvage effort.

Constantly test if your decision is wrong, and give yourself options to alter your decision as new information becomes available. We like to think that once we made a decision, we were right and we should get on with getting the decision accomplished. The decision to proceed to attempt to fix the late, overrun, and not working system should have been accompanied with options to stop with criteria to execute those options. Instead, once the decision to proceed was made, the possibility of failure was not seriously considered, as surely all that was needed was for some good engineers to replace the bad engineers who had been working on it so far. What then results is a "death march" for which there was no escape except a success that was not achievable. If hard work was the only ingredient needed to fix the problems, those charged with trying to save the system gave all there was to give.

What is good for a system is not necessarily good for an architecture. Since an architecture is a combination of systems, it does not follow that what is good for a system is good for an architecture. The faulty system was marginally beneficial for the architecture, so all the hard work fixing the system was at best marginally beneficial to the architecture. The same resources, devoted to other systems, could have tremendously improved the architecture. No one responsible for individual systems is likely to discern this, so it is critical that someone watch out for the architecture and make system decisions for the benefit of the architecture.

The following case study illustrates what happens when conflicting motivations add high levels of emotion to the decision-making process.

Case Study 3.3: Launch?

Background

After the loss of Space Shuttle Mission 51L, a massive effort to recertify the entire system was undertaken that took years. When the day to resume launches finally arrived, the manager responsible for the contractor support for launch and for on-orbit and entry mission planning and support is called upon to give an advisory recommendation to NASA whether or not it is safe to do each aspect of the mission. As was explained in Chapter 2, this involved simulating the trajectory through measured winds and atmosphere to determine both if sufficient propellant was on board to get the payload to orbit and that loads on the wings would be below tolerances. Unfortunately, for the trial 24 hours before launch, numerous instances of wing load levels exceeded allowable limits, and the responsible contractor supervisor advised that a launch scrub may be needed. When this was reported to the president of the division, that supervisor, his lead analyst, the manager, the head of the loads group, the lead load analyst, and two VPs were called to the division president's office for a data review. The supervisor presented plots showing the loads exceeding limits, and the manager expected that would be that. To his growing astonishment, the president started ordering the supervisor to reverse his recommendation. At first, the manager was dumbfounded; he presumed the president had misconstrued the data presented, so the manager, through a series of questions, tried to determine what part of the assessment the president found faulty or otherwise that would lead him to be comfortable to order a launch go recommendation when the data clearly showed the reverse. Rather than helping, these questions induced a red-faced, spittle-producing, profanity-infused rage, during which the manager was told in no uncertain terms his career was at stake if he didn't change the recommendation. The manager sat there for a moment completely uncertain what to do. He looked at his supervisor and his lead analyst, and they were literally shaking. He looked at his peer from the structures department and their lead analyst, and they appeared to be willing to comply—which really puzzled the manager as it was their wings he was trying to protect from failing. The manager knew the loads guys tended to be conservative, so much so that usually when their analyses showed load indicators exceeding limits, they tended to huddle and announce things were OK after all. And he looked at the Integration and Orbiter VPs for some hint of their opinion on the subject, and saw blank faces staring back. Then the manager thought of why 51L was lost and people died, and all the work the past year and a half putting in place a process for precisely this situation, and said something to the effect of "Sir, I will personally check every aspect of this analysis, but until such time as a flaw is found, I back the advisory we provided and will not change it, and if you continue to pressure me to do so I will bring this situation to the attention of the NASA Inspector General." At that, the president ordered everyone out of the room with a few expletives.

What Happened

An error was found in the analysis that introduced transient, higher than needed system dispersion protections, effectively overstating the possible loads. Coincidentally, other aspects of the Shuttle had problems that first launch attempt, and the launch was postponed anyway. By the next try, the system dispersion calculation corrections were made, and the winds of the day were much closer to those the trajectory was designed for, and there were no wing load issues and so a go advisory was issued. Was the president's behavior based on his experience, anticipating such a finding and trying to avoid an unnecessary flight cancellation? Or was his behavior an emotional reaction to having his order disobeyed? Why did all the other participants disengage from the decision? Was their silence support for the president's position, uncertainty on what to do, or fear of directly confronting his anger?

Lessons Learned

Do your work so you trust your work. If you are going to go to all the trouble to put in place a means to make a rational decision, then use that decision-making system you put into place. Don't confuse what you want to happen with what your decision-making system is suggesting you should do. The manager made the right decision, even though the source data turned out to be incorrect. Until the work was carefully checked and the error found, the only appropriate decision was to issue an advisory consistent with the data available. You can argue that it is prudent to use intuition or a hunch to make the opposite decision, and of course people do that all the time. Making decisions based on hunches is leaving your fate to random events or, in plain English, luck. Such behavior is certainly human, and is even admired by some people, but it is not rational decision making. We tend to celebrate the good outcomes from intuitive decision making as evidence of a superior intellect, but intuitive decision making precedes many individuals' biggest failures.

Ironically, making the correct rational decision often requires irrational courage. Making rational decisions is the easy part. People seem fundamentally built to first ensure their own safety and comfort and their own survival. Try as we might, none of us can be completely sure when these instincts have trumped rationality. As rational decisions usually result in some "winners" and some "losers," amazingly, often implementing the rational decision means going against the choice of the majority or at least a very motivated minority. In these situations, despite all one's efforts to be rational, it takes trust in one's rationality to proceed as chosen, while fully realizing there is the distinct possibility one could be wrong. Trust is, of course, just another word for belief or faith, an irrational response.

This chapter first addresses what it means to do the best we can with respect to the seven decision-making questions. Then we demonstrate formal methods to help us implement each approach.

3.1 MAKE GOOD DECISIONS

3.1.1 Determine What to Decide

One may need to do the following:

1. Assess the goodness of a diagnosis so one can determine what actions to take (or not take).
2. Make a sequence of decisions to optimize a return (or minimizes a loss).
3. Give people a fair method to choose from a static, finite set of options.
4. Allocate a static, finite set of resources so that the recipients feel equally satisfied.
5. Evolve options to best satisfy multiple criteria.
6. Select from a static, finite set of options that one which best achieves multiple dissimilar criteria.
7. Select from a static, finite set of options those which maximize benefits while minimizing detriments.
8. Do the optional endeavor with the best uncertain financial return.
9. Select a portfolio of investments that maximize return and minimize risk.
10. Choose the independent variable values that optimize an index of performance, perhaps subject to constraints, when the index of performance is inexpensive to determine.
11. Choose the independent variable values that optimize an index of performance, perhaps subject to constraints, when the index of performance is expensive to determine.
12. Define a dynamic control law to optimize a dynamics index of performance.
13. Define an optimal configuration.
14. Define the best strategy relative to a competitor.
15. Determine how best to address identified risks.

The first job is to determine which of these types of decisions is to be made, as the best means to make each of these decisions is different. Formal methods are presented to make all these kinds of decisions.

3.1.2 Determine When to Make the Decision

The decision must be made leaving sufficient time for it to be implemented and the results obtained prior to the consequences being of importance. If the situation is subject to change, then to effect the situation, decisions needs to be made with a frequency at least twice that of the rate of change.

3.1.3 Determine Who Should Make the Decision

Human societies implement hierarchies one way or another, with each level of the hierarchy usually defined by the decisions people are empowered to make.

Acknowledging this reality, four heuristics are useful to attempt to increase the odds of making a good decision.

First, people unaffected by the result of the decision will most likely make the best decision. Obviously, judges and juries, referees and umpires, are all manifestations of this heuristic.

Second, to the extent that implementing a decision requires the acceptance and cooperation of multiple people, the more the affected people participate in making the decision, the more likely the selected result will be implemented. All people, to some extent, can be commanded, but all people commanded against their will eventually revolt, even if only by passivity. On any complicated issue, there is unlikely to be a clear overwhelming majority agreement on what to do, so how to include the nonmajority in the implementation is critical to the decision-making process.

Third, the more people involved in the decision-making process, the more likely a good decision will be reached. This heuristic recognizes first, as Deming made famous, that "new knowledge always comes from outside." As individuals, our brains tend to lock onto a set of known things. To get a new idea into our heads requires interaction with someone for which this new thing is locked into their heads as something established. Also, studies have repeatedly shown that the mean assessment of many people has less error than the assessment of individual people. This so-called wisdom of crowds is often very useful to make good decisions.[2] But carefully note the conclusion: the mean prediction of a large group of people has less error than the mean of individual opinions. That does not mean the crowd was right, only generally less wrong. So the heuristic does not infer blindly following the crowd.

Fourth, we cannot predict the truly revolutionary events, so we must react wisely when they occur. These unanticipated, presumed to be impossible, but then proven to be very real events were popularized by Taleb[3] as "black swans." Black swan events sweep aside the consequences of any collection of incremental decisions we may make. Fortunately for most of us, though black swan events are inevitable, for long periods of time we can muddle along on our more pedestrian way, making localized decisions within the context of the current environment, until are lives are inexorably altered. All of us need to recognize the reality of the phenomenon, for which one must adopt a decision-making philosophy best described in Taleb's book. Taleb, in our vernacular, provides heuristics for a control law based on the state of the system to achieve an optimal result. In his framework, the status quo, rather than assumed constant, is expected to change in an unpredictable manner, and when it does, one optimizes to the new state, not the prior state. For example, if the price of something is inexorably rising, the status quo calls for decisions to be made presuming continuing price rises. Tabel's guidance is such a state of uninterrupted price increases must eventually change, and at an unknowable time, and further the prices will likely plummet back to unknowable but lower levels. So one should now be making decision anticipating rapid and deep price cuts, rather than the perceived certainty of continual prices increases. Taleb's guidance is that what appears to be predictable almost never is, stable situations will go unstable, apparently inexorable increases or decreases will stop and precipitously reverse, and abundance will turn into scarcity and vice versa. It will not be clear in advance what will cause these revolutions, but they will occur. So make decisions recognizing this reality.

TABLE 3.2
Leadership Style to Use when Making a Decision

Leadership Style to Use		Follower Acceptance Required to Implement Decision	
		Low	High
Number of correct solutions	One	Direct	Consult
	Many	Convenience	Consensus

When making decisions we need to adopt the appropriate leadership style and the appropriate method. The leadership style to use predominantly depends on whether there is likely to be one right solution or many acceptable options, and how much follower acceptance is required for the decision to be implemented. Table 3.2 provides heuristics on what leadership style to use depending on those two considerations.

If there are many acceptable solutions and implementation needs essentially everyone to accept the chosen solution, then consensus is the best leadership style; otherwise, any method to solicit and select an acceptable approach is acceptable. *Consensus* is defined here to mean that all can live with the decision, though none may be particularly thrilled by it. If there is probably one best solution, and to implement that solution follower acceptance is necessary, then one should consult with those affected, meaning seek guidance with respect to how to make the decision. If there is likely a single best solution and it can be achieved without a high level of follower buy-in, then one can direct the decision-making effort. Most of us have a default leadership style: we tend to direct others, consult, seek to achieve consensus, or adopt a "let them do what they want" approach. Notice, though, that to make the right decision it is critical to adopt the leadership style for the situation with respect to follower acceptance required and the likelihood that one solution is truly the best. Notice the very limited use of consensus leadership: "many correct solutions; follower acceptance is critical." Similarly, notice the limited use of direct leadership style: "one right decision; large amounts of follower acceptance are not required."

How to make the decision depends predominantly on two other considerations: time available and the follower's expertise with respect to the issue at hand. Table 3.3 provides guidance for the decision-making method with respect to these two considerations.

TABLE 3.3
Decision-Making Method to Use

Decision-Making Method		Follower Expertise	
		High	Low
Time available	Ample	Collaborate	Sell
	Constrained	Delegate	Command

If a decision is time critical, and the followers are not well prepared to assess the situation but the leader is, then the leader should command the decision. Alternatively, if there is ample time to make the decision, the leader should explore options with the followers and provide evidence for the appropriateness of each option with the intent of enabling each follower to come to agree with the chosen option. If followers are well prepared to address the decision because they have more expertise than the leader, and the situation is time critical, then the leader should delegate the decision to those best prepared to make it. Alternatively, if there is ample time to make the decision, then the leader should cast a wide net for options and assessments, and collaborate with all parties that could both recommend a solution and be affected by the solution, before making the decision.

3.1.4 DETERMINE THE DECISION MEASURE OF GOODNESS

Options and measure of goodness go hand in hand. The options must be definable in a manner appropriate to the measures of goodness. What follows are suggested options and measures of goodness for each of the types of decision types identified in Section 3.1.1:

Type: Assess the goodness of a diagnosis so one can determine what actions to take (or not take).
Measure of goodness: Probability of a correct diagnosis given the realities of false positives and false negatives.
Type: Make a sequence of decisions to optimize a return (or minimize a loss).
Measure of goodness: Consequence of the actions.
Type: Give people a fair method to choose from a static, finite set of options.
Measure of goodness: Number or percentile of people accepting the choice.
Type: Allocate a static, finite set of items so that the recipients feel equally satisfied.
Measure of goodness: Allocating in proportion to what the receivers are willing to pay.
Type: Evolve options to best satisfy multiple criteria.
Measure of goodness: Pugh[4] studied this for a long time and recommends the goodness measure be "better than (+)," "equal to (0)," or "inferior to (–)." He offers profound arguments why to avoid graduated scales.
Type: Select from a static, finite set of options that one which best achieves multiple dissimilar criteria.
Measure of goodness: Use fuzzy mathematics to combine the assessments of how well each option meets each criterion with fuzzy weights for the relative importance of each criterion.
Type: Select from a static, finite set of options those which maximize benefits while minimizing detriments.
Measure of goodness: On or near the efficient frontier of benefit versus detriment. The options that constitute the efficient frontier are those for which all other options have lower benefits with similar detriments.
Type: Do the optional endeavor with options with the best financial return.
Measure of goodness: Whatever the endeavor seeks to achieve.
Type: Select a portfolio of investments which maximize return and minimize risk.

Measure of goodness: On or very near the efficient frontier or return versus risk.

Type: Choose the independent variable values that optimize an index of performance, perhaps subject to constraints, when the index of performance criteria is inexpensive to determine.

Measure of goodness: The index of performance value augmented with a penalty times the extent each constraint is not satisfied.

Type: Choose the independent variable values that optimize an index of performance, perhaps subject to constraints, when the index of performance criteria is expensive to determine.

Measure of goodness: Taguchi quality indices.[4]

Type: Define a dynamic control law to optimize a dynamic index of performance.

Measure of goodness: The index of performance value.

Type: Define the optimal configuration.

Measure of goodness: What the configured item is to achieve.

Type: Define the best strategy relative to a competitor.

Measure of goodness: What the strategy seeks to maximize or minimize.

Type: Determine how best to address risks.

Measure of goodness: First minimize the potential uncertainty in the result, then drive the deviation from the desired result and the possible result to zero.

3.1.5 Ensure Option Evaluation Data Are Valid

Human history is full of spectacularly wrong decisions being made due to expertly applied approaches based on incorrect data or presumptions. Prior to implementing any decision-making process, it is critical that those empowered to make the decision, or those trusted to provide information to those who will make the decision, first perceive all data associated with the decision-making process to be completely and totally wrong until substantial evidence is presented to prove otherwise. Clearly, this adds a time delay to the process, which is probably the primary reason this critical activity is rarely done. Since people are smart, they can anticipate what conclusion the chosen decision-making process will reach given the input information, so then people who will benefit from a particular conclusion will do all they can to make sure the input data are those needed to reach their desired conclusion. Hence the recommendation in Section 3.1.3 that whenever possible, the decision should be made by people not affected by the decision to purposely minimize the likelihood of bias. Of course, one person's bias is another person's fact. So just what is bias and how do you recognize it? The equation

$$F = m * a$$

is not bias, though it may not always be correct. Definitions; assumptions; or screening of larger data sets for selected values of an F, m, or both, which results in a's that benefit a group of people at the expense of another group of people, is bias. To avoid bias, avoid making decisions based primarily on expert opinion, or accepting decisions made by people who will directly benefit from the decision. Though in neither case is it absolutely certain that bias tainted the decision, it may have, and it is wiser to presume a false conclusion. To put it another way, for every decision to be made by people with a vested interest in the outcome, there is always a way for the decision to be made by people without a vested interest, so use the method least subject to bias.

3.1.6 ASSESS THE ROBUSTNESS OF THE DECISION

When making a decision, it is wise to know how different the inputs can be for the decision to be the same, so you know when the decision is no longer valid and you need to make a new decision. In this case "different" means that for all key inputs to the decision-making process, if the "true" input was different than the presumed input, the same decision would be made. Certainly if the "difference" is large, it would be expected that a different conclusion would be reached. Another way to define robustness is to declare the range of the inputs for which the decision is valid: the larger the range, the more robust the decision. So, one needs to attempt to know the variability associated with each input, in either a probabilistic or fuzzy sense, and make decisions that explicitly acknowledge this variability. Some of the methods we next discuss inherently include such potential variability; some of the methods don't. If input uncertainty is not inherently part of the decision-making process, then you must make the decision repeatedly while purposefully varying the inputs over a wider and wider range until you understand under what conditions the decision is no longer valid.

3.1.7 DO THE RIGHT THING

All systems engineers and architects must realize that politics always trumps analysis. This is usually appropriate, since as we have already discussed, all analysis is inherently flawed at some level, so the political considerations may override the analytical conclusion. But there is a line that must be drawn. We each have an ethical responsibility to challenge decisions that are wrong. For example, if a credible analysis predicts something is highly unlikely to be accomplished, but the politically motivated decision purports a high likelihood of success, and thus results in the continuation of actions that may be harmful to people but benefit the decision makers, then those responsible for the analysis must, at the minimum, make the discrepancy known to as many people as possible. Truth is very hard to know. But we must be motivated to attempt to express it, and to attempt to recognize we may be wrong about what we believe most strongly. We must trust that a rational decision will, in the end, defeat a decision based on selfishness, if those who recognize this state of affairs will so declare.

3.2 MAKE GOOD DECISIONS BY SPECIFIC MEANS

3.2.1 ASSESS THE GOODNESS OF A DIAGNOSIS

The problem statement is as follows:

> A test has a probability for false positives (the test indicates a condition is true, but it isn't) of **Fp** and a probability of false negatives (the test indicates the absence of a condition that does exist) of **Fn**. What is the probability the condition really exists if the test returns positive?

To assess this situation, we use Bayes' rule, with **A** denoting event condition is True and **B** denoting event test indicates condition is True, as follows:

Event	Prior Probability	Conditional Probability	Joint Probability
Condition is true	$P(A)$	* $P(B|A) = 1 - Fn$	$= P(B$ intersect $A)$
Condition is false	$P(not\ A) = 1 - P(A)$	* $P(B|not\ A) = Fp$	$= P(B$ intersect not $A)$
Total	1		$P(B)$

So then:

$$P(A|B) = P(B\ \text{intersect}\ A)/P(B)$$

ProbTrueGivenPosTest determines $P(A|B)$ for provide $P(A)$, **Fp**, and **Fn**.

By treating $P(A)$ as a variable, we can plot the result from 0 to 1, for an **Fp** of 2% and an **Fn** of 1%.

In[146]:= **Plot[ProbTrueGivenPosTest[PA, .02, .01], {PA, 0, 1},**
PlotRange → All, AxesLabel → {"Prob. Condition", "Prob. Test Correct"}]

Out[146]=
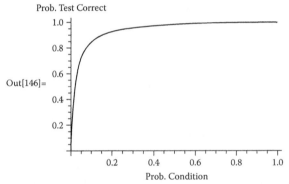

Note that unless the condition has a likelihood of at least 20%, the probability a positive test is truly indicating a positive result is diminishing small. Here's what happens for a probability of false positives of 0.1% and false negatives of 0.2%:

In[147]:= **Plot[ProbTrueGivenPosTest[PA, .001, .002], {PA, 0, 1},**
PlotRange → All, AxesLabel → {"Prob. Condition", "Prob. Test Correct"}]

Out[147]=
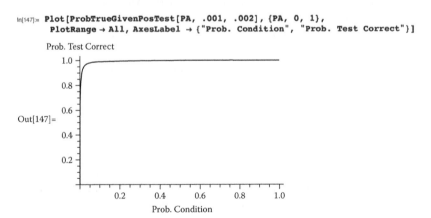

So very accurate tests are needed for a test to be useful for conditions that occur less than 5% of the time.

Alternatively, if we can establish the probability density function for the population without the condition with respect to a test value, and the separate probability density function for the population with the condition with respect to the same test value, we can construct a "receiver operating characteristic" curve that relates the true positive probability (which equals 1 minus the false negative probability) to the false positive probability (which equals 1 minus the true negative probability). A straight line on this plot denotes a 50% probability the test is predicting the result at the stipulated threshold.

Here is an example. Suppose the population not exhibiting the condition has a normal distribution of mean 20 with a standard deviation of 10, while the population exhibiting the condition has a normal distribution of mean 35 with a standard deviation of 10. Then the receiver operating characteristic curve looks like the following.

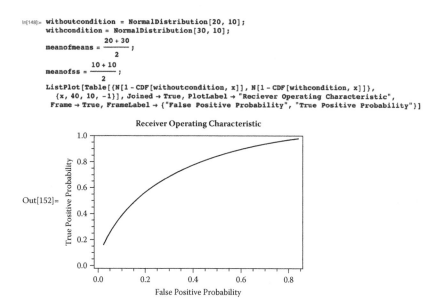

A lenient threshold has a high true positive probability but also a high false positive probability, while a strict threshold has the reverse. Selective thresholds can be explicitly noted on the plot as follows:

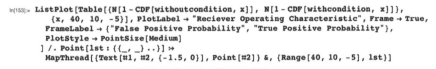

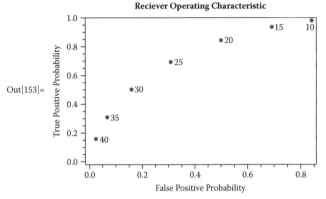

3.2.2 MAKE A SEQUENCE OF DECISIONS TO OPTIMIZE A RETURN

The problem statements are as follows:

1. At a node **N**, there are **M** possible outcomes each with known likelihood and consequence; what is the aggregate result?
2. At a node **N**, there are many possible choices with known benefits and costs; which is the choice that maximizes the net benefits?

The decision-making process requires the construction of a decision tree. The decision tree starts with a decision (often termed a *node* or *gate*) for which there are at least two options (each often termed a *leaf*); each option may have another decision with at least two options, to whatever number of decisions are to be made.

There is no formal method to guarantee correct identification of every option that needs to be considered, all the possibilities of those options occurring, and all the consequences. To visualize the problem, a network is drawn, with nodes used to denote the decision/result and arcs used to denote the options, typically labeled with the probability of occurrence or cost and benefit. What can be formally required is the methodology to solve the two problem statements. The formal requirements *probDecGateEval* address the first problem statement returning the most likely net result for all the options at a node.

As an example, suppose a node has three potential options with probabilities:

P1 = 0.1
P2 = 0.2
P3 = 0.7

with the returns:

R1 = {−1, 0, 1}
R2 = {0, 5, 10}
R3 = {−4, 1, 2}

This is evaluated as follows:

```
In[797]:= nodeinput = {{.1, {-1, 0, 1}}, {.2, {0, 5, 10}}, {.7, {-4, 1, 2}}};
         probDecGateEval[nodeinput]
Out[798]= {-2.9, 1.7, 3.5}
```

So option 3 is preferred. Trees with more arcs are evaluated by concatenating. For example, suppose the decision tree again starts with three options with the same probabilities as before, but this time the first option itself splits into the same 3 options. Then this tree is evaluated as follows:

```
In[156]:= treeinput = { {.1, probDecGateEval[nodeinput]}, {.2, {0, 5, 10}}, {.7, {-4, 1, 2}}};
         probDecGateEval[treeinput]
Out[157]= {-3.09, 1.87, 3.75}
```

So option 3 is still preferred. The solution to the second problem is to determine which branches to follow to maximize return. *fuzzDecisionNode* determines the net benefit to move from one node to another, when the cost and benefits for each transfer are known. In this case, both the cost and benefit are modeled as triangular fuzzy numbers.

To illustrate, suppose our choice is either to do a plant expansion or not to do the plant expansion. Suppose the fuzzy cost to do the expansion and benefits of doing an expansion are found to be

```
In[158]:= expansioncost = {-3, -2.2, -1.5};
         expansionbenefit = {.42, 3.568, 4.};
```

Since the cost to not expand is 0, and the benefit is found to be

```
In[160]:= noexpansioncost = {0, 0, 0};
         noexpansionbenefit = {.322, 1.87, 2.5};
```

we create a tree representing these branches as follows:

```
In[162]:= firsttree = {{expansioncost, expansionbenefit}, {noexpansioncost, noexpansionbenefit}};
```

fuzzyDecisionNode returns

```
In[163]:= fuzzDecisionNode[firsttree]
Out[163]= {{{-2.58, 1.368, 2.5}, 0.429333}, {{0.322, 1.87, 2.5}, 1.564}}
```

The defuzzified result for expansion is 0.43. The defuzzified result for no expansion is 1.564. So the decision should be to not expand.

To evaluate concatenated branches, use the *fuzzDecisionNode* to evaluate each individual node, and then sum the results for each unique node path. For example, if the expand branch had a second option to expand again or not expand with same values, then there are now three potential outcomes:

1. Expand and expand again.
2. Expand once.
3. No expansion.

This tree is modeled as follows using components of the **firsttree**:

```
In[164]:= secondtree = {firsttree[[1]] + firsttree[[1]], firsttree[[1]], firsttree[[2]]};
          fuzzDecisionNode[secondtree]
Out[165]= {{{-5.16, 2.736, 5.}, 0.858667}, {{-2.58, 1.368, 2.5}, 0.429333}, {{0.322, 1.87, 2.5}, 1.564}}
```

No expansion continues to be the best option.

3.2.3 GIVE PEOPLE A FAIR METHOD TO CHOOSE FROM A STATIC FINITE SET OF OPTIONS

The problem statement is as follows:

There are **C** candidates (an **M** × 1 array) for **N** people to choose from; which candidate is preferred?

When there are only two candidates, and each voter has one vote, the candidate who receives the majority of the votes is the candidate the majority of the people prefer. But if there are more than two candidates, it is necessary to allow people to vote for an alternative choice as well as their primary choice to help ensure the candidate who is selected is preferred by the most people. The Borda count is one of the methods for **N** people to select from three or more candidates the one who they collectively most prefer. The formal requirements *BordaCount* implement the methodology.

As an example, suppose there are four options (or candidates) **A**, **B**, **C**, and **D**, and six voters who cast their ballots as follows for each of the candidates:

Voter1: Doesn't have a preference, so casts one each for **A**, **B**, **C** and **D**.
Voter2: Is only happy if **A** is chosen, so casts all four for **A**.
Voter3: Is only happy if **B** is chosen, so casts all four for **B**.
Voter4: Prefers **A** or **B** equally, so casts two each for **A** and **B**.
Voter5: Prefers **A** and **C** equally, so casts two each for **A** and **C**.
Voter6: Can accept **B** but prefers **C**, so casts one for **B** and three for **C**.

Then input to *BordaCount* and results are

```
In[166]:=   votesforcandidates =
              { {1, 1, 1, 1},
                {4, 0, 0, 0},
                {0, 4, 0, 0},
                {2, 2, 0, 0},
                {2, 0, 2, 0},
                {0, 1, 0, 3}};
            BordaCount[{"A", "B", "C", "D"}, votesforcandidates]
```

Candidate	Total Votes
A	9
B	8
C	3
D	4

The result narrowly supports selection of candidate or option **A**. Your experience with dual-candidate, single-vote elections may result in your wanting to sponsor a second election between candidates **A** and **B** since neither obtained the majority of the 24 votes cast. But the Borda count methodology avoids the need for a second election. This is what should logically happen if everyone votes their stated preference, but the candidates are reduced to **A** and **B**. In theory, **A** will obtain the backing of **Voter2** and **Voter5**, while **B** obtains the support of **Voter3** and **Voter6**. So **Voter1** and **Voter4** will determine the result. But based on their initial preferences, both **Voter1** and **Voter4** are equally enamored with candidates **A** and **B**. Given the opportunity to vote again, **Voter1** and **Voter4** will determine the election based on an explicit comparison of just candidates **A** and **B**. **Voter1** and **Voter4** may not vote, since they do not care whether **A** or **B** wins. Or **Voter1** and **Voter4** may each toss a coin, so there is a 50% chance the runoff election ends in a tie, and a 25% chance either candidate **A** or **B** will obtain both **Voter1** and **Voter4** support. The original Borda count result provides a useful solution without the need for a runoff election.

The most important consideration when making a decision by voting is what will the people do when the result is obtained? Only make decisions by voting when you are certain the participants will abide by the result.

3.2.4   ALLOCATE A STATIC FINITE SET OF RESOURCES SO THAT THE RECIPIENTS FEEL EQUALLY SATISFIED

The problem statements are as follows:

1. Maximize the return of providing **N** identical items to those who may want them.
2. Maximize the return of providing **N** dissimilar items to those who may want them.
3. Distribute **N** dissimilar items between **M** people in a manner that gives equal opportunity for each to obtain each item.

The best way to allocate finite resources to those desiring them is to conduct an auction. A useful auction methodology for multiple items is the Dutch auction. In a Dutch auction, each bidder is given a number of tokens equal to the number of items to be chosen. Each bidder may allocate none to all of their tokens on each item. Each item goes to the bidder who allocated the most tokens to that item.

The formal requirements *DutchAuctionforSameItems* enable the bidders to make dissimilar bids for the same items. The formal requirements *DutchAuctionforDifferentItems* enable bidders to choose from dissimilar items.

For example, suppose selector **A** is willing to pay 100 for 10 items and 90 for 9 of the items, while selector **B** is willing to pay 200 for 8, and selector **C** is willing to pay 50 for 20, then 100 for 20, and finally 75 for 20.

Then *DutchAuctionforSameItems* returns

```
In[168]:= nitems = 50;
        bidsperitem =
          {{"A", { {100, 10}, {90, 9}}},
           {"B", {{200, 8}}},
           {"C", {{50, 20}, {100, 20}, {75, 20}}}};
        DutchAuctionforSameItems[nitems, bidsperitem]

        Total sale: 5635

        Total items sold: 50
```

Selector	Total cost	Total items
A	1810	19
B	1600	8
C	2225	23

As an example of an auction for different items, suppose there are three selectors, who each may want one or more of a house, car, plane, and painting. One approach is to achieve the maximum return to the seller. In that case, each selector's bid is unconstrained. Suppose

- Selector **A** is willing to pay 1000 for house, 10 for car, 1000 for plane, and 1 for painting.
- Selector **B** is willing to pay 900 for house, 20 for car, 0 for plane, and 5 for painting.
- Selector **C** is willing to pay 100 for house, 20 for car, 2000 for plane, and 5 for painting.

Then the input to and output from *DutchAuctionforDifferentItems* are

```
In[171]:= items = {"house", "car", "plane", "painting"};
        bidsperitem =
          {{"A", {1000, 10, 1000, 1}},
           {"B", {900, 20, 0, 5}},
           {"C", {800, 15, 2000, 3}}}; DutchAuctionforDifferentItems[items, bidsperitem]

        Item: house was awarded to: A with a bid of: 1000

        Item: car was awarded to: B with a bid of: 20

        Item: plane was awarded to: C with a bid of: 2000

        Item: painting was awarded to: B with a bid of: 5


        Total spent on items: 3025
```

Selector	Total spent
A	1000
B	25
C	2000

Should one or more of the bidders make the same maximum bid for an item, the input and output would be

```
In[173]:=  bidsperitem =
      {{"A", {1000, 10, 1000, 5}},
       {"B", {900, 20, 0, 5}},
       {"C", {100, 20, 2000, 3}}}; DutchAuctionforDifferentItems[items, bidsperitem]
```

Item: house was awarded to: A with a bid of: 1000

Selectors {B, C} offered same max value of 20 for item car

Item: plane was awarded to: C with a bid of: 2000

Selectors {A, B} offered same max value of 5 for item painting

Total spent on items: 3000

Selector	Total spent
A	1000
B	0
C	2000

Alternatively, if the objective is to offer equal opportunity for each selector to obtain any item, then the selectors are granted the same number of tokens to bid, and the items go to the selector who offers the most tokens for each item. Suppose each selector is given 1000 tokens, and

- Selector **A** offers 495 for house, 10 for car, and 495 for plane.
- Selector **B** offers 950 for house, 40 for car, and 10 for painting.
- Selector **C** offers 345 for house, and 10 for car, and 645 for plane.

Then the input and output for the routine are

```
In[174]:=  bidsperitem = {{"A", {495, 10, 495, 0}},
                          {"B", {950, 40, 0, 10}},
                          {"C", {345, 10, 645, 0}}};
       DutchAuctionforDifferentItems[items, bidsperitem]
```

Item: house was awarded to: B with a bid of: 950

Item: car was awarded to: B with a bid of: 40

Item: plane was awarded to: C with a bid of: 645

Item: painting was awarded to: B with a bid of: 10

Total spent on items: 1645

Selector	Total spent
A	0
B	1000
C	645

3.2.5 Evolve Options That Best Satisfy Multiple Criteria

The problem statement is as follows:

> Select a finite set of features for an item that collectively best satisfy a finite set of criteria.

Whenever making a decision to select from a finite number of options, broadly define those options, for clearly if you fail to identify the optimal option, the evaluation process does not find it for you. Human nature tends to limit the options considered to those that worked in the past, or those known to fail. To counter this human limitation, command the definition of at least six options prior to doing an assessment. Most people can quickly identify two to four options to do anything: one or two they won't want, and one or two they know might work from prior experience. Demanding at least six credible options compels an attempt to think more broadly and be innovative. We've all noticed some people are more innovative than others. How the more innovative of us achieve this capacity is more art than science. But some have thought deeply about how to mechanically aid the innovation process and have documented their opinions.[6,7]

Pugh[3] studied the creative process extensively and established principles for making decisions to select a finite set of features, which collectively best satisfy a finite set of criteria. In the Pugh process, the number of options and the number of criteria can dynamically vary. Pugh's process compares all options against a baseline, one criterion at a time. Pugh emphatically rejects numerically weighting the importance of each criterion. His rationale is that if a criterion is important to making a decision, then it must be utilized to make the decision. By default, if options compare equally with respect to that criterion, by definition that criterion is not important to make the decision. Pugh rejects both ranking and rating how well each option satisfies each criterion. Rather, for each criterion, each option is analyzed to determine if it is inferior, equal, or superior to how the baseline satisfies that criterion. Pugh's primary purpose is to synthesize an option that incorporates as much as practical the best features of any of the currently identified options. Though Pugh does not explicitly use this language, his approach mimics natural selection. Each option consists of features that can be perceived to be part of that option's genetic basis. Pugh perceived the decision process to be one of searching for and accepting beneficial features, while rejecting negative features, as determined by the criteria.

The formal requirements *PughComparison* enable one to see how options compare with respect to evaluation criteria and synthesize new options that may be superior to those initially identified.

As an example, suppose the baseline case is called **Status Quo**; we have three alternative options identified **O1**, **O2**, and **O3**; and we conduct a Pugh comparison with respect to four criteria, **C1**, **C2**, **C3**, and **C4**. Let the Pugh comparison of the options to the **Status Quo** be stipulated as shown in Table 3.4.

Examining the Pugh comparison, all the options are superior to **Status Quo** with respect to **C1**. With respect to criteria **C2**, **O1** is inferior to **Status Quo** and **O3** is superior. With respect to criteria **C3**, option **O1** is superior to **Status Quo** and option **O3** is inferior. With respect to **C4**, all the options are inferior to **Status Quo**. So, the solution

---

TABLE 3.4

Sample Pugh Comparison of Options to a Baseline

Status Quo	Option O1	Option O2	Option O3
Criteria **C1**	1	1	1
Criteria **C2**	−1	0	1
Criteria **C3**	1	0	−1
Criteria **C4**	−1	−1	−1

---

has any of the option features associated with **C1**, the features of **O3** for **C2**, and the features of **O1** for **C3** and of **Status Quo** for **C4**. The inputs for *PughComparison* are

```
In[176]:= baseline = "Status Quo";
          options = { "O1", "O2", "O3"};
          criteria = {"C1", "C2", "C3", "C4"};
          pughcompareOtoC = {{1, 1, 1}, {-1, 0, 1}, {1, 0, -1}, {-1, -1, -1}};
```

which, when provided to *PughComparison*, results in the following output:

```
In[180]:= PughComparision[baseline, options, criteria, pughcompareOtoC]
```

```
Status Quo  O1  O2  O3
    C1       1   1   1
    C2      -1   0   1
    C3       1   0  -1
    C4      -1  -1  -1
  Total      0   0   0
```

```
To synthesize a potentially better option:

consider substituting the feature of O1 for baseline, that provides criteria C1

consider substituting the feature of O2 for baseline, that provides criteria C1

consider substituting the feature of O3 for baseline, that provides criteria C1

consider substituting the feature of O3 for baseline, that provides criteria C2

consider substituting the feature of O1 for baseline, that provides criteria C3
```

3.2.6 SELECT FROM A FINITE SET OF STATIC ALTERNATIVES THE ONE THAT BEST ACHIEVED MULTIPLE DISSIMILAR CRITERIA

The problem statement is as follows:

> Which option **O** (chosen from an **n** × 1 array) best satisfies goals **G** (an **m** × 1 array), where the relative importance of each goal may need to be determined?

When selecting from a finite set of options with respect to multiple criteria, the relative importance of each criterion must be known. This is often conveniently done by assigning weights to each criterion so the sum of the weights is one. But when different people are involved in the decision, they may disagree as to the relative importance of each decision criterion. Also, people may find it easier to rate the importance of each criterion on a common scale, say, from 9 for the most important to 1 for the least important. "Importance" is inherently fuzzy, so on a 1–9 scale a person might judge a criterion to be a fuzzy triplet of <4, 5, 6>, indicating "not completely sure," but certainly not less than 4, not more than 6, and maybe 5. The formal requirements *GoalImportance* determine a composite importance for an arbitrary number of goals on a scale from 1 (for least) to 9 (for most), given multiple opinions.

As an example, suppose there are three goals or criteria and for which three different people are providing an indication of the relative importance for each goal, as follows:

- **Person1** gives **goal1** a fuzzy importance of <1, 2, 3>, **goal2** a fuzzy importance of <3, 4, 5>, and **goal3** a fuzzy importance of <6, 7, 8>.
- **Person2** gives **goal1** a fuzzy importance of <0.5, 1.5, 3>, **goal2** a fuzzy importance of <4, 4, 4>, and **goal3** a fuzzy importance of <7.5, 8, 8.5>.
- **Person3** gives **goal1** a fuzzy importance of <9, 9, 9>, **goal2** a fuzzy importance of <1, 1, 1>, and **goal3** a fuzzy importance of <1, 1, 1>.

For these inputs, *GoalImportance* returns

```
In[181]:= N[GoalImportance[
            {{{1, 2, 3}, {3, 4, 5}, {6, 7, 8}},
             {{.5, 1.5, 3}, {4, 4, 4}, {7.5, 8, 8.5}},
             {{9, 9, 9}, {1, 1, 1}, {1, 1, 1}}}]]
Out[181]= {3.10723, 2.51984, 3.82586}
```

On a scale from 1 to 9, **goal1** has a collective importance of 3.1, **goal2**'s importance is 2.5, and **goal3** has an importance of 3.8.

The formal requirements *AlternativeGoalSatisfaction* calculate how well each option or alternative satisfies the weighted sum of the goals or criteria.

An example, suppose there are two alternatives to be evaluated against two goals, and there are again three people making potentially different evaluations, as follows:

- **Evaluator1** determines **alternative1** satisfies criterion **A** by <1, 2, 3> and satisfies criterion **B** by <3, 4, 5>, while **alternative2** satisfies criterion **A** by <4, 5, 6> and criterion **B** by <7, 8, 9>.
- **Evaluator2** determines **alternative1** satisfies criterion **A** by <0.5, 1.5, 2.5> and satisfies criterion **B** by <3.5, 4.5, 5.5>, while **alternative2** satisfies criterion by **A** <4.5, 5.5, 6.5> and satisfies criterion **B** by <7.5, 8.5, 9>.
- **Evaluator3** determines **alternative1** satisfies criterion **A** by <1.5, 2.5, 3.5> and satisfies criterion **B** by <4, 5, 6>, while **alternative2** satisfies criterion **A** by <3.5, 4.5, 5.5> and satisfies criterion **B** by <6.5, 7.5, 8.5>.

Then *AlternativeGoalSatisfaction* returns

```
In[182]:= N[AlternativeGoalSatisfaction[ {{{{1, 2, 3}, {3, 4, 5}}, {{4, 5, 6}, {7, 8, 8}}},
            {{{0.5, 1.5, 2.5}, {3.5, 4.5, 5.5}}, {{4.5, 5.5, 6.5}, {7.5, 8.5, 9}}},
            {{{1.5, 2.5, 3.5}, {4, 5, 6}}, {{3.5, 4.5, 5.5}, {6.5, 7.5, 8.5}}}}]]
Out[182]= {{1.95743, 4.4814}, {4.98328, 7.8252}}
```

The output indicates **alternative1** scored about 2.0 against criterion **A** and about 4.5 against criterion **B**, while **alternative2** scored 5.0 against criterion **A** and 7.8 against criterion **B**.

The formal requirements *FuzzyDecisionMaker* input goals, the relative goal importance (calculated by *GoalImportance*), alternatives, and indications of how well each alternative satisfies each goal (calculated by *AlternativeGoalSatisfaction*) and returns net values between 0 and 1 for each alternative (that sum to 1); the higher the number, the better the alternative achieves the weighted goals.

As an example, suppose we want to choose one of four possible cars (sedan, SUV, midsize, or convertible) with respect to five criteria (medium cost, powerful engine, good gas mileage, large carrying capacity, and stylish interior), for which the relative importance has already been determined to be 5, 8, 2, 3, and 6, respectively, and we assess each option with respect to each criterion on a 1–9 scale as follows:

- For the sedan: medium cost 2, powerful engine 8, good gas mileage 4, large carrying capacity 6, and stylish interior 9
- For the SUV: medium cost 5, powerful engine 8, good gas mileage 2, large carrying capacity 8, and stylish interior 6
- For the midsize: medium cost 8, powerful engine 1, good gas mileage 9, large carrying capacity 2, and stylish interior 2
- For the convertible: medium cost 5, powerful engine 9, good gas mileage, 6, large carrying capacity 1, and stylish interior 9

Then *FuzzyDecisionMaker* returns

```
In[183]:= FuzzyDecisionMaker[{"medium cost", "powerful engine",
    "good gas mileage", "large carrying capacity", "stylish interior"},
    {5, 8, 2, 3, 6}, {"sedan", "suv", "midsize", "convertible"},
    {{2, 8, 4, 6, 9}, {5, 8, 2, 8, 6}, {8, 1, 9, 2, 2}, {5, 9, 6, 1, 9}}]

    Alternative sedan score 0.25789

    Alternative suv score 0.228671

    Alternative midsize score 0.152177

    Alternative convertible score 0.361262
```

The convertible is the preferred option. The robustness of the decision can be assessed two ways. To assess how the decision might change if the importance criteria are changed, redo the analysis with each of the entries {5, 8, 2, 3, 6} increased and decreased by 1 unit. To assess how the decision result might change if the option assessments are changed, simply one by one lower each convertible evaluation assessment that is not 1, by 1, and redo the analysis to see if it is still selected. That is, redo the analysis with the convertible's evaluation vector sequentially set to {4, 9, 6, 1, 9}, then {5, 8, 6, 1, 9}, then {5, 9, 6, 1, 8} to determine if the convertible remains the chosen car. Here is that process illustrated:

```
In[184]:= FuzzyDecisionMaker[{"medium cost", "powerful engine",
    "good gas mileage", "large carrying capacity", "stylish interior"},
    {5, 8, 2, 3, 6}, {"sedan", "suv", "midsize", "convertible"},
    {{2, 8, 4, 6, 9}, {5, 8, 2, 8, 6}, {8, 1, 9, 2, 2}, {4, 9, 6, 1, 9}}]
    Alternative sedan score 0.258247

    Alternative suv score 0.23312

    Alternative midsize score 0.154877

    Alternative convertible score 0.353756
```

In[185]:= **FuzzyDecisionMaker[{"medium cost", "powerful engine",**
"good gas mileage", "large carrying capacity", "stylish interior"},
{5, 8, 2, 3, 6}, {"sedan", "suv", "midsize", "convertible"},
{{2, 8, 4, 6, 9}, {5, 8, 2, 8, 6}, {8, 1, 9, 2, 2}, {5, 8, 6, 1, 9}}]

Alternative sedan score 0.289226

Alternative suv score 0.260008

Alternative midsize score 0.153415

Alternative convertible score 0.297351

In[186]:= **FuzzyDecisionMaker[{"medium cost", "powerful engine",**
"good gas mileage", "large carrying capacity", "stylish interior"},
{5, 8, 2, 3, 6}, {"sedan", "suv", "midsize", "convertible"},
{{2, 8, 4, 6, 9}, {5, 8, 2, 8, 6}, {8, 1, 9, 2, 2}, {5, 9, 5, 1, 9}}]

Alternative sedan score 0.25836

Alternative suv score 0.228801

Alternative midsize score 0.153667

Alternative convertible score 0.359172

In[187]:= **FuzzyDecisionMaker[{"medium cost", "powerful engine",**
"good gas mileage", "large carrying capacity", "stylish interior"},
{5, 8, 2, 3, 6}, {"sedan", "suv", "midsize", "convertible"},
{{2, 8, 4, 6, 9}, {5, 8, 2, 8, 6}, {8, 1, 9, 2, 2}, {5, 9, 6, 1, 8}}]

Alternative sedan score 0.278032

Alternative suv score 0.230981

Alternative midsize score 0.152599

Alternative convertible score 0.338388

The convertible remains preferred for all these evaluation adjustments so it is a robust choice.

3.2.7 Select from a Finite Set of Static Options Those That Maximize Benefits while Minimizing Detriments

The problem statement is as follows:

> For a finite set of options with dissimilar benefit and detriment, which is the best to choose?

The situation is easiest to visualize in two dimensions; for a given option, the benefit provided corresponds to a vertical axis value (such as performance or return) and the detriment (such as cost or risk) corresponds to the horizontal axis value. The option to pick, all else being equal, is the one that returns the highest benefit at the least cost, but since the collective options present a range of detriment and a range of benefit, the best options lie on what is called the *efficient frontier*, the set of options for which there are no options offering more benefit at the same detriment. The option you choose is the one corresponding to the most detriment you can tolerate.

Let's assume our benefit is a performance level and our detriment is the cost, and we have 20 options to choose from as follows:

```
In[188]:= options = {
             {{0.83, 4.68}, "A"},
             {{0.34, 7.95}, "B"},
             {{0.59, 7.94}, "C"},
             {{0.37, 4.64}, "D"},
             {{0.45, 3.02}, "E"},
             {{0.49, 1.88}, "F"},
             {{0.09, 2.45}, "G"},
             {{0.78, 0.47}, "I"},
             {{0.58, 4.03}, "J"},
             {{0.72, 9.48}, "K"},
             {{0.29, 2.23}, "L"},
             {{0.60, 9.79}, "M"},
             {{0.19, 0.34}, "N"},
             {{0.67, 6.70}, "O"},
             {{0.77, 7.39}, "P"},
             {{0.41, 4.06}, "Q"},
             {{0.92, 5.97}, "R"},
             {{0.57, 8.65}, "S"},
             {{0.90, 1.70}, "T"},
             {{0.94, 1.99}, "U"}};
```

The formal requirements *EfficientFrontier* present the information to make it easy for us to identify the best options.

```
In[189]:= EfficientFrontier[options, "performance, bigger better", "Cost ($M)"]
```

Out[189]=

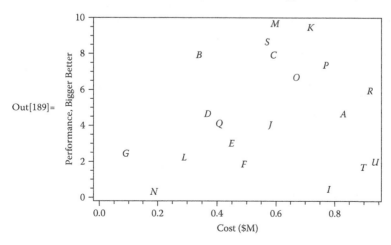

The efficient frontier consists of **G**, **B**, and **M**, because no other option offers superior performance at lower costs. Certainly, something other than a single performance criterion may be important to picking the option. For example, suppose option **M** is considered considerably riskier to achieve than option **S**; it is then wise to go with option **S**, sacrificing some performance for less risk.

3.2.8   DO THE OPTIONAL ENDEAVOR WITH THE BEST UNCERTAIN FINANCIAL RETURN

The problem statement is as follows:

> Multiple proposed endeavors require a time series of uncertain expenses with uncertain returns; which one is the best to do?

To make these types of decisions, first determine if an arbitrage situation is present. Such situations mean you can earn money without risk. An example is **BankA**, which is willing to pay 5% interest, and **BankB**, which is willing to loan money at 3% interest. Then borrow as much money as possible from **BankB** to invest with **BankA** to guarantee a 2% return as long as the condition exits. In competitive markets, arbitrage situations are rare, but when they exist the logical decision is to take advantage of them.

If an arbitrage situation does not exist, then we face a series of cash outflows followed by cash inflows. For example, what is the better investment: double your money in one year, or triple it in two years? Here are cash flows associated with the options:

Option1: –1, 2

Option2: –1, 0, 3

Two common methods to compare investment options with periods of negative and positive returns are net present value and internal rate of return. For the net present value method, the outflows and inflows of cash at fixed periods of time are discounted to the present using a chosen interest rate. The option with the larger value is preferred. Typically, the discount rate used for this method is chosen to be the risk-free return rate for the period of the cash flow. Though that is easy to say, it can be difficult to determine, but is often taken to be the current rate of the U.S. Treasury bonds for the period in question. For the internal rate of return method, the rate of return of the outflows and inflows is determined. The internal rate of return method solves a polynomial of order one less than the number of cash flow instances, for the interest rate that corresponds to the net present value of the cash flow to be 0.

The formal requirements *NPV* determine the net present value for a cash flow. The formal requirements *IRR* determine the internal rate of return for a cash flow. Here are net present values for the two illustrative cash flows, assuming the interest rate is 3%:

```
In[190]:= NPV[{-1, 2}, .03]
          NPV[{-1, 0, 3}, .03]
          IRR[{-1, 2}]
          IRR[{-1, 0, 3}]
```

Out[190]= 0.941748

Out[191]= 1.82779

Out[192]= {{rateofreturn → 1.}}

Out[193]= {{rateofreturn → 0.732051}}

The net present value evaluation method suggests selecting **Option2**, even though the payback is delayed a year, because the larger amount is worth the wait. The internal rate of return method suggests selecting **Option1** because it has the better rate of return. So which method should one use? If the intention is to repeat the investment, then using the internal rate of return is the more useful indicator. The internal rate of return is saying that, given a choice between doubling an investment every two years or tripling it every three years, it is preferable to double every year. If the intention is to make a one-time investment that cannot be repeated, then the net present value is the more appropriate measure of goodness.

Though useful and simple, neither net present value nor internal rate of return methods should be used to make decisions when the cash flows are uncertain, which is often the case. As we discussed in Chapter 2, one could model the cash flows as random variables or fuzzy numbers, and use Monte Carlo to determine probability density function or fuzzy logic to obtain the defuzzified net present value or internal rate of return. But often what we are looking for is a means to know when to do something, given that we at least understand our options. That is, what we want to know is, "Under what circumstances should we exercise our options?" Reference 8 shows several ways to do this.

As an example, suppose we can produce something of value up to 100,000 kilograms per year at a cost of $1000 per kilogram. The current price of the commodity is $2000 per kilogram. We have an option to design, build, and use a new machine to increase the production rate to 150,000 kilograms per year with a nonrecurring cost of $100M and a new operating cost of $1500 per kilogram. The risk-free rate of return is 5%. Our financial decision is to determine, "Under what conditions, if any, should the new machine be introduced?" To make this decision, we need to obtain a value for the operation, with and without the option being invoked at various times, explicitly modeling the reality that the future price of the commodity could vary quite considerably. To do this, we first establish the value of a 10-year lease for the as-is mine operation for different future prices. The price of the commodity is modeled as a binomial lattice such that each year the price will either increase by 1.15 or decrease by a factor of 0.9. We do not need to know the relative probability the price can increase or decrease. The formal requirements *BimonialLattice* routine builds such a lattice.

So for the commodity at a current price of $2000 per kilogram, with the potential each year to increase by 15% or decrease by 10%, the potential prices for 10 years are as follows:

In[194]:= **price = BinomialLattice[2000, 1.15, .9, 10]**

Out[194]= {{2000.}, {1800., 2300.}, {1620., 2070., 2645.}, {1458., 1863., 2380.5, 3041.75},
{1312.2, 1676.7, 2142.45, 2737.57, 3498.01}, {1180.98, 1509.03, 1928.21, 2463.82, 3148.21, 4022.71},
{1062.88, 1358.13, 1735.38, 2217.44, 2833.39, 3620.44, 4626.12},
{956.594, 1222.31, 1561.85, 1995.69, 2550.05, 3258.4, 4163.51, 5320.04},
{860.934, 1100.08, 1405.66, 1796.12, 2295.05, 2932.56, 3747.16, 4788.04, 6118.05},
{774.841, 990.075, 1265.1, 1616.51, 2065.54, 2639.3, 3372.44, 4309.23, 5506.24, 7035.75},
{697.357, 891.067, 1138.59, 1454.86, 1858.99, 2375.37, 3035., 3878.31, 4955.62, 6332.18, 8091.12}}

Which we can visualize a bit easier if we use the *Mathematica*® routines *MatrixForm* to make each row distinct and *NumberForm* to set the number of digits to display:

In[195]:= **NumberForm[MatrixForm[price], 4]**

Out[195]//NumberForm=

$$
\begin{pmatrix}
\{2000.\} \\
\{1800., 2300.\} \\
\{1620., 2070., 2645.\} \\
\{1458., 1863., 2381., 3042.\} \\
\{1312., 1677., 2142., 2738., 3498.\} \\
\{1181., 1509., 1928., 2464., 3148., 4023.\} \\
\{1063., 1358., 1735., 2217., 2833., 3620., 4626.\} \\
\{956.6, 1222., 1562., 1996., 2550., 3258., 4164., 5320.\} \\
\{860.9, 1100., 1406., 1796., 2295., 2933., 3747., 4788., 6118.\} \\
\{774.8, 990.1, 1265., 1617., 2066., 2639., 3372., 4309., 5506., 7036.\} \\
\{697.4, 891.1, 1139., 1455., 1859., 2375., 3035., 3878., 4956., 6332., 8091.\}
\end{pmatrix}
$$

The first row shows the current **price**. The second row shows the possible range of prices for one year from now. Each subsequent row shows a growing finite set of possible prices from the lowest to the highest for that year, given the possibility of the assumed potential maximum and minimum price changes for each price possibility in the prior year.

To value the venture, we note that it can be viewed as a lease that is valueless to us in the last year, since we must return operations to the owners, and work backward to determine the potential value of the lease for each price of the commodity, which is the profit that can be made that year, discounted by a risk-free interest rate, plus a risk-neutral expected value of the lease in the next period. Reference 7 shows the value of such a lease to be as follows:

$$(1 / 1 + \textbf{riskfreereturn}) * (\textbf{q} * \textbf{Cu} + (1 - \textbf{q}) * \textbf{Cd})$$

where **q** is the risk-neutral probability that equals

$$(1 + \textbf{riskfreereturn} - \textbf{down})/(\textbf{up} - \textbf{down})$$

and

$$\textbf{Cu} = Max[0, \textbf{up} * \textbf{value}]$$

and

$$\textbf{Cd} = Max[0, \textbf{down} * \textbf{value}]$$

The formal requirements *LeaseValues* determine the lease values lattice for the price lattice derived above in [*MatrixForm*[**price**], 4].

Using the *Mathematica* routine *ScientificForm* to set the displayed digits to 3, the lease values that correspond to the possible commodity prices are as follows:

```
In[196]:= leasevalue = LeaseValues[price, 0.05, 1.15, 0.9, 100000, 1000];
          ScientificForm[MatrixForm[leasevalue], 3]
Out[197]//ScientificForm=
```

$$\begin{pmatrix}
\{1.13 \times 10^9\} \\
\{8.32 \times 10^8, 1.26 \times 10^9\} \\
\{5.88 \times 10^8, 9.31 \times 10^8, 1.37 \times 10^9\} \\
\{3.94 \times 10^8, 6.63 \times 10^8, 1.01 \times 10^9, 1.45 \times 10^9\} \\
\{2.43 \times 10^8, 4.51 \times 10^8, 7.17 \times 10^8, 1.06 \times 10^9, 1.49 \times 10^9\} \\
\{1.31 \times 10^8, 2.86 \times 10^8, 4.85 \times 10^8, 7.4 \times 10^8, 1.07 \times 10^9, 1.48 \times 10^9\} \\
\{5.37 \times 10^7, 1.63 \times 10^8, 3.07 \times 10^8, 4.9 \times 10^8, 7.25 \times 10^8, 1.02 \times 10^9, 1.41 \times 10^9\} \\
\{9.56 \times 10^6, 7.7 \times 10^7, 1.74 \times 10^8, 2.98 \times 10^8, 4.56 \times 10^8, 6.59 \times 10^8, 9.17 \times 10^8, 1.25 \times 10^9\} \\
\{0, 2.4 \times 10^7, 8.18 \times 10^7, 1.56 \times 10^8, 2.51 \times 10^8, 3.73 \times 10^8, 5.28 \times 10^8, 7.26 \times 10^8, 9.79 \times 10^8\} \\
\{0, 0, 2.52 \times 10^7, 5.87 \times 10^7, 1.01 \times 10^8, 1.56 \times 10^8, 2.26 \times 10^8, 3.15 \times 10^8, 4.29 \times 10^8, 5.75 \times 10^8\} \\
\{0, 0, 0, 0, 0, 0, 0, 0, 0, 0, 0\}
\end{pmatrix}$$

This shows the lease **price** value for each year from now for 10 years; for the corresponding range of commodity prices each year, the current lease value is about $1.1 billion for the 10-year period.

To determine the value of the production enhancement option if implemented immediately, we repeat the above process using the new productivity and costs:

```
In[198]:= leasevaluewithoption = LeaseValues[price, 0.05, 1.15, 0.9, 150000, 1500];
          ScientificForm[MatrixForm[leasevaluewithoption], 3]
Out[199]//ScientificForm=
```

$$\begin{pmatrix}
\{1.12 \times 10^9\} \\
\{7.2 \times 10^8, 1.36 \times 10^9\} \\
\{4.09 \times 10^8, 9.13 \times 10^8, 1.57 \times 10^9\} \\
\{1.82 \times 10^8, 5.64 \times 10^8, 1.08 \times 10^9, 1.74 \times 10^9\} \\
\{4.14 \times 10^7, 3.02 \times 10^8, 6.95 \times 10^8, 1.2 \times 10^9, 1.86 \times 10^9\} \\
\{0, 1.19 \times 10^8, 4.04 \times 10^8, 7.86 \times 10^8, 1.27 \times 10^9, 1.9 \times 10^9\} \\
\{0, 1.57 \times 10^7, 1.96 \times 10^8, 4.69 \times 10^8, 8.21 \times 10^8, 1.27 \times 10^9, 1.85 \times 10^9\} \\
\{0, 0, 6.3 \times 10^7, 2.43 \times 10^8, 4.8 \times 10^8, 7.84 \times 10^8, 1.17 \times 10^9, 1.67 \times 10^9\} \\
\{0, 0, 0, 9.48 \times 10^7, 2.37 \times 10^8, 4.2 \times 10^8, 6.52 \times 10^8, 9.5 \times 10^8, 1.33 \times 10^9\} \\
\{0, 0, 0, 1.66 \times 10^7, 8.08 \times 10^7, 1.63 \times 10^8, 2.67 \times 10^8, 4.01 \times 10^8, 5.72 \times 10^8, 7.91 \times 10^8\} \\
\{0, 0, 0, 0, 0, 0, 0, 0, 0, 0, 0\}
\end{pmatrix}$$

The nonrecurring cost of $100 million needs to be subtracted from the optional $1.120 billion lease value if the option is to be implemented immediately, so the current lease value for the optional improved production is just a little over $1 billion, less than the $1.130 billion determined without the optional improvement, so it makes no sense to implement the improvement now. However, we can implement the higher productivity option in any year we care to, particularly if the commodity prices increase enough to favor it. The formal requirements *LeaseValuewithOption* determine when it's favorable to implement the new machine, by calculating a new lattice, where for each entry in the lattice, the value obtained with the option in place minus the cost to implement the option is compared with the value we had in the original lease lattice without the option. If the lattice entry value with the option is more than without the option, the option value is recorded for that lattice entry; otherwise a 0 is recorded.

Here is the result:

```
In[200]:= leasewithoption = LeaseValueswithOption[
            leasevaluewithoption, price, .05, 1.15, .9, 100000, 1000, 100 + 10^6];
        ScientificForm[MatrixForm[leasewithoption], 5]
```

```
Out[201]//ScientificForm=
```

$$
\begin{pmatrix}
\{1.1834 \times 10^9\} \\
\{8.4599 \times 10^8, \ 1.3402 \times 10^9\} \\
\{5.8957 \times 10^8, \ 9.541 \times 10^8, \ 1.4927 \times 10^9\} \\
\{3.9356 \times 10^8, \ 6.6604 \times 10^8, \ 1.0473 \times 10^9, \ 1.6398 \times 10^9\} \\
\{2.4277 \times 10^8, \ 4.5055 \times 10^8, \ 7.2137 \times 10^8, \ 1.1218 \times 10^9, \ 1.7563 \times 10^9\} \\
\{1.3073 \times 10^8, \ 2.8566 \times 10^8, \ 4.8525 \times 10^8, \ 7.4849 \times 10^8, \ 1.1746 \times 10^9, \ 1.7992 \times 10^9\} \\
\{5.3655 \times 10^7, \ 1.6284 \times 10^8, \ 3.065 \times 10^8, \ 4.9014 \times 10^8, \ 7.3912 \times 10^8, \ 1.171 \times 10^9, \ 1.7457 \times 10^9\} \\
\{9.5567 \times 10^6, \ 7.7045 \times 10^7, \ 1.7392 \times 10^8, \ 2.9787 \times 10^8, \ 4.5626 \times 10^8, \ 6.8373 \times 10^8, \ 1.0716 \times 10^9, \ 1.5673 \times 10^9\} \\
\{0, \ 2.3959 \times 10^7, \ 8.1804 \times 10^7, \ 1.5618 \times 10^8, \ 2.5121 \times 10^8, \ 3.7264 \times 10^8, \ 5.5225 \times 10^8, \ 8.4964 \times 10^8, \ 1.2296 \times 10^9\} \\
\{0, \ 0, \ 2.5247 \times 10^7, \ 5.8715 \times 10^7, \ 1.0148 \times 10^8, \ 1.5612 \times 10^8, \ 2.2595 \times 10^8, \ 3.1516 \times 10^8, \ 4.7232 \times 10^8, \ 6.9082 \times 10^8\} \\
\{0, \ 0, \ 0, \ 0, \ 0, \ 0, \ 0, \ 0, \ 0, \ 0, \ 0\}
\end{pmatrix}
$$

To compare the nonoption with option cases, we simply subtract the two lattices:

```
In[202]:= MatrixForm[leasewithoption - leasevalue]
```

```
Out[202]//MatrixForm=
```

$$
\begin{pmatrix}
\{5.07711 \times 10^7\} \\
\{1.38851 \times 10^7, \ 7.95926 \times 10^7\} \\
\{1.52802 \times 10^6, \ 2.32803 \times 10^7, \ 1.23767 \times 10^8\} \\
\{0., \ 2.67403 \times 10^6, \ 3.89579 \times 10^7, \ 1.9062 \times 10^8\} \\
\{0., \ 0., \ 4.67956 \times 10^6, \ 6.50566 \times 10^7, \ 2.64971 \times 10^8\} \\
\{0., \ 0., \ 0., \ 8.18923 \times 10^6, \ 1.0839 \times 10^8, \ 3.16605 \times 10^8\} \\
\{0., \ 0., \ 0., \ 0., \ 1.43312 \times 10^7, \ 1.46364 \times 10^8, \ 3.37922 \times 10^8\} \\
\{0., \ 0., \ 0., \ 0., \ 0., \ 2.50795 \times 10^7, \ 1.54381 \times 10^8, \ 3.196 \times 10^8\} \\
\{0, \ 0., \ 0., \ 0., \ 0., \ 0., \ 2.44459 \times 10^7, \ 1.23577 \times 10^8, \ 2.50245 \times 10^8\} \\
\{0, \ 0, \ 0., \ 0., \ 0., \ 0., \ 0., \ 0., \ 4.31543 \times 10^7, \ 1.15988 \times 10^8\} \\
\{0, \ 0, \ 0, \ 0, \ 0, \ 0, \ 0, \ 0, \ 0, \ 0, \ 0\}
\end{pmatrix}
$$

The results are indicating when and with what commodity price it makes sense to implement the option. For example, in two years, if the commodity price is at or above 2645, the highest value in the second year (which is shown as the third entry in the third row of price), we can make an extra $123.7 million (third entry in the third row above) if we implement the improvement. For the year and commodity price (row and column) for which the denoted profit is judged too small relative to the $100 million investment, the productivity option should not be invoked. Similarly, for the year and price for which the denoted profit is judged attractive relative to the investment, the productivity option should be invoked. So this type of analysis provides information as to when and under what manifestation of uncertain circumstances an option should be taken, or not.

3.2.9 SELECT A PORTFOLIO OF INVESTMENTS THAT MAXIMIZE RETURN AND MINIMIZE RISK

The problem statement is as follows:

How much to invest in a finite number of dissimilar investment opportunities with uncertain returns so that the likely return is maximized while the potential loss is minimized?

If an uncertain investment has a mean return and a variance, and considering any observable correlation that might exist between the investment option returns, various combinations of investment options can be plotted with respect to total mean return (goodness) and total variance (badness), which will create an efficient frontier. The investment portfolio to go with is the one on the frontier that offers the maximum mean return at the risk level (variance) that is palatable. Given also that one can periodically modify the portfolio, one can improve the cumulative return by rebalancing the level of investment in each option in the portfolio. That is, at the end of a period, one buys more of the investment option that performed the least well and one buys less of, or even sells, the investment option that did the best, so that each investment option is approximately the same value again. To fully benefit from this strategy, the individual investments must be independent or at least not strongly correlated. If the investment options are perfectly correlated, they will all achieve the same value at the end of the period so rebalancing is not possible.

Here are examples to illustrate the strategy. There are two investment options, one with a 50% probability of paying twice the amount bet, and the other a 50% probability of paying half the amount bet. *ProbReturn* calculates the probabilistic return and *CertReturn* models the certain return of invested funds, no more, no less (which is the equivalent of holding cash).

In[203]:= **ProbReturn[x_] := If$\left[\text{RandomReal[]} \leq 0.5, 2\,x, \dfrac{x}{2}\right]$;**

CertReturn[x_] := x;

The formal requirements *TotalReturn* returns the final amount after **n** periods, where at the start of each period, a proportion of available money is invested in each option.

Since the returns are random, a Monte Carlo simulation is conducted.

Suppose we start with $100 and make investments for 10 periods; here are the results for 10,000 Monte Carlo simulations if we put all our money in the certain return option:

In[205]:= **TotalReturn[100, 10, 10000, 0, 1, True]**

Mean amount at end of nperiods is 100.

The mean rate of return is 0.

Maximum amount at end of nperiods is 100.

Minimum amount at end of nperiods is 100.

The ratio of the return standard deviation to mean is 0.

Probability end up with less money than started with is 0.

Out[205]= {0., 0.}

Here are the results if we put all the money in the risky return:

In[206]:= **TotalReturn[100, 10, 10 000, 1, 0, True]**

Mean amount at end of nperiods is 906.731

The mean rate of return is 0.220468

Maximum amount at end of nperiods is 102 400.

Minimum amount at end of nperiods is 0.0976563

The ratio of the return standard deviation to mean is 8561.11

Probability end up with less money than started with is 0.3772

Out[206]= {0.3772, 0.220468}

We see that, on average, we net about nine times our starting money, for a mean rate of return of about 22%, but there is just under a 40% chance we will end up with less than our initial $100. Here is what happens if put 75% of our money in the certain return and 25% in the risky return:

In[209]:= **TotalReturn[100, 10, 10 000, .75, .25, True]**

Mean amount at end of nperiods is 546.331

The mean rate of return is 0.169805

Maximum amount at end of nperiods is 26 938.9

Minimum amount at end of nperiods is 0.909495

The ratio of the return standard deviation to mean is 1608.77

Probability end up with less money than started with is 0.372

Out[209]= {0.372, 0.169805}

The mean return is now 17%, and the risk of ending up with less money than we started is still about 40%.

To get a feel for the potential return versus potentially ending up with less money than we started with, we produce an efficient frontier plot for various percentages of investment in the risky and certain return options.

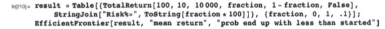

In[210]:= **result = Table[{TotalReturn[100, 10, 10 000, fraction, 1 - fraction, False],
    StringJoin["Risk%=", ToString[fraction * 100]]}, {fraction, 0, 1, .1}];
EfficientFrontier[result, "mean return", "prob end up with less than started"]**

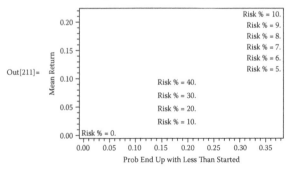

Clearly the sufficient frontier consists of 0% in the risky return, 40% in the risky return, and 100% in the risky return. Depending on one's risk tolerance, one should invest either 40% or 100% in the risky return.

Does a longer investment period improve our chances? First, increase the period from 10 to 20:

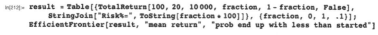

```
In[212]:= result = Table[{TotalReturn[100, 20, 10000, fraction, 1 - fraction, False],
         StringJoin["Risk%=", ToString[fraction * 100]]}, {fraction, 0, 1, .1}];
       EfficientFrontier[result, "mean return", "prob end up with less than started"]
```

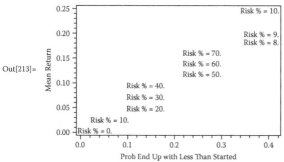

Out[213]=

Clearly, the longer investment period has reduced risk while retaining the potential returns. Here is the result for 40 periods:

```
In[214]:= result = Table[{TotalReturn[100, 40, 10000, fraction, 1 - fraction, False],
         StringJoin["Risk%=", ToString[fraction * 100]]}, {fraction, 0, 1, .1}];
       EfficientFrontier[result, "mean return", "prob end up with less than started"]
```

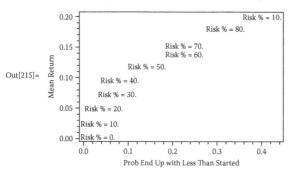

Out[215]=

Again, the more opportunities to rebalance the investments, the less risk for the same mean return.

Finally, what happens if we have two uncorrelated risky investments? The formal requirements *TotalReturnUCRisk* investigate this option.

Here are some cases for investing equally in two risky investments with the same probability of return as previously defined, but with the returns totally independent of each other:

In[216]:= **TotalReturnUCRisk[100, 10, 10000, .5, .5, True];**

Mean amount at end of nperiods is 924.095

The mean rate of return is 0.222364

Maximum amount at end of nperiods is 64 000.

Minimum amount at end of nperiods is 0.244141

The ratio of the return standard deviation to mean is 2058.02

Probability end up with less money than started with is 0.278

This is an amazing result, as we obtain a better return than the single pure-risk play. Notice the mean rate of return of 22%, but this time the odds of losing money are reduced from near 40% to near 30%. And time is again on our side; here are results for 20 and 40 periods rather than 10:

In[217]:= **TotalReturnUCRisk[100, 20, 10000, .5, .5, True];**

Mean amount at end of nperiods is 9003.1

The mean rate of return is 0.225008

Maximum amount at end of nperiods is 1.5625×10^6

Minimum amount at end of nperiods is 0.145519

The ratio of the return standard deviation to mean is 116 913.

Probability end up with less money than started with is 0.1766

In[218]:= **TotalReturnUCRisk[100, 40, 10000, .5, .5, True];**

Mean amount at end of nperiods is 628 057.

The mean rate of return is 0.21863

Maximum amount at end of nperiods is 2.27374×10^8

Minimum amount at end of nperiods is 0.0847033

The ratio of the return standard deviation to mean is 2.60941×10^7

Probability end up with less money than started with is 0.0878

The mean rate of return holds at over 20%, but the probability of ending up with less money than we started with keeps dropping.

From these examples, investment strategy heuristics are quite clear:

1. *Presume the possibility of what could happen, not what did happen.* Many people are very good at finding patterns in data for things that already occurred. One can examine the daily fluctuations for a stock over a period of time and create a probability density function that fits the data. But the

stock prices may or may not behave in the future as they have behaved in the past. When evaluating an investment opportunity, model the potential for what may happen, not just what has happened.

2. *Buy low, sell high.* The rebalancing strategy does this automatically.
3. *If the likely result is positive, maximize the number of opportunities.*
4. *Diversify*, and note that two independent risky investments provide a better return than one risky investment.
5. *Invest an amount in proportion to what one can lose.* All investment decisions must be made in full cognizance of the level of risk that can be tolerated.

3.2.10 Choose the Independent Variable Values That Optimize an Index of Performance Subject to Constraints When the Index of Performance Is Inexpensive to Determine

The problems statement is of the form:

Find the $\mathbf{x}^*$ (an array of n values) such that $J(\mathbf{x})$ is minimized, while $H(\mathbf{x}) = 0$ (an array of $\mathbf{m}$ equations) and $G(\mathbf{x}) < 0$ (an array of $\mathbf{p}$ equations).

Recall that finding the minimum of J is the same as finding the maximum of $-J$, so any maximization problem can be recast as a minimization problem.

There are many algorithms available to find solutions to the stated problem. Some of these algorithms require explicit or numerical derivatives of the index of performance, which are often inconvenient to obtain. I prefer methods that do not require derivatives. Some methods attempt to find the maximum while explicitly honoring the constraints. This is straightforward if the constraint involves only limits on the independent variables; simply do not attempt any trial solution with values exceeding the limits. But for large numbers of constraints, the permissible space for solution can be quite complicated, and perhaps not even convex, so I prefer methods that append the constraints to the index of performance as follows:

$$IP(\mathbf{x}) = J(\mathbf{x}) + \mathbf{W}(H.H + g.g)$$

where $g[[\mathbf{i}]] = 0$ if $G[[\mathbf{i}]]$ is < 0 and $g[[\mathbf{i}]] = G[[\mathbf{i}]]$ if $G[[\mathbf{i}]] > 0$, and $\mathbf{W}$ is a large but arbitrary value.

Then, to minimize *IP*, the algorithm will seek to drive each H to 0 and keep each G less than 0.

Mathematica comes with routines that enable determining the minimum of functions (*NMinimum*, *FindMinimum*, etc.), and as *Mathematica* evolves, the robustness of these routines will improve, so I urge the reader to first attempt optimization using these routines. The *Mathematica* numerical optimization routines are tested against 35 problems, with results shown in Appendix 3A.

Always check the obtained $\mathbf{x}^*$ for robustness before using. To do so, first check J in the neighborhood of $\mathbf{x}$. If small changes in any of the $\mathbf{x}^*$ components result in small changes in J, one can be reasonably assured the $\mathbf{x}^*$ obtained will be useful.

However, the solution to constrained optimization problems is often on the boundaries of one or more constraints. So also explicitly examine the constraints that are in effect at **x***. Determine to what extent these constraint equations may be wrong or approximate. Determine if it is better to be conservative, and use for the optimal solution an **x** that is a bit further away from the constraint than **x*** to make sure no constraint is violated, even though the resulting *J* may be larger.

3.2.11 Choose the Independent Variable Values That Optimize an Index of Performance Subject to Constraints When the Index of Performance Is Expensive to Determine

The problem statement is the same as in Section 3.2.10, but now the cost to evaluate either *J*, or the constraints, is very expensive. The "expense" may be that the index of performance or the constraints are so complicated that even the fastest computer needs a long time to provide the necessary evaluations so that the iterative search for the solution will take longer than we have to make the decision. Or the index of performance values need to be obtained from the actual output of a system that costs time and money to operate each time a different trial set of independent variables is used in attempt to find the minimum. So, we must sacrifice perfection for efficiency. We need to select and try the smallest possible sample of trial independent variables to evaluate the function to be minimized, and identify an approximate best set of independent variables from those few trials.

The Taguchi method[5] is a way to both select the trial cases and utilize the results to find an approximate optimum answer that is robust with respect to uncertainty in controlled factors and noise factors. The Taguchi method seeks to minimize a quality loss function that is presumed to be a quadratic with respect to the independent variables of interest. So the method finds control variables that minimize the variation from a target value. This is accomplished by calculating a signal-to-noise ratio, and finding the control variables that maximize the ratio. To achieve the desired mean response, a control variable is reserved to use as an adjustment factor, purposely selected to have the least impact on the signal-to-noise ratio, but a significant impact on the mean response.

The Taguchi method consists of six steps as follows:

Step 1. Select the appropriate quality loss function to optimize, which will be one of the following:
If smaller is better, use $-10 * Log[Sum[\mathbf{y}[[\mathbf{i}]]\hat{}2, \{\mathbf{i}, 1, \mathbf{n}] / \mathbf{n}]$.
If larger is better, use $-10 * Log[Sum[1 / \mathbf{y}[[\mathbf{i}]]\hat{}2, \{\mathbf{i}, 1, \mathbf{n}\} / \mathbf{n}]$.
If nominal is best, use $-10 * Log[(1 / \mathbf{n}) (\mathbf{T} - \mathbf{V}) / \mathbf{V})]$, where:

$$\mathbf{T} = Sum[\mathbf{y}[[\mathbf{i}]], \{\mathbf{i}, 1, \mathbf{n}\}]\hat{}2$$

$$\mathbf{V} = (Sum[\mathbf{y}[[\mathbf{i}]]\hat{}2, \{\mathbf{i}, 1, \mathbf{n}\}] - \mathbf{Total}) / (\mathbf{n} - 1)$$

Or use

$$-10 * Log[\mathbf{V}]$$

when $\mathbf{y}[[\mathbf{i}]]$ may be both negative and positive.

Step 2. Identify the **m** control factors that will be selected to achieve the optimum result, and the n noise factors, or things not explicitly controlled that still influence quality. Each control factor is presumed to affect quality independently, so if two control factors may affect quality as a consequence of their combined action, then that interaction is itself a separate control factor. When selecting both control and noise factors, it is important to be able to achieve selected test values, so one must also select these test values. A minimum of two values needs to be selected, but if there is any hint of nonlinear behavior in the quality loss function with respect to control or noise, then at least three values need to tested. The key to value selection is to bound the quality factor, for the approximate optimum will be determined within the interior ranges of these variables. So, it may be necessary to conduct preliminary tests to establish the appropriate bounding values. The more control factors and noise factors, the more tests that will need to be run. So, given the expense of the trials, limit the number of each to keep within cost constraints.

Step 3. Select an orthogonal array to use that specifies the control value combinations for each test. For each control value combination, the test is repeated for the full set of noise factors possible. For example, if there are two noise factors at three levels, then each control variable combination is repeated $3^2 = 9$ times for the nine different combinations of three things taken two at a time. Reference 5 documents orthogonal arrays for two levels, two and three levels, four levels, two and four levels, and five levels. Along with the arrays are line graphs, which help one determine which column of a given array should be used for any controls that are needed to model interactions between two controls. The orthogonal arrays are constructed so those factors that are hardest to manipulate (change) should be assigned to the first few columns.

Step 4. Conduct the test runs as stipulated by the orthogonal array to obtain the quality loss values.

Step 5. Perform the calculations necessary to ascertain the most significant control factors and the best values for each. This requires determining the mean value for each experiment, the mean signal-to-noise ratio (quality loss function) for each experiment, the mean signal-to-noise ratio of each control factor at each test level, the effect of each factor level on the signal-to-noise ratio (the difference between the largest and smallest values achieved), the identification of those control factors with the largest quality loss function effect (roughly the half of the control factors with the largest signal-to-noise ratio effect), the mean response for each control factor level, the effect of each factor on the mean response, identifying the roughly half of the control factors that influence the mean response, and finally selecting the control factor values that predict the minimal quality loss function.

Step 6. Conduct a test with the chosen control variable values to confirm the predicted quality loss achieved.

Here is an example analysis.

Step 1. Quality loss function: Nominal is best, with a target value of 35.75.

Step 2. Eight control factors are identified, one with two levels, seven with three levels, and initially no interactions presumed significant.

Step 3. The **L18** orthogonal array is found to be appropriate as it enables stipulating 1 two-level variable and up to 7 three-level variables. (If we had instead only four control factors, all of which were to be assessed at three levels, then we would have used the **L9** orthogonal array. Alternatively, if we had as many as 13 control factors, to be assessed at three levels, we would have used the **L27** array.) The **L18** array specifies the level to use for 18 experiments. The **L18** array is as follows. Each row is an experiment, to perform the 1, 2, or 3 entry specifying the level to set the control variable to for that experiment:

```
In[219]:= L18 =
          {{1, 1, 1, 1, 1, 1, 1, 1},
           {1, 1, 2, 2, 2, 2, 2, 2},
           {1, 1, 3, 3, 3, 3, 3, 3},
           {1, 2, 1, 1, 2, 2, 3, 3},
           {1, 2, 2, 2, 3, 3, 1, 1},
           {1, 2, 3, 3, 1, 1, 2, 2},
           {1, 3, 1, 2, 1, 3, 2, 3},
           {1, 3, 2, 3, 2, 1, 3, 1},
           {1, 3, 3, 1, 3, 2, 1, 2},
           {2, 1, 1, 3, 3, 2, 2, 1},
           {2, 1, 2, 1, 1, 3, 3, 2},
           {2, 1, 3, 2, 2, 1, 1, 3},
           {2, 2, 1, 2, 3, 1, 3, 2},
           {2, 2, 2, 3, 1, 2, 1, 3},
           {2, 2, 3, 1, 2, 3, 2, 1},
           {2, 3, 1, 3, 2, 3, 1, 2},
           {2, 3, 2, 1, 3, 1, 2, 3},
           {2, 3, 3, 2, 1, 2, 3, 1}};
```

Step 4. There are two critical noise factors, one to be tested at three levels and one to be tested at two levels, for a total of six combinations. Thus, for each 18 variations of control factors, we run six tests at the different noise factors, for a total of 84 experiments. Let's say the 18 tests at six different settings for noise factors give the 84 results as follows:

```
In[220]:= TestResults =
          {{34.10, 32.95, 35.65, 35.05, 39.30, 40.20},
           {35.00, 34.35, 35.75, 35.40, 36.50, 36.80},
           {35.00, 34.50, 36.30, 36.55, 37.60, 38.50},
           {34.60, 33.95, 35.55, 35.15, 36.50, 37.15},
           {32.70, 31.45, 35.55, 34.80, 38.55, 39.60},
           {34.15, 33.35, 36.00, 36.50, 37.85, 38.70},
           {32.00, 30.75, 35.25, 34.50, 38.65, 40.45},
           {32.00, 30.75, 35.15, 34.40, 38.50, 39.50},
           {33.45, 32.30, 35.70, 35.00, 38.00, 38.80},
           {35.35, 35.00, 35.70, 35.50, 36.05, 35.65},
           {33.75, 32.30, 35.80, 36.20, 37.90, 38.80},
           {34.50, 33.85, 36.25, 36.50, 38.00, 39.10},
           {33.10, 32.15, 35.50, 34.90, 37.95, 39.05},
           {35.60, 35.45, 35.55, 35.40, 35.50, 35.60},
           {33.20, 32.10, 36.00, 36.70, 38.95, 39.85},
           {32.50, 31.55, 35.25, 34.65, 38.10, 39.10},
           {33.50, 32.70, 35.20, 34.70, 37.05, 37.85},
           {33.25, 32.10, 35.70, 35.00, 38.20, 38.65}};
```

Step 5. The calculations that are needed are as follows:
1. Mean experiment value (i.e., the mean of the result for the noise cases for each run), **ymean**
2. Mean of the signal-to-noise ratio for each run, **ston**
3. The signal-to-noise ratio for each factor at each level, **stonfactor[[level]]**
4. The signal-to-noise ratio response table (which shows the **stonfactor** for each control factor at each level and the difference between the maximum and minimum signal-to-noise achieved at any level)
5. The mean response level for each factor at each level, **meanresp**
6. The mean response table which shows the **meanresp** for each control factor at each level and the difference between the maximum mean response and minimum mean response at any level

These calculated results are used to first identify which factors should be chosen to maximize the signal-to-noise ratio (and thus minimize the variability in output) and what value each should be set to. Then we identify the control variables that contribute most to the mean response. We look for the control variable that does not also affect the signal-to-noise ratio to use to move the mean response to the target. The formal requirements *TaguchiResponseTable* perform the necessary calculations for a provided orthogonal array, test results, and an indication of the number of control variables (which are presumed to be assigned in order to the columns of the orthogonal array).

The results obtained are as follows:

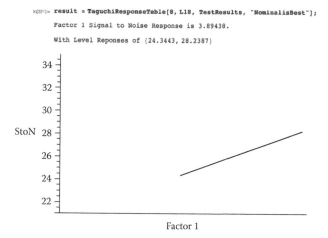

In[22:1]:= **result = TaguchiResponseTable[8, L18, TestResults, "NominalisBest"];**

Factor 1 Signal to Noise Response is 3.89438.

With Level Reponses of {24.3443, 28.2387}

Factor 2 Signal to Noise Response is 6.71763
. With Level Reponses of {28.3014, 28.6454, 21.9277}

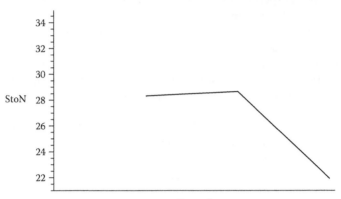

Factor 2

Factor 3 Signal to Noise Response is 4.8442
. With Level Reponses of {25.7947, 28.962, 24.1178}

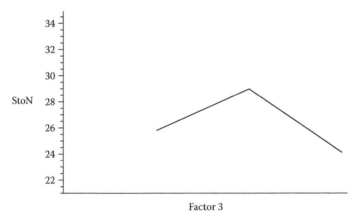

Factor 3

Factor 4 Signal to Noise Response is 7.39364
. With Level Reponses of {24.0047, 23.7381, 31.1317}

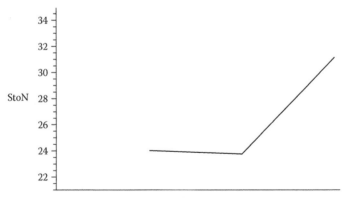

Factor 4

Factor 5 Signal to Noise Response is
2.5879. With Level Reponses of {27.49, 24.9021, 26.4825}

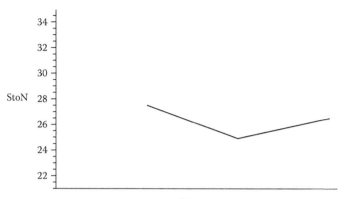

Factor 5

Factor 6 Signal to Noise Response is 10.9227
. With Level Reponses of {23.2333, 33.2819, 22.3592}

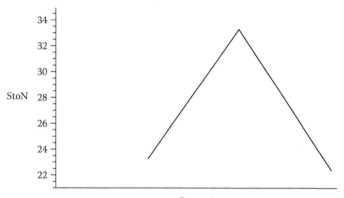

Factor 6

Factor 7 Signal to Noise Response is 3.26453
. With Level Reponses of {27.5297, 27.0797, 24.2651}

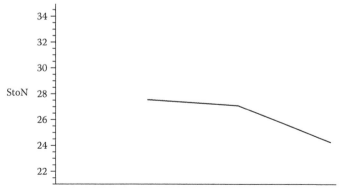

Factor 7

Factor 8 Signal to Noise Response is 5.46633
. With Level Reponses of {24.5055, 24.4513, 29.9177}

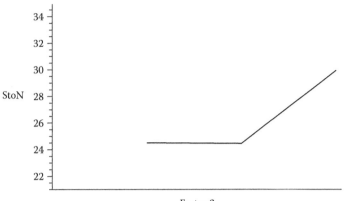

Factor 2

Factor 1 Mean Response is 0.0546296. With Level Reponses of {35.6806, 35.6259}

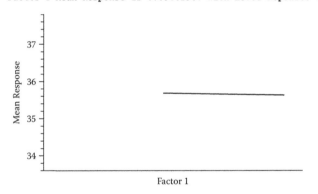

Factor 1

Factor 2 Mean Response is 0.708333. With Level Reponses of {35.9917, 35.6847, 35.2833}

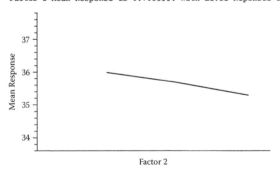

Factor 2

Factor 3 Mean Response is 0.570833. With Level Reponses of {35.5222, 35.4333, 36.0042}

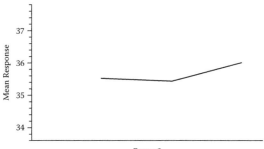

Factor 4 Mean Response is 0.115278. With Level Reponses of {35.7208, 35.6056, 35.6333}

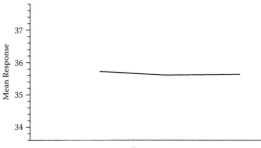

Factor 5 Mean Response is 0.136111. With Level Reponses of {35.7264, 35.6431, 35.5903}

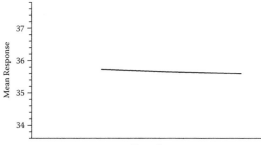

Factor 6 Mean Response is 0.1875. With Level Reponses of {35.7208, 35.5333, 35.7056}

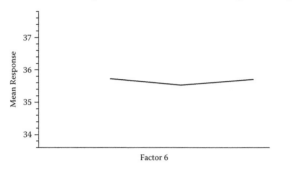

Factor 7 Mean Response is 0.101389. With Level Reponses of {35.7111, 35.6389, 35.6097}

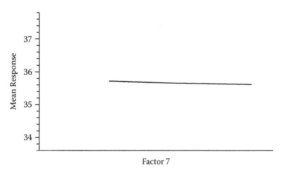

Factor 8 Mean Response is 0.0861111. With Level Reponses of {35.6431, 35.6153, 35.7014}

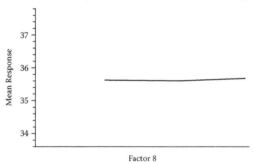

The results above indicate that the control factors that most impact the signal-to-noise ratio (variability in result) are 6, 4, and 2, with respective responses of 10.9, 7.4, and 6.7. A Taguchi method heuristic is to use approximately half of the control factors that have the most impact on the signal-to-noise ratio. Since there are eight control factors, we would nominally select four. But a second Taguchi method heuristic is to stop selecting control factors when the difference in response results between it and its predecessor first increases. In this case, the difference between the largest and second largest is

In[222]:= **10.92 - 7.39**

Out[222]= **3.53**

while the difference between the second and third largest is

In[223]:= **7.39 - 6.72**

Out[223]= **0.67**

while the difference between the third and fourth largest is

In[224]:= **6.72 - 5.47**

Out[224]= **1.25**

and, lastly, the difference between the fourth and fifth largest is

In[225]:= **5.47 - 4.84**

Out[225]= **0.63**

Notice these differences first increased after decreasing for the third and fourth largest, so we stop our selection for control factors to control the signal-to-noise ratio to the three largest.

So the difference in response results first increases after the third highest control variable, so we stop our selection with control factor 2 which had a 6.72 response impact on the signal-to-noise ratio.

Examining the signal-to-noise ratio plots, the control values to choose to maximize the signal-to-noise ratio and therefore minimize the variability in the output are

For control factor 2, at level 2
For control factor 4, at level 3
For control factor 6, at level 2

as these are the levels with the maximum values in respective response plots.

Each orthogonal array has unique properties. The **L18** array implicitly contains the interaction of factor 1 and 2, even though this interaction was not assigned to an array dimension. Since we choose factor 2 to obtain the approximate optimal solution, it is important to then also find what value of control factor 1 to utilize since it inherently has interacted with control factor 2 in the experiments, due to the construct of the orthogonal array, even though we did not explicitly intend it to do so. All that is required of us is to use the experiment data to calculate the following means:

Control factor 1 at level 1 and control factor 2 at levels 1, 2, and 3

and

Control factor 1 at level 2 with control factor 2 at levels 1, 2, and 3

The formal requirements *InteractionResponse* report the interaction mean response and signal-to-noise ratio, which provides the following information:

In[226]:= **InteractionResponse[result, L18, 1, 2]**

Signal to Noise for Interaction of Control Factor 1 at level
1 and Control Factor 2 at level 1 is: 27.0985

Signal to Noise for Interaction of Control Factor 1 at level
2 and Control Factor 2 at level 1 is: 29.5042

Signal to Noise for Interaction of Control Factor 1 at level
1 and Control Factor 2 at level 2 is: 25.0852

Signal to Noise for Interaction of Control Factor 1 at level
2 and Control Factor 2 at level 2 is: 32.2056

Signal to Noise for Interaction of Control Factor 1 at level
1 and Control Factor 2 at level 3 is: 20.8492

Signal to Noise for Interaction of Control Factor 1 at level
2 and Control Factor 2 at level 3 is: 23.0063

Mean response for Interaction of Control Factor 1 at level
1 and Control Factor 2 at level 1 is: 36.0833

Mean response for Interaction of Control Factor 1 at level
2 and Control Factor 2 at level 1 is: 35.9

Mean response for Interaction of Control Factor 1 at level
1 and Control Factor 2 at level 2 is: 35.6722

Mean response for Interaction of Control Factor 1 at level
2 and Control Factor 2 at level 2 is: 35.6972

Mean response for Interaction of Control Factor 1 at level
1 and Control Factor 2 at level 3 is: 35.2861

Mean response for Interaction of Control Factor 1 at level
2 and Control Factor 2 at level 3 is: 35.2806

Since will use control factor 2 at level 2, we see that the signal-to-noise ratio is maximized for control factor 1 also set to level 2 (32.2056 > 25.08); similarly control factor 1 set to level 2 also, just barely, provides the maximum mean response.

Revisiting the output from *TaguchiResponseTable*, the control factors that most influence the mean response are 2 and 3, with respective values of 0.71 and 0.57. Again, even though we normally would be looking to set about half, or in this case four, of the control variables, we quit after selecting the two control variables because the difference in the mean responses obtained increase after the second selection.

The difference in response between control factors 2 and 3 is

In[227]:= **0.71 - 0.57**

Out[227]= 0.14

While the difference in response between control factors 3 and 6 is

In[228]:= **0.57 - 0.19**

Out[228]= 0.38

Step 6. Select a value for control factor 3, estimate the mean result expected, and run a confirmation experiment at the selected values to see if the result is as predicted.

Since the signal-to-noise maximization assessment determined control factor 2 should be part of the set to maximize the signal-to-noise ratio (and therefore minimize the variability in the response), we use it for that purpose, and use control factor 3 as the factor to adjust the mean closer to the desired value. We find the desired value for control factor 3 by predicting the mean response based on the selected control factors and their selected values:

```
In[229]:= ymean = Mean[TestResults];
          Tbar = Mean[ymean]
          meanCF2at2 = 35.6842;
          target = 35.75;
          Solve[target == Tbar + (meanCF2at2 - Tbar) + (meanCF3 - Tbar), {meanCF3}]
Out[230]= 35.6532
Out[233]= {{meanCF3 → 35.719}}
```

Notice that the mean response of control factor 3 at level 1 is 35.52, the mean response of control factor 3 at level 2 is 35.42, and the mean response of control factor 3 at level 3 is 36.00, so to move the mean to the target with the chosen values for control factor 1, 2, 4, and 6, we need to set control factor 3 to a value intermediate between our chosen level 2 and level 3.

To test our analysis results, run one more experiment with the control factors set as stipulated, for the six noise levels to determine if the mean response is as predicted.

3.2.12 DEFINE A DYNAMIC CONTROL LAW TO OPTIMIZE A DYNAMIC INDEX OF PERFORMANCE

The problems statement is of the form:

Find an $\mathbf{m} \times 1$ vector of control variables that are a function of the state of the system and an independent variable, $\mathbf{u}(\mathbf{x}, \mathbf{t})$, such that for the $\mathbf{n} \times 1$ state vector $\mathbf{x}$,

$$d\mathbf{x}/dt = f(\mathbf{x}, \mathbf{u}, \mathbf{t})$$

with

$$\mathbf{x}(0) \text{ usually known}$$

and

$$\text{minimize } J = g(\mathbf{x}, \mathbf{u}, \mathbf{t}),$$

There are two ways to offer a solution. The easier solution finds, $\mathbf{u}$, the "controls" to be provided as a function of independent variable, $\mathbf{t}$. Though often easy to do, when we attempt to apply the control in the real world, it will usually perform badly because the solution is optimal only for the precise equations specified. Any errors in the mathematical model's representation of reality, or the inability to precisely provide the stipulated control levels, will manifest themselves in the system drifting off the desired optimum. This is because, unless the control law includes the current values of $\mathbf{x}$, there is no way for $\mathbf{u}$ to adjust its value accordingly. So what we really seek is a control law, a function of the independent variables and the state variables. Granted, the solution obtained remains only truly optimal for exactly the situation modeled, but since we can model noise in the measurement of the states, and noise in the application of the control, we have a much better chance of implementing a solution that will actually work. Unfortunately, finding such solutions are amongst the most difficult numerical processes known. Still, very smart people have been at work on this for a long time and their progress can be harvested.

For the case:

$$f[\mathbf{x}, \mathbf{u}, \mathbf{t}] = \mathbf{A} * \mathbf{x} + \mathbf{B} * \mathbf{u},$$

with $\mathbf{x}(0)$ and final time $(\mathbf{tf})$ specified and

$$\mathbf{F} * \mathbf{x}[\mathbf{tf}] = \mathbf{Fo},$$

where:

$\mathbf{A}$ is an $\mathbf{n} \times \mathbf{n}$ matrix of known functions of time,
$\mathbf{B}$ is an $\mathbf{n} \times \mathbf{m}$ matrix of known functions of time,
$\mathbf{F}$ is a $\mathbf{p} \times \mathbf{n}$ matrix of known constants, and
$\mathbf{Fo}$ is a $\mathbf{p} \times 1$ matrix of known constants.

If we seek to minimize

$J = 0.5 * Transpose[\mathbf{x[tf]}] * \mathbf{G} * \mathbf{x[tf]} + 0.5 * Integrate[Transpose[\mathbf{x[t]}] * \mathbf{W1} * \mathbf{x[t]} + 2 * Transpose[\mathbf{x[t]}] * \mathbf{W2} * \mathbf{u[t]} + Transpose[\mathbf{u[t]}] * \mathbf{W3} * \mathbf{u[t]}$, for $\{\mathbf{t, 0, tf}\}]$,

where:

G is an **n** × × **n** symmetric matrix of known constants,
W1 is an **n** × **m** symmetric matrix of known functions of time,
W2 is an **n** × **m** matrix of known functions of time, and
W3 is an **m** × **m** symmetric matrix of known functions of time and is taken to
  be positive definite.

Then the control law is known to be[9]

$\mathbf{u[x, t]} = -Inverse[\mathbf{W3}] * Transpose[\mathbf{B}] * \mathbf{R[t]} * Inverse[\mathbf{Q}] * \mathbf{C} * \mathbf{x[tf]} - Inverse[\mathbf{W3}]$
$* (Transpose[\mathbf{W2}] + Transpose[\mathbf{B}] * (\mathbf{P[t]} - \mathbf{R[t]} * Inverse[\mathbf{Q}] * Transpose[\mathbf{R}])) * \mathbf{x[t]}$

where:

$d\mathbf{P}/dt = -\mathbf{P} * \mathbf{A} - Transpose[\mathbf{A}] * \mathbf{P} - \mathbf{W1} + (\mathbf{P} * \mathbf{B} + \mathbf{W2}) * Inverse[\mathbf{W3}] *$
$(Transpose[\mathbf{W2}] + Transpose[\mathbf{B}] * \mathbf{P})$,

with $\mathbf{P[tf]} = \mathbf{G}$. Note **P** is an **n** × **n** matrix of functions of time determined by integrating backward from $\mathbf{P[tf]}$.

$d\mathbf{R}/dt = (\mathbf{P} * \mathbf{B} * Inverse[\mathbf{W3}] * Transpose[\mathbf{B}] - Transpose[\mathbf{A}] + \mathbf{W2} * Inverse[\mathbf{W3}]$
$* Transpose[\mathbf{B}]) * \mathbf{R}$, with $\mathbf{R[tf]} = Transpose[\mathbf{F}]$.

Note that **R** is an **n** × **p** matrix of functions of time determined by integrating backward from $\mathbf{R[tf]}$.

$d\mathbf{Q}/dt = Transpose[\mathbf{R}] * \mathbf{B} * Inverse[\mathbf{W3}] * Transpose[\mathbf{B}] * \mathbf{R}$, with $\mathbf{Q[tf]} = 0$.

Note **Q** is a **p** × **p** matrix of functions of time determine by integrating backward from $\mathbf{Q[tf]}$).

Similar results can be quoted with the introduction of noise or uncertainty in the equations of motion, measurement of state variables, and application of controls. For the case of **A**, **B** constant, **G** and **W3** zero matrices, and **tf** taken to be infinity, the solution simplifies considerably to

$$\mathbf{u[x, t]} = -\mathbf{K} * \mathbf{x},$$

where

$$\mathbf{K} = Inverse[\mathbf{W3}] * Transpose[\mathbf{B}] * \mathbf{P}$$

and

$$Transpose[A] * P + P * A - P * B * Inverse[W3] * Transpose[B] + \mathbf{W1} = 0$$

The formal requirements *ContFunctforLinConstEOM* provide solutions to this type of problem. Here is a simple example:

```
In[151]:= A = {{0, 1}, {0, 0}};
B = {0, 1};
W1 = {{1, 0}, {0, c}};
W3 = {1};
ContFunctforLinConstEOM[A, B, W1, W3]
```

Candidate solution: 1 $P = \begin{pmatrix} \sqrt{-2+c} & -1 \\ -1 & -\sqrt{-2+c} \end{pmatrix}$ $K = \begin{pmatrix} -1 \\ -\sqrt{-2+c} \end{pmatrix}$

Candidate solution: 2 $P = \begin{pmatrix} -\sqrt{-2+c} & -1 \\ -1 & \sqrt{-2+c} \end{pmatrix}$ $K = \begin{pmatrix} -1 \\ \sqrt{-2+c} \end{pmatrix}$

Candidate solution: 3 $P = \begin{pmatrix} -\sqrt{2+c} & 1 \\ 1 & -\sqrt{2+c} \end{pmatrix}$ $K = \begin{pmatrix} 1 \\ -\sqrt{2+c} \end{pmatrix}$

Candidate solution: 4 $P = \begin{pmatrix} \sqrt{2+c} & 1 \\ 1 & \sqrt{2+c} \end{pmatrix}$ $K = \begin{pmatrix} 1 \\ \sqrt{2+c} \end{pmatrix}$

A stable solution is obtained only for **P**'s that are positive definite, which is the case for candidate solution 4, provided **c** > –2. Notice the optimal control law requires measurement of all the states, this is a universal characteristic of such solution, so clearly then, unless the states can be measured or estimated, the optimal control cannot be implemented.

Let's examine the solution for the presumed equations of motion. The resulting differential equations are

```
In[156]:= A.{x1[t], x2[t]} + B * (-x1[t] - Sqrt[c + 2] * x2[t])
```

Out[156]= $\left\{ x2[t], -x1[t] - \sqrt{2+c}\, x2[t] \right\}$

Assuming **c** = 1 and **x1**[0] = 1 and **x2**[0] = 0, the index of performance is trying to drive both states to 0 as soon as practical:

```
In[157]:= c = 1;
          result = NDSolve[{x1'[t] == x2[t], x2'[t] == -x1[t] - Sqrt[c + 2] * x2[t], x1[0] == 1, x2[0] == 0},
            {x1, x2}, {t, 0, 1000}]
          Plot[Evaluate[x1[t] /. result], {t, 0, 10}, PlotLabel → "x1 vs time"]
          Plot[Evaluate[x2[t] /. result], {t, 0, 10}, PlotLabel → "x2 vs time"]
          Plot[Evaluate[-x1[t] - Sqrt[c + 2] * x2[t]] /. result,
            {t, 0, 10}, PlotLabel -> "u vs time", PlotRange → All]
Out[158]= {{x1 → InterpolatingFunction[{{0., 1000.}}, <>],
            x2 → InterpolatingFunction[{{0., 1000.}}, <>]}}
```

Out[159]=

x1 vs time

Out[160]=

x2 vs time

Out[161]=

u vs time

So the obtained control law works pretty well. But suppose the equations of motion are slightly wrong, for example, the 1's in **A** and **B** should really be 1.1 and a small bias of 0.01 is unknowingly provided by the real-world control mechanisms:

```
In[162]:= c = 1;
         result = NDSolve[{x1'[t] == 1.1 * x2[t], x2'[t] == 1.1 * (-x1[t] - Sqrt[c + 2] * x2[t]) + .01,
            x1[0] == 1, x2[0] == 0}, {x1, x2}, {t, 0, 1000}]
         Plot[Evaluate[x1[t] /. result], {t, 0, 10}, PlotLabel → "x1 vs time"]
         Plot[Evaluate[x2[t] /. result], {t, 0, 10}, PlotLabel → "x2 vs time"]
         Plot[Evaluate[-x1[t] - Sqrt[c + 2] * x2[t]] /. result,
            {t, 0, 10}, PlotLabel -> "u vs time", PlotRange → All]

Out[163]= {{x1 → InterpolatingFunction[{{0., 1000.}}, <>],
          x2 → InterpolatingFunction[{{0., 1000.}}, <>]}}
```

Out[164]=

x1 vs time

Out[165]=

x2 vs time

Out[166]=

u vs time

The optimal control law is still working very well but no longer perfectly; you can discern a bit longer to get to 0 values and some oscillation about the 0 value with time. Now suppose neither state can be perfectly measured, that there will be random noise corrupting the **x1** and **x2** utilized by the controller, and suppose this noise uniformly distributed between −0.01 and 0.01, we can simulate this situation as follows:

```
In[893]:= c = 1;
result = NDSolve[{x1'[t] == 1.1` x2[t], x2'[t] ==
    1.1` (-x1[t] (1.01`-1.02` RandomReal[]) - √(c+2) x2[t] (1.01`-1.02` RandomReal[])) +
    0.01`, x1[0] == 1, x2[0] == 0}, {x1, x2}, {t, 0, 1000}]
Plot[Evaluate[x1[t] /.result], {t, 0, 10}, PlotLabel → "x1 vs time"]
Plot[Evaluate[x2[t] /.result], {t, 0, 10}, PlotLabel → "x2 vs time"]
Plot[Evaluate[-x1[t] - √(c+2) x2[t]] /.result,
    {t, 0, 10}, PlotLabel → "u vs time", PlotRange → All]
Out[894]= {{x1 → InterpolatingFunction[{{0., 1000.}}, <>],
    x2 → InterpolatingFunction[{{0., 1000.}}, <>]}}
```

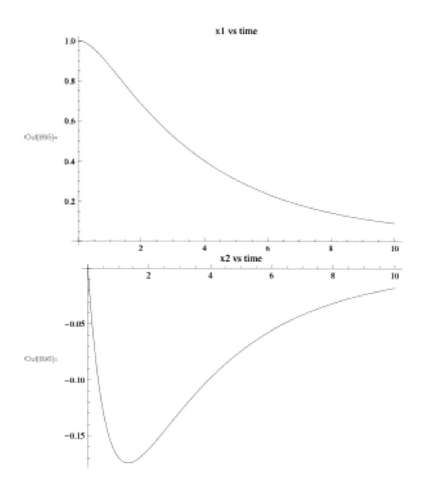

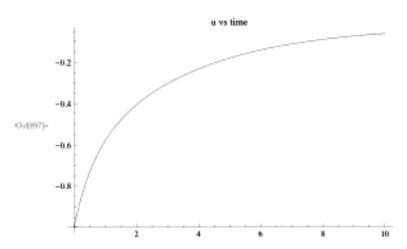

The optimal controller is still working pretty well, though we do see it takes longer to get to the 0 values for both states and we see the beginnings of some over- and undershoot on both variables. Still, the power of the method is evident; even with reality incorrectly modeled, biases in control application, and noise in the measurements, the controller is working.

For system models with terms that vary with time, or with nonlinear *f*'s, closed form solutions using calculus of variations are rarely derivable. The reader should be able to find developed application packages for these much more complicated situations, and sometimes they will work wonderfully.

My suggestion to the reader is that fuzzy logic offers a general purpose solution process, that though approximate, is quite robust in implementation. The process is as follows:

1. Establish the equations of motion with careful attention to what can really be accurately measured and accurately controlled in a timely fashion.
2. For these equations of motion, find by whatever means is possible, a near-optimum nominal control function.
3. Vary the key parameters that could affect the control function over the likely range of possible alternative values and determine two things—the new control law and the difference between that and the nominal control law.
4. Use the above sensitivity data to establish fuzzy logic rules for establishing a control based on fuzzy senses values.

We'll use the flat earth, no-atmosphere equations of motion introduced in Chapter 2 to illustrate the creation of and use of a fuzzy controller. The problem statement is

minimize **tfinal** with

$dx1/dt = x3$, **x1**(0) = 0, **x1(tfinal)** unspecified
$dx2/dt = x4$, **x2**(0) = 0, **x2(tfinal)** = 50,000 meters
$dx3/dt = A * Cos[$**Theta**$]$, **x3**(0) = 0, **x3(tfinal)** = 5444 meters/second
$dx4/dt = A * Sin[$**Theta**$] - g$, **x4**(0) =0, **x4(tfinal)** = 0 meters/second

where **g** is acceleration due to gravity, assumed to be a constant of 5.32 meters/second^2; **A** is thrust divided by mass and assumed to be a constant of 20.82 meters/second^2; and atmospheric affects such as lift and draft are assumed to be negligible.

Recall the open loop optimal steering law is given by

$$Tan[\mathbf{Theta}] = \mathbf{a} * \mathbf{t} + \mathbf{b}$$

where **a** and **b** are constants which are determined by solving three nonlinear equations in three unknowns of **a**, **b**, and **tfinal**. Despite the almost absurd simplifications, I assure you this "lintan" control function is remarkably robust for the exo-atmospheric phase of rocket motion, and provides a near optimal solution even with the realities of an oblong planet and gravity decreasing with altitude. The Space Shuttle used this control law. Thanks to a math trick I won't bother you with, we can find equations of motion that let us use on-board computers of modest power to solve for a new **a** and **b** every few seconds; since the new solution is dependent on the current state, a control law is thus achieved. But, though effective, the solution so obtained remains sensitive to modeling errors, and a less than optimal solution will be obtained in proportion to the size of the errors.

Alternatively, we can use fuzzy logic to approximate the control law. What follows is the step-by-step process for so doing. When we are done, we'll compare how well the fuzzy logic solution deals with modeling and control system implementation realities to utilizing the open-loop optimum provided by calculus of variations.

Knowing the solution is the tangent varying linearly with time, we can use the *Mathematica* routine *FindMinimum* to find the linear constants (**a** and **b**) that minimize the index of performance (**tfinal**). To do this, we must integrate the equations of motion with the trial control law and figure out when to stop the integration. This is a common issue in such problems. In this case, **x3** is a convenient variable to use to stop integration since for essentially any control law we try, we see from the equations **x3** will monotonically increase by **A** * *Cos*[**Theta**], eventually reaching the desired value of 5444. Once it does we stop integrating, and note what **tfinal** is, and what value **x2** got to, hoping it is near 50,000, while **x4[tf]** is near 0. So we ask *FindMinimum* to find **a** and **b** so that **tfinal** is minimized and the penalties (**x2[tf]**/50000–1)^2 and **x4[tf]**^2 are driven to zero.

First define the needed constants, which are the desired final altitude (**H**), the desired final horizontal velocity (**U**), the rocket's assumed constant acceleration (**A**), and the planet's assumed constant gravity acceleration (**g**):

```
In[172]:=  H = 50 000;
           U = 5444;
           A = 20.82;
           g = 5.32;
```

The formal requirements *FlatNoAtmConstTEOM* integrate and save the state and control values in arrays from time equals 0 to achieving the stopping condition (**x3** achieves its desired value), and report the corresponding **tfinal**, **x2[tfinal]** and **x4[tfinal]**.

The index of performance is defined to be

$$\text{minimize } J = \mathbf{tfinal} + \mathbf{w1} * (\mathbf{x2[tfinal]}) / \mathbf{H} - 1)^2 + \mathbf{w2} * (\mathbf{x3[tfial]})^2$$

where **w1** and **w2** are positive numbers used as weights on the penalty terms added to **tfinal** to achieve the terminal condition constraints. So, *J* is defined as follows:

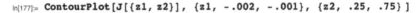

```
In[176]:= J[x_ ? (VectorQ[#, NumericQ] &)] := Module[{intresult, retresult},
     intresult = FlatNoAtmConstTEOM[x[[1]], x[[2]], A, g, U, 0];
     retresult = intresult[[1]] + 10 000 * (intresult[[2]] / H - 1) ^2 + 1 * (intresult[[3]]) ^2;
     Return[retresult]
     ]
```

Since there are only two independent variables, it is easy to produce a contour plot using **z1** to represent possible **a** values and **z2** to represent possible **b** values, to help visualize where the solutions may be

```
In[177]:= ContourPlot[J[{z1, z2}], {z1, -.002, -.001}, {z2, .25, .75}]
```

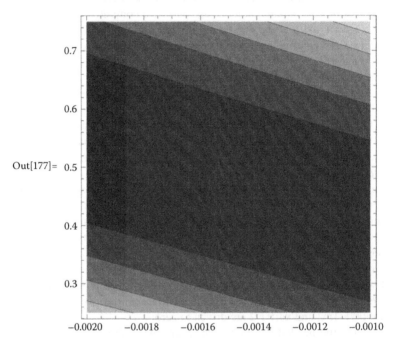

Out[177]=

The plot indicates a huge area in which *J* is very close to a minimum. This means the exact global optimum is going to be difficult to find numerically. If the solution space is approximately symmetric, the solution is somewhere near the center of the area which is **a** equal to −.0015 and **b** equal to .5. With some difficulty, *FindMinimum* finds an optimal set of steering constants and we check the trajectory achieved:

```
In[178]:= FindMinimum[J[{z1, z2}], {z1, -.0015, -.002, -.001}, {z2, .5, .25, .75}]
```

FindMinimum::lstol :
    The line search decreased the step size to within the tolerance specified by AccuracyGoal and PrecisionGoal
        but was unable to find a sufficient decrease in the function. You may need more
        than MachinePrecision digits of working precision to meet these tolerances. ≫

Out[178]= {272.5, {z1 → −0.00159672, z2 → 0.487634}}

To check how well the solution did, substitute **z1** for **a** and **z2** for **b** into *FlatNoAtmConstTEOM*:

In[179]:= **FlatNoAtmConstTEOM[-0.00159672, 0.487634, A, g, U, 0]**

Out[179]= {272.5, 50000.8, 0.0028251}

So we terminate thrust at 272.5 seconds, 0.8 above the desired altitude of 50,000, and just under 0.003 over the desired vertical velocity of 0.

Please note what the optimum steering law is: it is a complete time history for **theta**, from initial time to final time, that is dependent on the terminal conditions sought (**H**, **U**, and that the final vertical velocity is to be 0) as well as the system variables **A** and **g**, and the form of the equations of motion. Change any of these factors, and a totally different theta time history is optimum, not a deviation from the nominal optimum. Unfortunately then, should reality not match our model, the chosen steering decision will not be optimum. Since our model is at best an approximation, we can then be certain our so-called optimal steering law is in reality going to be suboptimum when it is applied in real life.

The goodness of our decision is dependent on the goodness of our model. We can estimate our sensitivity to the modeling inaccuracies by performing a Monte Carlo simulation, in which the potential alternative values for **A** and **g** are estimated by probability density functions and we can model to what extent the real-world control system may implement a **theta** value somewhat off the one we command. As an example, possible real-world values for **A**, **g**, and the steering bias are represented as uniform distributions between a minimum and maximum value with mean the value we assumed to determine the optimum steering law. So if the real **A** could be as much as 5% more or less than the assumed **A**, the real **g** plus or minus 1% the assumed **g**, and a bias of plus or minus 0.1 degree could manifest itself in the control mechanism, then we produce 500 random possibilities as follows:

In[180]:= **Apossibilities = RandomReal[{0.95 A, 1.05 A}, 500];**
gpossibilities = RandomReal[{0.99 g, 1.01 g}, 500];
thetabiaspossibilities = RandomReal[$\left\{\dfrac{1}{180}\,(-0.1)\,\pi,\ \dfrac{0.1\,\pi}{180}\right\}$, 500];
result = Table[FlatNoAtmConstTEOM[-0.00159672, 0.487634, Apossibilities[[i]],
gpossibilities[[i]], U, thetabiaspossibilities[[i]]], {i, 1, 500}];

So, at **tfinal**, the mean and standard deviation obtained are

In[184]:= **tfinalpossibilities = Table[result[[i, 1]], {i, 1, 500}];**
N[Mean[tfinalpossibilities]]
N[Sqrt[Variance[tfinalpossibilities]]]

Out[185]= 272.898

Out[186]= 7.86927

For the final altitude achieved, which is desired to be 50,000, the mean and standard deviation are

```
In[187]:= Hfinalpossibilities = Table[result[[i, 2]], {i, 1, 500}];
          N[Mean[Hfinalpossibilities]]
          Sqrt[Variance[Hfinalpossibilities]]

Out[188]= 49 638.4

Out[189]= 7655.13
```

For the final vertical velocity achieved, which is desired to be 0, the mean and standard deviation obtained are

```
In[190]:= X4finalpossibilities = Table[result[[i, 3]], {i, 1, 500}];
          N[Mean[X4finalpossibilities]]
          Sqrt[Variance[X4finalpossibilities]]

Out[191]= -2.15306

Out[192]= 77.8895
```

Our job is to determine **theta** so that we can keep the mean result for final altitude and vertical velocity closer to the goals of 50,000 and 0, while still minimizing the **tfinal**. Fuzzy logic controllers are usually superior to any alternative method because the fuzzy logic controller expects the sensed information to be uncertain, the applied control to be uncertain, and the underlying equations used to establish the control relationship uncertain. It is not surprising then that the resultant controller is very robust when confronted with all these uncertainties.

To develop the fuzzy controller for this problem, we note that the optimal control law is directly a result of the initial state, the desired final state, and the system variables, **A**, **g**, and any **theta** bias that may manifest itself. Let's assume all these are unknown but constant while we are flying, then the optimal linear tangent steering law still has the form of a linear tangent relationship, just with unknown constants. So we must first solve a system identification problem to ascertain, as easily as we can to facilitate real time determination, what combination of real **A**, **g**, and **theta** bias are we experiencing, then determine the best control law from that point forward. Alternatively, we can examine the measured state vectors and compare them to the values we would have if everything was nominal, and use the difference in the state vectors as a proxy indicating we must have different **A**, **g**, or **theta** bias than expected, and again, relate the state vector differences to new values for **a** and **b** we should use from that point on. For this problem, a body-mounted accelerometer enables dx3/dt and dx4/dt (vertical and horizontal acceleration) to be approximately measured at any time in the trajectory. So we could integrate this information onboard and estimate **x3** and **x4** (vertical and horizontal velocity) and relate the values obtained to the nominal values we would expect at a point in time that we'll call the *decision gate.*

To make the optimal control decisions, we need to get some data. Since we have three uncertainties and we can easily assume at least five trial values for each, we could evaluate 125 cases. But to minimize the work we'll take advantage of the Taguchi method and use an orthogonal array to reduce the number of need optimizations. Using the **L25(5x6)** array we can assess combinations of up to six variables at five levels with only 25 evaluations. The 25 optimization problems that were solved to get the training data are documented in Appendix 3B. For each of those optimization problems, we fly the nominal steering profile to 3 seconds, while the trial acceleration, gravity, and bias are set to specify off nominal values. Starting at 3 seconds, the optimization routine finds out what the steering law needs to be for those off nominal parameters. The 25 results are summarized in Table 3.5, which also show what the horizontal (**x3**) and vertical (**x4**) velocities were at 3 seconds.

TABLE 3.5

Training Data for Fuzzy Controller for Optimal Rocket Trajectory

Trial	A value	g value	bias value	x3 at 3s	x4 at 3s	a*	b*
1	1	1	1	53.4292	10.0104	−0.0014458	0.491329
2	1	2	2	53.4067	9.97719	−0.00143856	0.491326
3	1	3	3	53.3841	9.94399	−0.00143928	0.492063
4	1	4	4	53.3614	9.91076	−0.00144029	0.492805
5	1	5	5	53.3388	9.87752	−0.00144138	0.49356
6	2	1	2	54.8121	10.7374	−0.00151469	0.487227
7	2	2	3	54.7889	10.7055	−0.0015158	0.487933
8	2	3	4	54.7657	10.6735	−0.00151693	0.489337
9	2	4	5	54.7424	10.6415	−0.00151796	0.489337
10	2	5	1	54.8352	10.3704	−0.0015134	0.494674
11	3	1	3	56.1938	11.4670	−0.00159324	0.484222
12	3	2	4	56.1699	11.4362	−0.00159433	0.484879
13	3	3	5	56.1461	11.4054	−0.00159543	0.485538
14	3	4	1	56.2413	11.1294	−0.00159198	0.490727
15	3	5	2	56.2175	11.0987	−0.00159309	0.490277
16	4	1	4	57.5742	12.1989	−0.00167183	0.481433
17	4	2	5	57.5497	12.1693	−0.0016729	0.482044
18	4	3	1	57.6473	11.8885	−0.0067895	0.488112
19	4	4	2	57.6230	11.8590	−0.00167011	0.487771
20	4	5	3	57.5986	11.8294	−0.00167128	0.488401
21	5	1	5	58.9543	12.9333	−0.00174927	0.478731
22	5	2	1	59.0533	12.6475	−0.00175585	0.484796
23	5	3	2	59.0284	12.6192	−0.00175686	0.485362
24	5	4	3	59.0034	12.5909	−0.00175769	0.485902
25	5	5	4	58.9784	12.5626	−0.00175901	0.486511

From the table above, here are the training data to determine **a** as a function of **x3** and **x4**:

```
In[193]:= atrainingdata =
          {{53.43, 10.01, -0.00144580},
           {53.41, 9.98, -0.00143856},
           {53.38, 9.94, -0.00143928},
           {53.36, 9.91, -0.00144029},
           {53.34, 9.88, -0.00144138},
           {54.81, 10.74, -0.00151469},
           {54.79, 10.71, -0.00151580},
           {54.77, 10.68, -0.00151693},
           {54.74, 10.64, -0.00151796},
           {54.83, 10.37, -0.00151340},
           {56.19, 11.47, -0.00159324},
           {56.17, 11.44, -0.00159433},
           {56.15, 11.40, -0.00159543},
           {56.24, 11.13, -0.00159198},
           {56.22, 11.10, -0.00159309},
           {57.57, 12.20, -0.00167183},
           {57.55, 12.17, -0.00167290},
           {57.65, 11.89, -0.00167895},
           {57.62, 11.86, -0.00167011},
           {57.60, 11.83, -0.00167128},
           {58.95, 12.93, -0.00174927},
           {59.05, 12.65, -0.00175585},
           {59.03, 12.62, -0.00175686},
           {59.00, 12.59, -0.00175769},
           {58.98, 12.56, -0.00175901}
          };
```

Since we want the steering law to be a function of both the **x3** and **x4**, the fuzzy logic rules are of the form:

IF **x3** = **fuzzyX3[[i]]**and **x4** = **fuzzyX4[[i]]**THEN **a** = **A[[i]]**

for **i** equal 1 to the number of regions we care to subdivide the possible horizontal and vertical velocities into. Since we see from the training data **x3** values between 53 and 60, we elect to subdivide the region into eight fuzzy **x3** regions, as follows:

fuzzyX3[[1]] = x3_very_very_very_low = <53, 53, 54>

fuzzyX3[[2]] = x3_very_very_low = <53, 54, 55>,

fuzzyX3[[3]] = x3_very_low = <54, 55, 56>,

fuzzyX3[[4]] = x3_low = <55, 56, 57>,

fuzzyX3[[5]] = x3_high = <56., 57, 58>,

fuzzyX3[[6]] = x3_very_high = <57, 58, 59>,

fuzzyX3[[7]] = x3_very_very_high = <58, 59, 60>

fuzzyX3[[8]] = x3_very_very_very_high = <59, 60, 60>

So:

```
In[320]:= fuzzyx3 = {{53, 53, 54}, {53, 54, 55}, {54, 55, 56},
          {55, 56, 57}, {56, 57, 58}, {57, 58, 59}, {58, 59, 60}, {59, 60, 60}};
```

Similarly, since the training data shows **x4** values between 9.5 and 12.5, we define eight fuzzy **x4** regions as follows:

fuzzyX4[[1]] = x4_very_very_very_low = <9.5, 9.5, 10>,

fuzzyX4[[2]] = x4_very_very_low = <9.5, 10, 10.5>

fuzzyX4[[3]] = x4_very_low = <10, 10.5, 11>

fuzzyX4[[4]] = x4_low = <10.5, 11, 11.5>

fuzzyX4[[5]] = x4_high = <11, 11.5, 12>

fuzzyX4[[6]] = x4_very_high = <11.5, 12, 13>

fuzzyX4[[7]] = x4_very_very_high = <12, 12.5, 13>

fuzzyX4[[8]] = x4_very_very_very_high = <12.5, 13, 13>

So:

```
In[195]:= fuzzyx4 = {{9.5, 9.5, 10}, {9.5, 10, 10.5}, {10, 10.5, 11}, {10.5, 11, 11.5},
          {11, 11.5, 12}, {11.5, 12, 12.5}, {12, 12.5, 13}, {12.5, 13, 13}};
```

Since there are eight ranges for each input variables **x3** and **x4**, the total input range space is an 8×8 grid. We lack training data for every one of these 64 boxes because all our trial cases fall into a subset of the 64 boxes essentially near the diagonal of the range because small changes in the fuzzy input parameters result in small changes in the output parameters. This is a common occurrence for fuzzy rules associated with physical motion. We know the fuzzy rules will be of the form:

{{**x3range**[[i]], **x4range**[[j]]}, **output_value**}

Recall from Chapter 2, the **output_value** determined by the counting method is the mean of the minimum, mean, and maximum training data that fell into that fuzzy range. If a region has no data associated with it, there is no means to calculate **output_value**. We could just leave the output undefined, after all it is, but should our rocket end up in a state associated with one of these fuzzy zones, there would be no steering command to provide and things will only get worse. So, even though our training data suggests there is a low probability we need steering commands for all 64 fuzzy regions, it is prudent to define some output just in case. For our problem, we'll assume the open loop optimal steering command is the best we can recommend for these unlikely to reach fuzzy states.

FuzRulesIn1In2ForAbyCounting formulates fuzzy rules for two inputs and one deterministic output. Providing **atraining** data and **fuzzyX3** and **fuzzyX4** to

FuzRulesIn1In2ForAbyCounting yields the rules to determine **a*** as a function of **x3** and **x4**, which is called **astarrules:**

```
In[196]:= astarrules = FuzRulesIn1In2ForAbyCounting[atrainingdata, -0.00159672, fuzzyx3, fuzzyx4]

Out[196]= {{{53, 53, 54}, {9.5, 9.5, 10}, -0.00143994}, {{53, 53, 54}, {9.5, 10, 10.5}, -0.00144181},
          {{53, 53, 54}, {10, 10.5, 11}, -0.0014458}, {{53, 53, 54}, {10.5, 11, 11.5}, -0.00159672},
          {{53, 53, 54}, {11, 11.5, 12}, -0.00159672}, {{53, 53, 54}, {11.5, 12, 12.5}, -0.00159672},
          {{53, 53, 54}, {12, 12.5, 13}, -0.00159672}, {{53, 53, 54}, {12.5, 13, 13}, -0.00159672},
          {{53, 54, 55}, {9.5, 9.5, 10}, -0.00143994}, {{53, 54, 55}, {9.5, 10, 10.5}, -0.00146836},
          {{53, 54, 55}, {10, 10.5, 11}, -0.00148929}, {{53, 54, 55}, {10.5, 11, 11.5}, -0.00151633},
          {{53, 54, 55}, {11, 11.5, 12}, -0.00159672}, {{53, 54, 55}, {11.5, 12, 12.5}, -0.00159672},
          {{53, 54, 55}, {12, 12.5, 13}, -0.00159672}, {{53, 54, 55}, {12.5, 13, 13}, -0.00159672},
          {{54, 55, 56}, {9.5, 9.5, 10}, -0.00159672}, {{54, 55, 56}, {9.5, 10, 10.5}, -0.0015134},
          {{54, 55, 56}, {10, 10.5, 11}, -0.00151571}, {{54, 55, 56}, {10.5, 11, 11.5}, -0.00151633},
          {{54, 55, 56}, {11, 11.5, 12}, -0.00159672}, {{54, 55, 56}, {11.5, 12, 12.5}, -0.00159672},
          {{54, 55, 56}, {12, 12.5, 13}, -0.00159672}, {{54, 55, 56}, {12.5, 13, 13}, -0.00159672},
          {{55, 56, 57}, {9.5, 9.5, 10}, -0.00159672}, {{55, 56, 57}, {9.5, 10, 10.5}, -0.00159672},
          {{55, 56, 57}, {10, 10.5, 11}, -0.00159672}, {{55, 56, 57}, {10.5, 11, 11.5}, -0.00159367},
          {{55, 56, 57}, {11, 11.5, 12}, -0.00159367}, {{55, 56, 57}, {11.5, 12, 12.5}, -0.00159672},
          {{55, 56, 57}, {12, 12.5, 13}, -0.00159672}, {{55, 56, 57}, {12.5, 13, 13}, -0.00159672},
          {{56, 57, 58}, {9.5, 9.5, 10}, -0.00159672}, {{56, 57, 58}, {9.5, 10, 10.5}, -0.00159672},
          {{56, 57, 58}, {10, 10.5, 11}, -0.00159672}, {{56, 57, 58}, {10.5, 11, 11.5}, -0.00159367},
          {{56, 57, 58}, {11, 11.5, 12}, -0.00163149}, {{56, 57, 58}, {11.5, 12, 12.5}, -0.00167402},
          {{56, 57, 58}, {12, 12.5, 13}, -0.00167236}, {{56, 57, 58}, {12.5, 13, 13}, -0.00159672},
          {{57, 58, 59}, {9.5, 9.5, 10}, -0.00159672}, {{57, 58, 59}, {9.5, 10, 10.5}, -0.00159672},
          {{57, 58, 59}, {10, 10.5, 11}, -0.00159672}, {{57, 58, 59}, {10.5, 11, 11.5}, -0.00167402},
          {{57, 58, 59}, {11, 11.5, 12}, -0.00167417}, {{57, 58, 59}, {11.5, 12, 12.5}, -0.00167402},
          {{57, 58, 59}, {12, 12.5, 13}, -0.00171766}, {{57, 58, 59}, {12.5, 13, 13}, -0.00175453},
          {{58, 59, 60}, {9.5, 9.5, 10}, -0.00159672}, {{58, 59, 60}, {9.5, 10, 10.5}, -0.00159672},
          {{58, 59, 60}, {10, 10.5, 11}, -0.00159672}, {{58, 59, 60}, {10.5, 11, 11.5}, -0.00159672},
          {{58, 59, 60}, {11, 11.5, 12}, -0.00159672}, {{58, 59, 60}, {11.5, 12, 12.5}, -0.00159672},
          {{58, 59, 60}, {12, 12.5, 13}, -0.00175467}, {{58, 59, 60}, {12.5, 13, 13}, -0.00175467},
          {{59, 60, 60}, {9.5, 9.5, 10}, -0.00159672}, {{59, 60, 60}, {9.5, 10, 10.5}, -0.00159672},
          {{59, 60, 60}, {10, 10.5, 11}, -0.00159672}, {{59, 60, 60}, {10.5, 11, 11.5}, -0.00159672},
          {{59, 60, 60}, {11, 11.5, 12}, -0.00159672}, {{59, 60, 60}, {11.5, 12, 12.5}, -0.00159672},
          {{59, 60, 60}, {12, 12.5, 13}, -0.00175678}, {{59, 60, 60}, {12.5, 13, 13}, -0.00175678}}
```

Recall in Chapter 2 we used *EvalFuzRulesIFA1ANDA2THENBconst* to evaluate such rules. Here we do so again and show the results using *Plot3D*:

```
In[222]:= Plot3D[EvalFuzRulesIFA1ANDA2THENBconst[x3, x4, astarrules],
          {x3, 53, 60}, {x4, 9.5, 13}, AxesLabel → {x3, x4, a}, ColorFunction → Black]
```

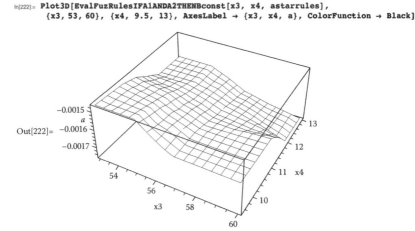

We can look at the training points relative to the fuzzy logic fit as follows:

```
In[223]:=  outplot = Plot3D[EvalFuzRulesIFA1ANDA2THENBconst[x3, x4, astarrules],
             {x3, 53, 60}, {x4, 9.5, 13}, AxesLabel → {x3, x4, a},
             ColorFunction → Black, DisplayFunction → Identity];
           Show[outplot, Graphics3D[{PointSize[0.02], Thread[Point[atrainingdata]]}],
             DisplayFunction → $DisplayFunction, ViewPoint → {1, 1, 1}]
```

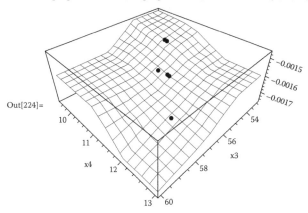

Out[224]=

Finally, we can calculate the error between what the fuzzy rules return and the training data input values and the training data–dependent value:

```
In[200]:=  Table[(EvalFuzRulesIFA1ANDA2THENBconst[atrainingdata[[i, 1]], atrainingdata[[i, 2]],
             astarrules] - atrainingdata[[i, 3]]) / atrainingdata[[i, 3]], {i, 1, 25}]
```

```
Out[200]=  {0.00551628, 0.00916814, 0.00715822, 0.00559007, 0.00410804, -0.0015304,
           -0.00245675, -0.00338419, -0.00431588, -0.00495593, 0.00429967, 0.00284103,
           0.00163492, 0.00491683, 0.00375835, 0.0068922, 0.00575034, -0.0067906, -0.00221135,
           -0.00349049, 0.00212282, -0.000561758, -0.00117749, -0.00171703, -0.00287231}
```

Next, repeat the process for **b***. First, get the training data from the table summarizing the 25 optimization problems that were solved:

```
In[201]:=  btrainingdata =
             {{53.43, 10.01, 0.491329},
              {53.41, 9.98, 0.491367},
              {53.38, 9.95, 0.492063},
              {53.36, 9.91, 0.492805},
              {53.34, 9.88, 0.493560},
              {54.81, 10.74, 0.487227},
              {54.79, 10.71, 0.487933},
              {54.77, 10.67, 0.488641},
              {54.74, 10.64, 0.489337},
              {54.83, 10.37, 0.494674},
              {56.19, 11.47, 0.484222},
              {56.17, 11.44, 0.484879},
              {56.14, 11.40, 0.485538},
              {56.24, 11.13, 0.490727},
              {56.22, 11.10, 0.491390},
              {57.57, 12.20, 0.481433},
              {57.55, 12.17, 0.482044},
              {57.65, 11.89, 0.488112},
              {57.62, 11.86, 0.487771},
              {57.60, 11.83, 0.488401},
              {58.95, 12.93,  0.478731},
              {59.05, 12.65, 0.484796},
              {59.03, 12.62, 0.485362},
              {59.00, 12.59, 0.485902},
              {58.98, 12.57, 0.486511}
             };
```

Then, this time use the open loop optimal **b*** values as the default output for when no training data exists, the rules, **bstarrules**, for establishing **b*** values based on measured **x3** and **x4** at **t** = 3 are then found to be

In[202]:= **bstarrules = FuzRulesIn1In2ForAbyCounting[btrainingdata, 0.487634, fuzzyx3, fuzzyx4]**

Out[202]= {{{53, 53, 54}, {9.5, 9.5, 10}, 0.492459}, {{53, 53, 54}, {9.5, 10, 10.5}, 0.492371},
{{53, 53, 54}, {10, 10.5, 11}, 0.491329}, {{53, 53, 54}, {10.5, 11, 11.5}, 0.487634},
{{53, 53, 54}, {11, 11.5, 12}, 0.487634}, {{53, 53, 54}, {11.5, 12, 12.5}, 0.487634},
{{53, 53, 54}, {12, 12.5, 13}, 0.487634}, {{53, 53, 54}, {12.5, 13, 13}, 0.487634},
{{53, 54, 55}, {9.5, 9.5, 10}, 0.492459}, {{53, 54, 55}, {9.5, 10, 10.5}, 0.492879},
{{53, 54, 55}, {10, 10.5, 11}, 0.490586}, {{53, 54, 55}, {10.5, 11, 11.5}, 0.488283},
{{53, 54, 55}, {11, 11.5, 12}, 0.487634}, {{53, 54, 55}, {11.5, 12, 12.5}, 0.487634},
{{53, 54, 55}, {12, 12.5, 13}, 0.487634}, {{53, 54, 55}, {12.5, 13, 13}, 0.487634},
{{54, 55, 56}, {9.5, 9.5, 10}, 0.487634}, {{54, 55, 56}, {9.5, 10, 10.5}, 0.494674},
{{54, 55, 56}, {10, 10.5, 11}, 0.490488}, {{54, 55, 56}, {10.5, 11, 11.5}, 0.488283},
{{54, 55, 56}, {11, 11.5, 12}, 0.487634}, {{54, 55, 56}, {11.5, 12, 12.5}, 0.487634},
{{54, 55, 56}, {12, 12.5, 13}, 0.487634}, {{54, 55, 56}, {12.5, 13, 13}, 0.487634},
{{55, 56, 57}, {9.5, 9.5, 10}, 0.487634}, {{55, 56, 57}, {9.5, 10, 10.5}, 0.487634},
{{55, 56, 57}, {10, 10.5, 11}, 0.487634}, {{55, 56, 57}, {10.5, 11, 11.5}, 0.487654},
{{55, 56, 57}, {11, 11.5, 12}, 0.487654}, {{55, 56, 57}, {11.5, 12, 12.5}, 0.487634},
{{55, 56, 57}, {12, 12.5, 13}, 0.487634}, {{55, 56, 57}, {12.5, 13, 13}, 0.487634},
{{56, 57, 58}, {9.5, 9.5, 10}, 0.487634}, {{56, 57, 58}, {9.5, 10, 10.5}, 0.487634},
{{56, 57, 58}, {10, 10.5, 11}, 0.487634}, {{56, 57, 58}, {10.5, 11, 11.5}, 0.487654},
{{56, 57, 58}, {11, 11.5, 12}, 0.487747}, {{56, 57, 58}, {11.5, 12, 12.5}, 0.485129},
{{56, 57, 58}, {12, 12.5, 13}, 0.481738}, {{56, 57, 58}, {12.5, 13, 13}, 0.487634},
{{57, 58, 59}, {9.5, 9.5, 10}, 0.487634}, {{57, 58, 59}, {9.5, 10, 10.5}, 0.487634},
{{57, 58, 59}, {10, 10.5, 11}, 0.487634}, {{57, 58, 59}, {10.5, 11, 11.5}, 0.487634},
{{57, 58, 59}, {11, 11.5, 12}, 0.488089}, {{57, 58, 59}, {11.5, 12, 12.5}, 0.485129},
{{57, 58, 59}, {12, 12.5, 13}, 0.482722}, {{57, 58, 59}, {12.5, 13, 13}, 0.482986},
{{58, 59, 60}, {9.5, 9.5, 10}, 0.487634}, {{58, 59, 60}, {9.5, 10, 10.5}, 0.487634},
{{58, 59, 60}, {10, 10.5, 11}, 0.487634}, {{58, 59, 60}, {10.5, 11, 11.5}, 0.487634},
{{58, 59, 60}, {11, 11.5, 12}, 0.487634}, {{58, 59, 60}, {11.5, 12, 12.5}, 0.487634},
{{58, 59, 60}, {12, 12.5, 13}, 0.483167}, {{58, 59, 60}, {12.5, 13, 13}, 0.483167},
{{59, 60, 60}, {9.5, 9.5, 10}, 0.487634}, {{59, 60, 60}, {9.5, 10, 10.5}, 0.487634},
{{59, 60, 60}, {10, 10.5, 11}, 0.487634}, {{59, 60, 60}, {10.5, 11, 11.5}, 0.487634},
{{59, 60, 60}, {11, 11.5, 12}, 0.487634}, {{59, 60, 60}, {11.5, 12, 12.5}, 0.487634},
{{59, 60, 60}, {12, 12.5, 13}, 0.48535}, {{59, 60, 60}, {12.5, 13, 13}, 0.48535}}

The resulting **b*** value variability with the input values is visualized using *Plot3D*:

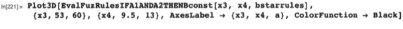

In[221]:= **Plot3D[EvalFuzRulesIFA1ANDA2THENBconst[x3, x4, bstarrules],**
{x3, 53, 60}, {x4, 9.5, 13}, AxesLabel → {x3, x4, a}, ColorFunction → Black]

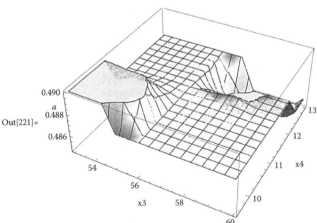

Now we can test the decision-making process. We again run a Monte Carlo analysis of the flights for randomly dispersed **A**, **g**, and **thetabias** values. We make the optimal controller's job rightfully more difficult by modeling **x3** and **x4** measurement errors by adding a random error to both **x3** and **x4** obtained at **t** = 3 seconds, then use the fuzzy rules just defined about to set a new **a** and **b** to use to fly that point forward. To do this, we define *FlatNoAtmConstTEOMopt* to simulate the trajectory to first generate the corrupted **x3** and **x4** values at **t** = 3, then to enable the derived fuzzy rules to be evaluated to obtain the steering constants.

For the Monte Carlo analysis we use the same 500 random inputs for **Apossibilities**, **gpossibilities** and **thetabiaspossibilities** as used before. The mean and standard deviation for the fuzzy logic controller are

```
In[204]:= result = Table[FlatNoAtmConstTEOMopt[Apossibilities[[i]],
         gpossibilities[[i]], U, thetabiaspossibilities[[i]]], {i, 1, 500}];
```

The mean and standard deviation for the flight time is

```
In[205]:= tfinalpossibilities = Table[result[[i, 1]], {i, 1, 500}];
         N[Mean[tfinalpossibilities]]
         N[Sqrt[Variance[tfinalpossibilities]]]

Out[206]= 272.945

Out[207]= 8.64586
```

The mean and standard deviation for final altitude, which is to be 50,000, are

```
In[208]:= Hfinalpossibilities = Table[result[[i, 2]], {i, 1, 500}];
         N[Mean[Hfinalpossibilities]]
         Sqrt[Variance[Hfinalpossibilities]]

Out[209]= 50 041.3

Out[210]= 940.402
```

The mean and standard deviation for the final vertical velocity, which is to be 0, are

```
In[211]:= X4finalpossibilities = Table[result[[i, 3]], {i, 1, 500}];
         N[Mean[X4finalpossibilities]]
         Sqrt[Variance[X4finalpossibilities]]

Out[212]= -0.0294951

Out[213]= 7.09454
```

Table 3.6 compares the open and fuzzy logic controller results.

TABLE 3.6

Terminal Conditions Achieved Using Open-Loop Optimal and 64-Rule Fuzzy Controller

Parameter	Open-Loop Optimal Control	64-Rule Fuzzy Control
Mean final time	272.9	272.9
Final time standard deviation	< 10	<10 but > 10% larger
Error in mean final altitude	100's	10's
Final altitude standard deviation	< 10000	<1000
Error in mean final vertical velocity	1's	0.1's
Final vertical velocity standard deviation	< 100	<10

The fuzzy controller improves the mean performance and reduces the standard deviation by roughly a factor of 10. Notice the mean final time is about the same but the fuzzy controller standard deviation is actually more than 10% larger than when using the open loop optimal controller. The longer the rocker must burn, the more fuel it must carry, and the less payload it can deliver. These results indicate the final orbit insertion accuracy can be improved substantially, but only if more propellant is held in reserve. So, as with nearly everything in life, to get one thing, you need to prepare to sacrifice something else.

We can run experiments to see the impact of other rule sets. Suppose for example, our on-board processor is a bit limited, and we'd like to reduce the rule set to 25, roughly for measured **x3** and **x4** with fuzzy values of very high, high, nominal, low or very low. To start, redefine **fuzzyx3** and **fuzzyx4** to have five potential fuzzy values:

```
In[94]:= fuzzyx3 = {{53, 53, 55}, {53, 55, 57}, {55, 56.5, 58}, {56, 58, 60}, {58, 60, 60}};
fuzzyx4 =
    {{9.5, 9.5, 10.5}, {9.5, 10.5, 11.5}, {10.5, 11.25, 12}, {11, 12, 13}, {12, 13, 13}};
```

Then update the rules to determine **a*** and **b*** using the same training data and default values, but the new fuzzy ranges. First, for **a***:

```
In[216]:= astarrules = FuzRulesInIn2ForAbyCounting[atrainingdata, -0.00159672, fuzzyx3, fuzzyx4]

Out[216]= {{{53, 53, 55}, {9.5, 9.5, 10.5}, -0.00146836}, {{53, 53, 55}, {9.5, 10.5, 11.5}, -0.00147831},
    {{53, 53, 55}, {10.5, 11.25, 12}, -0.00151633}, {{53, 53, 55}, {11, 12, 13}, -0.00159672},
    {{53, 53, 55}, {12, 13, 13}, -0.00159672}, {{53, 55, 57}, {9.5, 9.5, 10.5}, -0.00146836},
    {{53, 55, 57}, {9.5, 10.5, 11.5}, -0.00151693},
    {{53, 55, 57}, {10.5, 11.25, 12}, -0.00155646}, {{53, 55, 57}, {11, 12, 13}, -0.00159367},
    {{53, 55, 57}, {12, 13, 13}, -0.00159672}, {{55, 56.5, 58}, {9.5, 9.5, 10.5}, -0.00159672},
    {{55, 56.5, 58}, {9.5, 10.5, 11.5}, -0.00159367},
    {{55, 56.5, 58}, {10.5, 11.25, 12}, -0.00163149},
    {{55, 56.5, 58}, {11, 12, 13}, -0.00163475}, {{55, 56.5, 58}, {12, 13, 13}, -0.00167236},
    {{56, 58, 60}, {9.5, 9.5, 10.5}, -0.00159672},
    {{56, 58, 60}, {9.5, 10.5, 11.5}, -0.00159367},
    {{56, 58, 60}, {10.5, 11.25, 12}, -0.00163149}, {{56, 58, 60}, {11, 12, 13}, -0.00167504},
    {{56, 58, 60}, {12, 13, 13}, -0.00172092}, {{58, 60, 60}, {9.5, 9.5, 10.5}, -0.00159672},
    {{58, 60, 60}, {9.5, 10.5, 11.5}, -0.00159672},
    {{58, 60, 60}, {10.5, 11.25, 12}, -0.00159672},
    {{58, 60, 60}, {11, 12, 13}, -0.00175467}, {{58, 60, 60}, {12, 13, 13}, -0.00175467}}
```

Using *Plot3D* to visualize how the new rules translate the measured **x3** and **x4** into the steering parameter:

```
In[219]:= Plot3D[EvalFuzRulesIFA1ANDA2THENBconst[x3, x4, astarrules],
    {x3, 53, 60}, {x4, 9.5, 13}, AxesLabel → {x3, x4, a}, ColorFunction → Black]
```

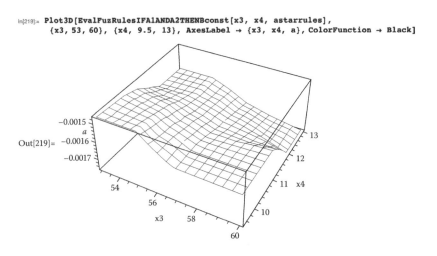

Then for **b***:

```
In[225]:= bstarrules = FuzRulesIn1In2ForAbyCounting[btrainingdata, 0.487634, fuzzyx3, fuzzyx4]
```

```
Out[225]= {{{53, 53, 55}, {9.5, 9.5, 10.5}, 0.492879},
   {{53, 53, 55}, {9.5, 10.5, 11.5}, 0.490932}, {{53, 53, 55}, {10.5, 11.25, 12}, 0.488283},
   {{53, 53, 55}, {11, 12, 13}, 0.487634}, {{53, 53, 55}, {12, 13, 13}, 0.487634},
   {{53, 55, 57}, {9.5, 9.5, 10.5}, 0.492879}, {{53, 55, 57}, {9.5, 10.5, 11.5}, 0.489536},
   {{53, 55, 57}, {10.5, 11.25, 12}, 0.487793}, {{53, 55, 57}, {11, 12, 13}, 0.487654},
   {{53, 55, 57}, {12, 13, 13}, 0.487634}, {{55, 56.5, 58}, {9.5, 9.5, 10.5}, 0.487634},
   {{55, 56.5, 58}, {9.5, 10.5, 11.5}, 0.487654}, {{55, 56.5, 58}, {10.5, 11.25, 12}, 0.487747},
   {{55, 56.5, 58}, {11, 12, 13}, 0.486425}, {{55, 56.5, 58}, {12, 13, 13}, 0.481738},
   {{56, 58, 60}, {9.5, 9.5, 10.5}, 0.487634}, {{56, 58, 60}, {9.5, 10.5, 11.5}, 0.487654},
   {{56, 58, 60}, {10.5, 11.25, 12}, 0.487747}, {{56, 58, 60}, {11, 12, 13}, 0.485281},
   {{56, 58, 60}, {12, 13, 13}, 0.482927}, {{58, 60, 60}, {9.5, 9.5, 10.5}, 0.487634},
   {{58, 60, 60}, {9.5, 10.5, 11.5}, 0.487634}, {{58, 60, 60}, {10.5, 11.25, 12}, 0.487634},
   {{58, 60, 60}, {11, 12, 13}, 0.483167}, {{58, 60, 60}, {12, 13, 13}, 0.483167}}
```

```
In[226]:= Plot3D[EvalFuzRulesIFA1ANDA2THENBconst[x3, x4, bstarrules],
       {x3, 53, 60}, {x4, 9.5, 13}, AxesLabel → {x3, x4, a}, ColorFunction → Black]
```

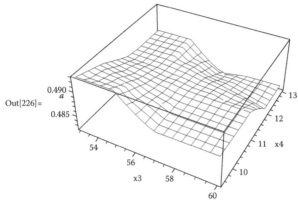

Out[226]=

Finally, redo the Monte Carlo analyses with this new fuzzy controller:

```
In[227]:= result = Table[FlatNoAtmConstTEOMopt[Apossibilities[[i]],
       gpossibilities[[i]], U, thetabiaspossibilities[[i]]], {i, 1, 500}];
```

The mean and standard deviation for the flight time are

```
In[228]:= tfinalpossibilities = Table[result[[i, 1]], {i, 1, 500}];
       N[Mean[tfinalpossibilities]]
       N[Sqrt[Variance[tfinalpossibilities]]]
```

```
Out[229]= 272.922
```

```
Out[230]= 8.57309
```

The mean and standard deviation for final altitude, which is to be 50,000, are

```
In[231]:= Hfinalpossibilities = Table[result[[i, 2]], {i, 1, 500}];
       N[Mean[Hfinalpossibilities]]
       Sqrt[Variance[Hfinalpossibilities]]
```

```
Out[232]= 49 773.7
```

```
Out[233]= 1269.86
```

The mean and standard deviation for the final vertical velocity, which is to be 0, are

```
X4finalpossibilities = Table[result[[i, 3]], {i, 1, 500}];
N[Mean[X4finalpossibilities]]
Sqrt[Variance[X4finalpossibilities]]

-2.85523

10.5159
```

Table 3.7 compares the three results.

The 25 rules case still provides considerable reduction in variability in the final altitude and vertical velocity, but the mean performance degrades considerably. As we showed in Chapter 2, more rules enables a better fit to the data, so controllers based on less rules are likely to be less "good." But there is only so much information in the data, and adding too many rules slows down the calculations to no benefit.

The basic process for establishing a fuzzy controller is as illustrated above, establish a mathematical model of the system behavior, and then formulate rules to provide controls based on some of the measured states. If mathematics enables finding a near optimal way to formulate these rules, then certainly take advantage of the situation. The Taguchi method can be used to minimize the trial cases needed to form the training data to establish the rules. Often, the rules can be formulated just as well using common sense. In such instances, each rule invokes a control in proportion to the measured state. A third approach is to structure the rules using the IF clauses, then select the THEN values by having each be the independent variables in an optimization problem.

3.2.13 Determine an Optimum Configuration

By *configuration*, I mean an arrangement of objects or activities. The area of a beam versus its length is a simple configuration. A wing span, angle, and taper is a more complicated configuration. The locations for sensors to track satellites, with different

---

TABLE 3.7

Terminal Conditions Achieved Using Open-Loop Optimal and 25- and 64-Rule Fuzzy Controllers

Parameter	Open-Loop Optimal Control	64-Rule Fuzzy Control	25-Rule Fuzzy Control
Mean final time	272.9	272.9	272.9
Final time standard deviation	< 10	<10	< 10 but > 10% larger
Error in mean final altitude	100's	10's	½ open loop
Final altitude standard deviation	< 10000	~1000	>1000
Error in mean final vertical velocity	1's	0.1's	>1
Final vertical velocity standard deviation	< 100	<10	~10

viewing and data communications capabilities around the globe, are a still more complicated configuration. The problem statement is as follows:

What is the optimal configuration to achieve a stipulated result?

Optimum configurations are sought for a wide variety of problems, such as the following:

1. How much of an item to move from multiple sources to multiple destinations, with different costs for transport between the sources and destinations? This is also known as the *transportation problem.*
2. Which order to visit nodes to minimize the cost of travel? This is also known as the *traveling salesman problem.*
3. What sequence of operations completes an assignment at minimum cost? This is also known as the *scheduling problem.*
4. How to assign multiple resources to accomplish specified activities during available times to maximize goals? This is also known as the *table problem.*
5. Fastest ways to group objects into a set of categories? This is also known as the *partitioning problem.*
6. What path should an object follow from point A to B where there are impassable obstacles or dangers to be avoided? This is also known as the *navigation problem.*

To illustrate determining optimal configurations, here is a problem documented in reference 10:

$$\text{Minimize } J = Sum[f[\mathbf{x}[[\mathbf{i},\mathbf{j}]]],$$

with

$$\mathbf{i} = 1 \text{ to } 3$$

and

$$\mathbf{j} = 1 \text{ to } 4$$

where

$$f[\mathbf{x}[[\mathbf{i},\mathbf{j}]]] = 0 \text{ if } \mathbf{x}[[\mathbf{i}, \mathbf{j}]] = 0, \text{ otherwise} = \mathbf{d} + \mathbf{c} * Sqrt[\mathbf{x}[[\mathbf{i}, \mathbf{j}]]]$$

with

$$\mathbf{d} = 5$$

$$\mathbf{c} = \{\{0, 21, 50, 62\}, \{21, 0, 17, 54\}, \{50, 17, 0, 60\}\},$$

and

$$\mathbf{x}[[1, 1]] + \mathbf{x}[[1, 2]] + \mathbf{x}[[1, 3]] + \mathbf{x}[[1, 4]] = 10$$

$$\mathbf{x}[[2, 1]] + \mathbf{x}[[2, 2]] + \mathbf{x}[[2, 3]] + \mathbf{x}[[2, 4]] = 15$$

$$\mathbf{x}[[3, 1]] + \mathbf{x}[[3, 2]] + \mathbf{x}[[3, 3]] + \mathbf{x}[[3, 4]] = 20$$

$$\mathbf{x}[[1, 1]] + \mathbf{x}[[2, 1]] + \mathbf{x}[[3, 1]] = 3$$

$$\mathbf{x}[[1, 2]] + \mathbf{x}[[2, 2]] + \mathbf{x}[[3, 2]] = 20$$

$$\mathbf{x}[[1, 3]] + \mathbf{x}[[2, 3]] + \mathbf{x}[[3, 3]] = 5$$

$$\mathbf{x}[[1, 4]] + \mathbf{x}[[2, 4]] + \mathbf{x}[[3, 4]] = 17$$

The last equation is actually redundant as the prior six equations uniquely determine the answer.

This is a 12-variable transportation problem to get $\{10, 15, 20\}$ items from three sources to four places needing $\{3, 20, 5, 17\}$ items, respectively.

TransportProblem represents the index of performance for which the constraints are added as penalty functions:

```
In[146]:= TransportProblem[x_] := Module[{c, d},
    d = 5;
    c = {0, 21, 50, 62, 21, 0, 17, 54, 50, 17, 0, 60};
    Module[{i}, N[Sum[If[x[[i]] == 0, 0, d + c[[i]] * Sqrt[x[[i]]]], {i, 1, 12}]] +
      500 * (
        (x[[1]] + x[[2]] + x[[3]] + x[[4]] - 10) ^2 +
        (x[[5]] + x[[6]] + x[[7]] + x[[8]] - 15) ^2 +
        (x[[9]] + x[[10]] + x[[11]] + x[[12]] - 20) ^2 +
        (x[[1]] + x[[5]] + x[[9]] - 3) ^2 +
        (x[[2]] + x[[6]] + x[[10]] - 20) ^2 +
        (x[[3]] + x[[7]] + x[[11]] - 5) ^2)]]
```

Simple configurations can often be found using the algorithms described in previous sections, but both *FindMinimim* and *NMinimize* fail to solve this optimization problem. Our best clue to find a solution is to mimic the way nature evolves a life form, which is, after all, a profoundly complicated configuration. To do so, we manufacture a method to denote a configuration uniquely as a series of numbers, perhaps as a long binary string, where finite lengths of the string correspond to parts of the configuration description. In essence, we create a set of genes to represent the configuration, where each number in the gene, whether we use just ones and zeros, or real numbers, roughly corresponds to chromosomes. We randomly create many gene sequences, so we have a population of diverse individual test cases. For each member of the population we assess goodness against our criteria. We then select two members from the population, perhaps randomly, perhaps with the odds in proportion to their goodness, and we manufacture a means to create one or more new members of the population based on the chromosomes associated with the chosen parents. There are numerous ways to do this. We could specify an arbitrary point in the number sequence, and up to that point, we could copy numbers from the first parent, and after that point we could copy numbers from the second parent. This is called *crossover*. We could do this at an arbitrary number of places in the sequence, for arbitrary quantity of numbers, to the limit of each chromosome's value; this is called *multiple crossover*. We could compare the two numbers from each parent and select an arbitrary number between the two values, or even extrapolated beyond the two variables. This is called arithmetic combination. In addition, at a low probability, we could mutate one or more of the numbers to a value that has no correspondence to either parent. We can make any number of children by all these means. We can assess all their respective goodness. We can keep for the next population just the children, or a finite set of those that score the best, and repeat the process. What will gradually happen is all members of the population become essentially the same as represented by their numerical genes. We will see very little difference in goodness, and we can select the one that has the best goodness as our answer. In a nutshell, this is a description of the genetic programming algorithm. Though we can utilize such an algorithm to solve any optimization problem, I have found that usually this method is best suited for configuration problems. The tricky part is to figure out an efficient means to represent the configuration so goodness can be easily and rapidly calculated. "Rapid" is necessary, because to find the optimum and not get stuck on a local optimum, these methods, like evolution, need a sizable population with genetic diversity. A heuristic for the trial population size is at least 50 members for each parameter sought. Then, at least hundreds and usually thousands of generations must be produced, so J evaluation needs to be very efficient, or the method takes a long time to reach a conclusion. For this reason, it is wise to also monitor the solution obtained for incremental generations.

The very first genetic algorithms represented independent variable values in base two using binary strings of equal length and sufficient number of digits to span the permissible range for the independent variables. The gene string was the binary representation of the each independent variable, concatenated from first to last. You can see how crossover got invented, picking an arbitrary point along the total string of 1's and 0's, (most likely interior to the span associated with an independent variable) and swapping prior values with one parent's string and post values with another parent's string, changes the value for all independent variables. Also, mutation is easy to visualize: randomly change any or several 0's to a 1's or vice versa. Experiments soon showed the methodology converged faster if real numbers were used. Crossover became not only swapping numbers, but also arithmetically combining numbers, and mutation changed the value of one or more numbers to a new value between a permissible minimum and maximum. Basic ways to mimic genetic manipulations are

1. Uniform mutation: an existing population member has one randomly chosen element modified based on a uniform distribution between a minimum and maximum value
2. Boundary mutation: an existing population member has one randomly chosen element set to either its minimum or maximum value, usually with equal probability
3. Non-uniform mutation: an existing population member has one randomly chosen element set to a value randomly intermediate between its current value or its minimum or maximum value

4. Arithmetic crossover: two randomly chosen members of the population, with gene strings **p1** and **p2** respectively, create two children as follows:

$$\mathbf{r} * \mathbf{p1} + (1 - \mathbf{r}) * \mathbf{p2}$$

and

$$\mathbf{r} * \mathbf{p2} + (1 - \mathbf{r}) * \mathbf{p1}$$

where **r** is a uniform random variable between 0 and 1

5. Simple crossover: two randomly chosen members of the population switch elements starting at a randomly chosen element, to the end of the end of the last element
6. Heuristic crossover: one parent is chosen randomly, **p1**, and a second parent **p2** is chosen randomly from those with a better index of performance; the new population member is

$$\mathbf{r} * (\mathbf{p2} - \mathbf{p1}) + \mathbf{p2}$$

(Note that if the numerical values have limits, they need to be checked, and if exceeded, they need to be set to the appropriate limit.)

The formal requirements *FindMinGA* perform a genetic optimization.

To apply *FindMinGA* to optimize *TransportProblem*, first define the minimum and maximum values each independent variable can have, which are **l** and **u**, respectively, then run *FindMinGA* a number of times since the initial variable assignment and subsequent modifications are randomly determined. Here are 10 example runs with a population size of 100, 10 parents per generation, 5000 maximum generations, and a stopping condition set for when the difference between the best and worst index of performance for the population is less than 1%:

```
In[147]:= l = Table[0, {12}];
          u = {3, 20, 5, 17, 3, 20, 5, 17, 3, 20, 5, 17};
          Do[FindMinGA[TransportProblem, 0, 12, 100, 10, 5000, .01, 5000, l, u], {10}]
```

```
Stop on |(Jbest/Jworst -1)| <= 0.01

Generation: 36

Number of evaluations: 2496

Best value: 699.137

x at best value:

{1, 6, 2, 1, 1, 7, 1, 6, 1, 7, 2, 10}
```

```
Stop on |(Jbest/Jworst -1)| <= 0.01

Generation: 48

Number of evaluations: 3198

Best value: 702.295

x at best value:

{1, 5, 1, 3, 1, 8, 3, 3, 1, 7, 1, 11}


Stop on |(Jbest/Jworst -1)| <= 0.01

Generation: 50

Number of evaluations: 3396

Best value: 683.752

x at best value:

{1, 0, 2, 7, 0, 9, 1, 5, 2, 11, 2, 5}

Stop on |(Jbest/Jworst -1)| <= 0.01

Generation: 45

Number of evaluations: 3052

Best value: 649.312

x at best value:

{0, 7, 3, 0, 2, 2, 0, 11, 1, 11, 2, 6}


Stop on |(Jbest/Jworst -1)| <= 0.01

Generation: 43

Number of evaluations: 2946

Best value: 672.287

x at best value:

{1, 0, 1, 8, 1, 9, 2, 3, 1, 11, 2, 6}


Stop on |(Jbest/Jworst -1)| <= 0.01

Generation: 29

Number of evaluations: 2004

Best value: 1126.88

x at best value:

{1, 2, 0, 7, 0, 9, 0, 6, 2, 9, 4, 5}
```

```
Stop on |(Jbest/Jworst -1)| <= 0.01

Generation: 56

Number of evaluations: 3768

Best value: 652.944

x at best value:

{0, 8, 0, 2, 2, 6, 2, 5, 1, 6, 3, 10}

Stop on |(Jbest/Jworst -1)| <= 0.01

Generation: 51

Number of evaluations: 3430

Best value: 584.437

x at best value:

{3, 4, 1, 2, 0, 11, 2, 2, 0, 5, 2, 13}

Stop on |(Jbest/Jworst -1)| <= 0.01

Generation: 65

Number of evaluations: 4366

Best value: 623.565

x at best value:

{0, 9, 0, 1, 2, 7, 4, 2, 1, 4, 1, 14}

Stop on |(Jbest/Jworst -1)| <= 0.01

Generation: 53

Number of evaluations: 3574

Best value: 729.628

x at best value:

{1, 3, 4, 2, 0, 6, 1, 8, 2, 11, 0, 7}
```

Notice all the independent variables are now integers, since we are assuming only whole things can be shipped from place to place. There is quite a range in results, but a few hint as to the optimum answer.

As genetic algorithm research matured, people soon realized the process was a specific implementation of an evolving strategy, where randomness could be used in a "controlled" way to mimic the same result. The simplest random search tries one new randomly selected set of independent variables. The more effective random search methods use an annealing function which provides a number less than 1 that slowly decreases in value as the search proceeds. The annealing function value is the probability a worse trial solution is accepted rather than a better trial solution, to help prevent premature termination at a local minimum. In evolving strategies, randomness is used differently, rather than to directly select a new trial point, randomness is used to create solutions similar to existing solutions with the

rationale that good characteristics can be retained and bad characteristics gradually eliminated.

A typical evolving strategy algorithm is as follows:

$$\text{To min } J[\mathbf{x}],$$

with

$$\mathbf{x} \text{ an array of dimension } \mathbf{n},$$

and

$$\mathbf{L}[[i]] <= \mathbf{x}[[\mathbf{i}]] <= \mathbf{U}[[\mathbf{i}]] \text{ for } \mathbf{i} = 1 \text{ to } \mathbf{n},$$

also define

$$\mathbf{sigma}[[\mathbf{i}]] = (\mathbf{U}[[\mathbf{i}]] - \mathbf{L}[[\mathbf{i}]])/6 \text{ for } \mathbf{i} = 1 \text{ to } \mathbf{n}$$

Step 1. Form a population of **P** **x**'s by randomly assigning values between **L**[[**i**]]
and **U**[[**i**]] to each **x**[[**i**]], **i** = 1 to **n**.

Step 2. Do the following at least **P** times: Randomly select 2 individuals from the
population, **x1** and **x2** along with their associated **sigma1** and **sigma2**,
with a 50% probability, create a trial member **y** by selecting with 50%
probability each element from **x1** or **x2** and **sigma1** and **sigma2**, with
50% probability, create a trial member:

$$\mathbf{y} = \mathbf{alpha} * \mathbf{x1} + (1 - \mathbf{alpha}) * \mathbf{x2}$$

and

$$\mathbf{sigmay} = \mathbf{alpha} * \mathbf{sigma1} + (1 - \mathbf{alpha}) * \mathbf{sigma2}$$

with **alpha** randomly selected between 0 and 1. To each member of **sigmay** apply a
mutation as follows

$$\mathbf{sigmaymutated} = \mathbf{sigmay} * Exp[NormalDistribution(0, \mathbf{delta})]$$

(where *NormalDistribution*(**m**, **s**) is a normal probability density function with mean
m and standard deviation **s**. Note, for small **delta**, this step will increase or decrease
sigmay a small amount). To each member of **y** apply a mutation as follows:

$$\mathbf{ymutated} = \mathbf{y} + NormalDistribution(0, \mathbf{sigmaymutated}).$$

Evaluate the index of performance for **y**.

Step 3. Of the **P** trials, let **ybest** be the **ymutated** that provides the best *J*, denoted
Jbest, and **yworst** be the **ymutated** that provides the worst *J*, denoted **Jworst**.

$$\text{If } |\mathbf{Jbest} - \mathbf{Jworst}| \text{ is less than a } \mathbf{tolerance}, \text{ stop.}$$

Otherwise, replace the population **x** with **ymutated**, and go to step 2. (Or keep the **P**
members from **x** or **ymutated** that provided the **P** best *J* values. Or keep the **x** or **ymutated** which provides the best *J* values, and replace the remaining **P** − 1 members of the
population with the all the **ymutated**, except the one which generates the worst *J* value.)

The formal requirements *MakeChildrenSigma* do precisely that, and make evolving trial solutions and their corresponding **sigma** values.

The formal requirements *FindMinES* minimizes an index of performance using the evolving strategy outlined above. Here are five trial runs of *FindMinES* for the *TransportProblem*:

```
In[150]:= l = Table[0, {12}];
          u = {3, 20, 5, 17, 3, 20, 5, 17, 3, 20, 5, 17};
          Do[FindMinES[TransportProblem, 0, 12, 100, 5000, .01, 5000, 1, u], {5}]
```

Stop on |(Jbest - Jworst)/Jworst| <= 0.01

Generation: 381

Number of evaluations: 38 200

Best value: 577.157

x at best vaule:

{2, 0, 0, 8, 1, 7, 4, 3, 0, 13, 1, 6}

Stop on |(Jbest - Jworst)/Jworst| <= 0.01

Generation: 548

Number of evaluations: 54 900

Best value: 589.379

x at best vaule:

{2, 2, 0, 6, 1, 9, 3, 2, 0, 9, 2, 9}

Stop on |(Jbest - Jworst)/Jworst| <= 0.01

Generation: 289

Number of evaluations: 29 000

Best value: 605.487

x at best vaule:

{3, 1, 2, 4, 0, 9, 0, 6, 0, 10, 3, 7}

Stop on |(Jbest - Jworst)/Jworst| <= 0.01

Generation: 397

Number of evaluations: 39 800

Best value: 564.583

x at best vaule:

{0, 5, 0, 5, 2, 12, 1, 0, 1, 3, 4, 12}

Stop on |(Jbest - Jworst)/Jworst| <= 0.01

Generation: 343

Number of evaluations: 34 400

Best value: 604.037

x at best vaule:

{2, 1, 0, 7, 0, 10, 4, 1, 1, 9, 1, 9}

Notice we are again getting a range of answers, with slightly better results than obtained using the genetic algorithm, that require substantially more index of performance evaluations.

Lastly, we examine what can happen if we spend the time to produce a genetic algorithm that is tailored to the problem. In both the proceeding attempts we constructed a vector to represent the solution. But the natural construct is clearly a matrix that relates how much to move from source **i** to destination **j**. Our primary difficulty is to find a way to assign values to the matrix that satisfy the constraints. Reference 10 explains a very clever solution, which is implemented here in *Mathematica* by *InitialTransport1*.

The formal requirements *InitialTransport1* can be used as many times as we want to create an initial population, all members of which satisfy the constraints. Here is one instantiation:

```
In[153]:= Sources = {15, 25, 5};
         Demands = {5, 15, 15, 10};
         v = InitialTransport1[Sources, Demands]
Out[155]= {{0, 0, 15, 0}, {5, 15, 0, 5}, {0, 0, 0, 5}}
```

Next, we need a way to mutate a population member, in a manner that again assures the mutation satisfies the constraints. To do so, take a random matrix (at least 2×2) and apply the above routine to that matrix, then insert the modified result back into the elements of the total matrix. This is accomplished by the formal requirements *TransportMutateType1*. Here is an example:

```
In[156]:= v
         TransportMutateType1[v]
Out[156]= {{0, 0, 15, 0}, {5, 15, 0, 5}, {0, 0, 0, 5}}

Out[157]= {{0, 15, 0, 0}, {5, 0, 15, 5}, {0, 0, 0, 5}}
```

The above allows us to rearrange elements of the matrix. The following allows us to actually change element values. The *InitialTransport2* is identical to *InitialTransport1*, except a conditional is introduced, such that up to assigning the last mutated element, the predecessor elements can have randomly assigned values up to the lower of the source or destination limits.

The formal requirements *InitialTransport2* are used in *TransportMutateType2* to change values in matrices that are permissible solutions. Here is an example output:

```
In[158]:= v
         TransportMutateType2[v]
Out[158]= {{0, 0, 15, 0}, {5, 15, 0, 5}, {0, 0, 0, 5}}

Out[159]= {{1.05894, 4.29071, 0.176775, 0.225275},
          {0.443389, 6.67488, 13.9426, 3.34014}, {3.49676, 0.490637, 0.736648, 0.275951}}
```

For crossover, an arithmetic combination works just fine, since we are certain the child will still honor the constraints. (In reference 10, a fixed combination term was used; here we use a random term.) The formal requirements *TransportCrossOver* implement this. An example crossover output:

```
In[160]:= v1 = {{1, 0, 0, 7, 0}, {0, 4, 0, 0, 0}, {2, 1, 4, 0, 5}, {0, 0, 6, 0, 0}};
          v2 = {{0, 0, 5, 0, 3}, {0, 4, 0, 0, 0}, {0, 0, 5, 7, 0}, {3, 1, 0, 0, 2}};
          result = TransportCrossOver[v1, v2];
          result[[1]]
          result[[2]]

Out[163]= {{0.99836, 0., 0.00819856, 6.98852, 0.00491914},
           {0., 4., 0., 0., 0.}, {1.99672, 0.99836, 4.00164, 0.011478, 4.9918},
           {0.00491914, 0.00163971, 5.99016, 0., 0.00327943}}

Out[164]= {{0.00163971, 0., 4.9918, 0.011478, 2.99508}, {0., 4., 0., 0., 0.},
           {0.00327943, 0.00163971, 4.99836, 6.98852, 0.00819856},
           {2.99508, 0.99836, 0.00983828, 0., 1.99672}}
```

The formal requirements *TransportGA* solve the general (nonlinear, that is the solution can be real, not all integers) transport problem for a specified index of performance denoting the cost of transportation from sources to destinations. The requirements first make a population of a user-specified size, using *InitialTransport1*. Then for each member of the population, for a specified probability, it may combine with a randomly selected second member of the population to create two children by crossover, or with a specified probability, create a child by either *TransportMutate1* or *TransportMutate2*, with equal probability. So this is a very explicit example of applying genetic algorithm to a problem in exactly the form of the problem.

To utilize these formal requirements, we need to modify the index of performance to make its calculation based on a matrix input, rather than a vector, as follows:

```
In[165]:= J[m_] := Module[{},
          N[If[m[[1, 1]] == 0, 0, 5] +
            If[m[[1, 2]] == 0, 0, 5 + 21 * Sqrt[m[[1, 2]]]] +
            If[m[[1, 3]] == 0, 0, 5 + 50 * Sqrt[m[[1, 3]]]] +
            If[m[[1, 4]] == 0, 0, 5 + 62 * Sqrt[m[[1, 4]]]] +
            If[m[[2, 1]] == 0, 0, 5 + 21 * Sqrt[m[[2, 1]]]] +
            If[m[[2, 2]] == 0, 0, 5] +
            If[m[[2, 3]] == 0, 0, 5 + 17 * Sqrt[m[[2, 3]]]] +
            If[m[[2, 4]] == 0, 0, 5 + 54 * Sqrt[m[[2, 4]]]] +
            If[m[[3, 1]] == 0, 0, 5 + 50 * Sqrt[m[[3, 1]]]] +
            If[m[[3, 2]] == 0, 0, 5 + 17 * Sqrt[m[[3, 2]]]] +
            If[m[[3, 3]] == 0, 0, 5] +
            If[m[[3, 4]] == 0, 0, 5 + 60 * Sqrt[m[[3, 4]]]] +
            500 * (
              (m[[1, 1]] + m[[1, 2]] + m[[1, 3]] + m[[1, 4]] - 10)^2 +
              (m[[2, 1]] + m[[2, 2]] + m[[2, 3]] + m[[2, 4]] - 15)^2 +
              (m[[3, 1]] + m[[3, 2]] + m[[3, 3]] + m[[3, 4]] - 20)^2 +
              (m[[1, 1]] + m[[2, 1]] + m[[3, 1]] - 3)^2 +
              (m[[1, 2]] + m[[2, 2]] + m[[3, 2]] - 20)^2 +
              (m[[1, 3]] + m[[2, 3]] + m[[3, 3]] - 5)^2)]]
```

Now to apply the approach to the transport problem:

```
In[166]:= sources = {10, 15, 20};
         demands = {3, 20, 5, 17};
         Do[TransportGA[sources, demands, 100, .05, .2, .01, 5000], {5}]
```

Stop on |(Jbest - Jworst)/Jworst| <= 0.01

Generation: 66

Number of evaluations: 2111

Best value: 356.989

x at best vaule:

{{3, 7, 0, 0}, {0., 13, 2, 0.}, {0., 0, 3, 17.}}

Stop on |(Jbest - Jworst)/Jworst| <= 0.01

Generation: 62

Number of evaluations: 1996

Best value: 356.989

x at best vaule:

{{3., 7., 0., 0.}, {0., 13., 2., 0.}, {0., 0., 3., 17.}}

Stop on |(Jbest - Jworst)/Jworst| <= 0.01

Generation: 68

Number of evaluations: 2111

Best value: 356.989

x at best vaule:

{{3., 7., 0., 0.}, {0., 13., 2., 0.}, {0., 0., 3., 17.}}

Stop on |(Jbest - Jworst)/Jworst| <= 0.01

Generation: 64

Number of evaluations: 1975

Best value: 356.989

x at best vaule:

{{3, 7, 0, 0}, {0., 13, 2, 0.}, {0., 0, 3, 17.}}

Stop on |(Jbest - Jworst)/Jworst| <= 0.01

Generation: 95

Number of evaluations: 2881

Best value: 356.989

x at best vaule:

{{3., 7., 0., 0.}, {0., 13., 2., 0.}, {0., 0., 3., 17.}}

Clearly, this genetic search provides better solutions with fewer index of performance evaluations. Hence the heuristic, for configuration problems, use genetic algorithms coded in the context of the problem.

3.2.14   Determine the Best Strategy Relative to a Competitor

The problem statement is

What strategy will maximize my payoff versus a competitor?

This problem is always extremely difficult to solve since the options to choose from quickly grow so huge they cannot be exhaustively assessed in any reasonable time. Nevertheless, progress can be made: witness the success of software programmed to complete against humans at checkers, backgammon, and chess. To achieve this success took considerable effort over many years, fast computers, and the discovery of clever approximations to help guide strategy selection, rather than brute force tabulation of every possible move and counter move.

Here, we limit our attention to two competitors, **C1** and **C2**. **C1** can make one of **n** decisions; **C2** can make one of **m** decisions. The payoff is known for all possible decisions and represented by an **n** by **m** matrix so that payoff [[**i**, **j**]] declares the amount **C1** pays **C2** if **C1** invokes strategy **i** and **C2** invokes strategy **j**. Some care is recommended in producing the payoff matrix. For example, there is no reason to include a row that has all its values less than another row; if you did, there no is reason to ever pick the strategy corresponding to that row. Similarly, there is no reason to have a column with every element less than some other column.

For some payoff matrices, there is a mutual strategy selection that maximizes the payoff for both players relative to all other strategy options. This [[**i**, **j**]] is called a *saddle point*. If the value of the payoff is not 0, it will not be "fair" since if positive, **C2** keeps getting more at expense of **C1**, and if it is negative, **C1** keeps gaining at the expense of **C2**, but neither player can alter their strategy without risking larger loss. The formal requirements *SaddlePoint* determine if a saddle point exists and the corresponding payout.

P1 and **P2** as defined below are payoff matrices, with saddle points.

```
In[169]:= P1 = {{1, 1, 0, 2}, {1, 1, 0, 1}, {1, -1, 1, 0}, {3, 2, 1, 1}};
          SaddlePoint[P1]

Out[170]= {{4, 3}, 1}

In[171]:= P2 = {{10, 4, 3, 3}, {3, 0, 3, 7}, {7, 5, 9, 6}, {5, 3, 5, 4}, {4, 3, 8, 5}};
          SaddlePoint[P2]

Out[172]= {{3, 2}, 5}
```

So for **P1**, if **C1** always invokes strategy 4 and **C2** always invokes strategy 3, **C2** always wins 1. Though not nice for **C1**, any other strategy could result in a worse payoff. There are several other payoff cells with the same value, namely, [[1,1]], [[1,2]], [[2,1]], [[2,2]], [[2,3]], [[3,1]], [[3,3,]], and [[4,4,]]. For all these other strategy combinations, there is a counterstrategy that pays off better for one competitor. For example, if **C1** tries strategy 1, the risk of loss is now 2, if **C2** plays strategy 4. If **C1** plays strategy 2, he will only be better off if **C2** plays strategy 3, so **C2** will not, and **C1** continues to lose 1. If **C1** plays strategy 3, though he would be better off if **C2** plays strategy 2 or 4, **C2** will not do so, so again **C1** continues to lose 1. A similar review of **C2** options will show though there are options that might provide a bigger payoff than 1, each can be countered by a strategy option for **C1** that limits the payoff to 1. For **P2**, **C1** has numerous larger winning possibilities, but they consistently play strategy 3 and **C2** consistently plays strategy 2, and there is no better combined outcome for the two competitors, even though the payoff always favors **C2**.

Here is an example without a saddle point:

In[173]:= **P2 = {{0, 1}, {1, 0}};**
SaddlePoint[P2]

No saddle point

If there is no saddle point in the payoff matrix, then, given no insight into the Competitor's intentions, assuming the competition will last several rounds, one seeks to randomly pick a strategy with the probability of selecting each strategy set to maximize one's likely payoff over the multiple competitions. For a square payoff matrix (**m = n**) with no negative payoffs (this can be assured by adding the same constant to each element so no element is less than 1), the probabilities are found by solving the following optimization problem:

For **Competitor2**, let **c2[[j]]** be the probability to invoke strategy **j**, 1<= **j** <= **m**. Then:

$$max[Sum[x[[i]],\{i, 1, \textbf{number\_strategies}\}]]$$

subject to the constraints:

$$\textbf{P.x} <= 1.$$

Then:

$$c2[[i]] = x[[i]] \,/\, determinent[P]$$

For **Competitor1**, apply the same algorithm on *Transpose*[**P**].

The formal requirements *StrategyProbabilities* determines the probabilities. Here is the strategy for **C2** for the payoff matrix **Pm**:

In[175]:= **Pm = {{2, 5, 4}, {5, 2, 3}, {7, 1, -3}};**
C2strat = StrategyProbabilities[Pm]

Out[176]= {0.472222, 0.444444, 0.0833333}

Competitor1 strategies are

In[177]:= **C1strat = StrategyProbabilities[Transpose[Pm]]**

Out[177]= {0.5, 0.5, 0.}

Here is a simulation of the two competitors playing their respective strategies 10 times:

```
In[178]:= Competitor2result = {};
          cumcompetitor2result = 0;
          Do[
            ran2 = RandomReal[];
            Competitor2strategy = If[ran2 ≤ 0.472, 1, If[0.472 < ran2 && ran2 ≤ 0.916, 2, 3]];
            ran1 = RandomReal[];
            Competitor1strategy = If[ran1 ≤ 0.5, 1, 2];
            payoff = Pm[[Competitor1strategy, Competitor2strategy]];
            Print["Competitor1 strategy ", Competitor1strategy,
              " Competitor2 strategy ", Competitor2strategy, " payoff ", payoff];
            cumcompetitor2result = cumcompetitor2result + payoff;
            Competitor2result = Append[Competitor2result, cumcompetitor2result]
            , {i, 1, 10}];
          Print["Cum Competitor 2 gain ", Competitor2result[[10]]]
          ListPlot[Competitor2result, AxesLabel → {"Competitions", "Competitor 2 Gain"}]

          Competitor1 strategy 2 Competitor2 strategy 1 payoff 5

          Competitor1 strategy 1 Competitor2 strategy 2 payoff 5

          Competitor1 strategy 1 Competitor2 strategy 2 payoff 5

          Competitor1 strategy 2 Competitor2 strategy 1 payoff 5

          Competitor1 strategy 1 Competitor2 strategy 2 payoff 5

          Competitor1 strategy 1 Competitor2 strategy 2 payoff 5

          Competitor1 strategy 1 Competitor2 strategy 1 payoff 2

          Competitor1 strategy 1 Competitor2 strategy 2 payoff 5

          Competitor1 strategy 1 Competitor2 strategy 1 payoff 2

          Competitor1 strategy 2 Competitor2 strategy 2 payoff 2

          Cum Competitor 2 gain 41
```

Out[182]=

The formal requirements *GameSim* executes 10 games, with each competitor playing an optimum random strategy 1000 times. Here is the result for 1000 trials of the optimum strategy:

```
In[264]:= GameSim[Pm, C1strat, C2strat]

          Minimum payoff for Competitor 2 for 1000 trials of 10 games: 22.

          Mean payoff for Competitor 2 for 1000 trials of 10 games: 34.938

          Maximum payoff for Competitor 2 for 1000 trials of 10 games: 48.

          Variance in payoff for Competitor 2 for 1000 trials of 10 games: 20.3225
```

The formal requirements *GameSimStrategy3* simulates 1000 trials of 10 games in which **Competitor1** knows **Competitor2** will invoke such a strategy, so then **Competitor1** picks strategy 3 every time since that is the only strategy for which **Competitor1** gets a return. Here is the result for 1000 trials of the optimum strategy:

```
In[266]:= GameSimStrategy3[Pm, C2strat]

          Minimum payoff for Competitor 2 for 1000 trials of 10 games: 2.

          Mean payoff for Competitor 2 for 1000 trials of 10 games: 35.08

          Maximum payoff for Competitor 2 for 1000 trials of 10 games: 70.

          Variance in payoff for Competitor 2 for 1000 trials of 10 games: 123.445
```

Although the mean result is about the same, the variance is about six times larger. By invoking a non-optimum strategy, **Competitor1** does create the possibility of less loss (the minimum payoff for **Competitor2** is much about 10 times less), but also the possibility of much worse losses (the maximum payoff for **Competitor2** is substantially higher). You cannot beat an optimum strategy unless your intelligence is very specific, in this case, you have to know precisely what **Competitor2**'s strategy will be just before invoked; only then could you formulate a strategy that will necessarily be toward your advantage. But then again, if luck is all you have, you might as well depend on it.

Even if real-world competition complexities are beyond practical analysis, one can usually approximately analyze the strategic options and consequences by the means illustrated above. The heuristics that should be heeded are

1. If the payoff matrix value is in your favor, then compete, if the worse possible loss can be tolerated. Remember, it is very rare you are absolutely guaranteed to win. Your competitor is probably just as smart as you. They will recognize a situation that shows they are unlikely to win, and so most likely they will avoid competition. This is potentially to your advantage.
2. Should the payoff matrix show you will lose, then don't compete. It is amazing how often people fail to heed this advice. Rather, seek to change the situation so the payoff matrix has at least a saddle point of 0 value or with too low a value for your competitor to be interested in competing.
3. If cooperation either minimizes your maximum loss or maximizes your minimum gain, then cooperate.
4. If your competitor is executing an optimum strategy, and you can discern this, intelligence is of value only if it can determine exactly what the competitor will do immediately prior to doing it.

3.2.15 Make Decisions Regarding Risks

A risk is something that can happen with detrimental consequences. Risk management is a program management function, but often the program manager asks the systems engineers or architects to support, so methodologies are presented to do decision making regarding identifiable and unidentifiable risks.

3.2.15.1   Make Decisions to Mitigate Identifiable Risks

For identifiable risks, the risk management process is

Step 1. Identify potential risks.
Step 2. Assess the likelihood and consequence.
Step 3. Prepare a plan to attempt to reduce the likelihood or make the risk consequence acceptable.
Step 4. Execute the risk mitigation plan.

The four primary ways to identify risks are to do the following:

1. Recall what bad things happened previously on similar endeavors.
2. Utilize thought prompting questions based on prior experience.
3. Challenge the veracity of assumptions made.
4. Identify any items not previously used before or any activities not previously accomplished successfully.

How the prior similar systems or architectures were designed, or analyzed, or how people used the prior similar system or architecture, may have resulted in undesirable consequences. Risk mitigation for these prior issues is straightforward; implement a solution that accommodates the bad thing happening to the fullest extent willing to allocate resources to do so. For example, if a prior building was destroyed due to a particularly energetic earthquake, then design and build future buildings to withstand such earthquakes. Of course, it is much easier to say this than do it. Considerable effort will need to be expended to analyze, simulate, and test ways to make earthquake tolerant buildings.

Many organizations utilize standard questions to try to institutionalize risk identification. A particularly good set of questions is documented in reference 11.

One never knows for certain how long it will take for technology to mature, nor even if it will ever perform as desired. Such risks are also easy to mitigate. First, minimize dependency on immature technology. In the limit, this could result in avoiding implementing a system solution, which will upset those who want the solution, but if the technology never matures, the system never comes in to existence so it seems imminently logical to avoid implementing systems with immature technology. Alternatively, there may be an acceptable, but a less well-performing system that could be fielded until the technology is matured. Under these conditions, the risk mitigation decision is when to switch to the more mature technology.

Any assumptions that turn out to be incorrect can manifest themselves in unintended consequences. Uncertainty risks are managed by modeling the variability or uncertainty to determine the amount of design margins to impose. These risk decisions involve setting the magnitude of the uncertainties to accommodate and the establishment of how much margin to impose.

It is currently fashionable to identify a risk as a mark on a 5×5 matrix, where rows depict different levels of probability of occurrence, and the columns depict different levels of consequence, as shown in Table 3.8.

TABLE 3.8
Typical Risk Register Currently in Vogue

Risk Level		Consequence				
		Very Low	**Low**	**Medium**	**High**	**Very High**
Probability	**Very high**	Low risk	Low risk	Medium risk	High risk	High risk
	High	Low risk	Low risk	Medium risk	Medium risk	High risk
	Medium	Low risk	Low risk	Medium risk	Medium risk	Medium risk
	Low	Low risk	Low risk	Low risk	Low risk	Low risk
	Very low	Low risk	Low risk	Low risk	Low risk	Low risk

The rows and column headers may use relative words as shown above, or may specify the probability ranges numerically. The consequences columns may state ranges for dollars, days of delay, or physical units for some desired performance. Each cell is often color coded, from blue or green for those combinations of probability and consequence that are considered low risk, to yellow and then red for those cells that denote high risk. The risk register approach, though certainly better than nothing, has two serious weaknesses.

First, virtually nothing that can go wrong (what a risk is) has a single distinct probability and consequence, but is rather a continuum of possible results at either different probabilities, or a fuzzy number depicting the uncertainty associated with the result. For example, a risk might be as follows:

IF the structural mass exceeds its allocation, THEN the payload mass allocation will have to be reduced in proportion.

During system development, the mass is a guess, and has a wide range of possible values, many of which are under the launch vehicle's limit, a few of which may be over the limit. Rather than communicate the mass risk on a 5 × 5 table, show the mass estimate as a fuzzy number with respect to the launch vehicle capability.

Second, the state of the mitigation effort affects where the mark should be made. When using the risk register it is common to denote multiple marks for each risk, indicating where it is "now" (or "before mitigation") and where it will be moved to as the mitigation effort is completed. But what really happens as the mitigation effort proceeds is that the uncertainty in the result is hopefully reduced and the difference between the predicted value and the desired value hopefully shrinks. None of this is communicated by placing marks on the 5 × 5 matrix. Let's go back to the mass risk above. As the design matures the mass estimate uncertainty will undoubtedly improve: it may get heavier, it may get lighter, but the certainty will increase, and the consequence is partly a result of what the final number is for the entity, and equally what the final numbers are of all the other masses, relative to a requirement, which itself might change. If it is ever determined the mass estimate is going to cause a problem, then the mitigation effort may be to reduce the item's mass by removing material, or perhaps substituting material, or the mitigation effort might be to accept the mass increase in one item and reduce mass elsewhere, or to find a carrier that can tolerate the higher mass. None of these options is well documented on the 5 × 5 matrix. At best you put a mark in one of the yellow or red areas for the mass "as is," and you also put a mark in one of the green areas to denote that you predict the mass "will be" acceptable at some point. A better communication is shown in Chapter 2, plot the "was," "is" and "will be" mass estimates.

The following recommend ways to identify, assess, and mitigate identifiable risks.

3.2.15.1.1   To Mitigate Cost Risks Retain Financial Reserves

All developmental and operations activities have inherent cost risks, the less prior experience doing either activity, the more the risk. For developmental cost estimates, the situation is aggravated since the acquirer will often deliver the contract to the lowest bidder, while the provider also bids the cost low to obtain the job. This mutually destructive behavior is unavoidable.

Development cost risks are mitigated by the acquirer using the methodology illustrated in Chapter 2 to adjust the seller's proposed cost based on technology maturity and proposal risk (also known as estimation certainty.) Obviously, should the seller apply such a technique to their estimate, the acquirer should only adjust the factors necessary to remove possible underbidding adjustments.

Operations costs for new systems are equally uncertain. To the extent operations require expendables, for example, such consumables as fuel, future operations costs will fluctuate with these costs. Use the lattice technique illustrated earlier in this chapter to determine possible operations cost ranges, then formulate options to enable cost reductions that pay off in proportion to the consumable cost increases that may occur.

3.2.15.1.2   To Mitigate Schedule Risks Promise Late and Deliver Early

Model the activities as shown in Chapter 2, honoring dependencies, with fuzzy numbers. Plan on the de-fuzzified time to be the necessary time to complete the effort. Note the worst possible time, and allow some contingency for the difference between the defuzzified total time and the worst possible time.

3.2.15.1.3   To Mitigate Performance Risks Require the Entity to Tolerate Off Nominal Conditions and Predict Performance Acknowledging the Likelihood of Uncertainty

If an item is susceptible to less than perfect performance, then impose requirements to tolerate the uncertainty to some amount. A simple example is to require a structure be designed to accommodate a factor of safety times the most likely worst load, as mitigation for the risks of underestimating the load, or that the material properties are worse than assumed, or that the structure is utilized in a manner that will cause more loading than planned. Or, for example, if the item must operate in a high radiation environment, then require a level of radiation to be tolerated. Or, if the item may have an engine failure, but must still be able to conduct its mission, then require the capacity to do so. The art of this effort is to be able to name the possible things that can go wrong for which design tolerance must be included. Typical sources are as follows:

- Extremes in the operating environment (examples: loads, temperatures, humidity, dirt, pressure, radiation)
- Malfunction of components (examples: power failure, propulsion failure, structural failure, sensor failure, processing failure, communication failure)
- Purposeful damage by a specified means (examples: exposed to fire, hit by a car, shot with a bullet, attacked by a virus)

As was shown in Chapter 2, use either Monte Carlo or fuzzy mathematics to model the uncertainty and predict the likely performance, so the design allows for all components performing less than perfectly.

3.2.15.1.4 Avoid Reliance on Immature Technologies for Key System or Architecture Performance and Mature Technology with Exit Gates

Should a system or architecture be dependent on a yet to be demonstrated technology, then obviously that lack of maturity represents a risk. If the cost adjustment and schedule adjustment techniques mentioned above are used, then the some of the potential impact on the project are quantified. Sometimes backup alternatives for the low maturity technologies can be substituted, then explicit circumstances should be identified when a decision will be made for the backup to substitute for the immature technology.

Technology development should be planned in the following manner. The current maturity level should be explicitly stated (recall in Chapter 2 that we showed how to do this using fuzzy number to allow for the ambiguity that might exist as to the itemized current state). Activities should be described that explain what will be done to attempt to get the technology to the next higher maturity level, step by step. Each activity needs to consist of declared tasks, rather than a implicit call for a miracle to happen. The capability that is to be demonstrated by the end of the actions must be quantified. As with any plan, a time estimate and a cost estimate are to be provided. As you approach the end of the activity, make a decision: continue or stop? At this point the money and time are spent, so looking back does no good. Rather estimate forward, an inherently inaccurate thing to do, but as demonstrated in Chapter 2 and as shown using uncertainties lattices, we can make reasonable predictions. The primary motivation should be to kill the immature technology in favor of the proven technology, so to continue, extraordinary good evidence must exist that the cost to go and time to go and capability to go are very attractive relative to the money and time available.

Please understand, I am all for technology development; it is exactly how new systems and architecture are created to enhance our lives. But they are fundamentally gambles, and most do not pay off. Thank goodness though that both the fun of chase and the vision of the reward are tantalizing enough that people keep trying. My suggestion is, getting a complicated systems or architecture fielded is fraught with enough problems, so try to avoid the woes due to the unpredictable realities of technology development. System and architect engineers should communicate what their current systems do, and customers should communicate what they wish they could do, and the technology developer should try to create something better. System engineers and architects should avoid like the plague the conscious inclusion of new technologies into an operational system. If the system or architecture one envisions to achieve the customer and user needs is dependent on an envisioned technology, then one should concentrate effort initially and exclusively on maturing that technology. Since complicated systems and architectures often take a long time to come into being even if use existing technology, many organizations gamble that the technology maturation efforts can proceed in parallel with the more mature efforts. But this is a gamble that history has showed nearly everyone loses, at least in the sense of achieving original cost and schedule targets. So, recognize technology maturation for what it is, a gamble. And like any successful gambler, you must do your best to get the odds in your favor, or at least stipulate a loss level you do not go beyond. If you have decided you need the technology no matter what, then cost and schedule limits must be approached in a different way. If your need is so great, cost is not the prime consideration, you certainly don't want to behave that way or cost will grow unbounded. Useful heuristics are

1. *Compete the development.* Though one may think it is counterproductive to split one's resources, given human nature, it is almost always better to fund two different organizations to competitively deliver the needed technology.
2. *Utilize cost plus award fee contracts.* That is, recognizing the cost is essentially impossible to predict, be willing to reimburse the costs you approve, but keep the profit a judgment call on how well the effort is progressing.
3. *Parcel out the money in parts, consistent with the planned activities.* This makes it easier to save resources if you need to stop. And thus makes it easier to stop if you need to.
4. *Review the planned activities, and object to and remove all unnecessary expenses other than those needed to mature the technology.* Staffing levels should be low and overheads and other charges for expenses secondary to the technology maturation should be challenged and cut to the minimum level. Never let the ratio go below 20 active members for 1 administrative member.
5. *Make the decision whether to proceed based on estimate of time and money to spend, not spent.* Never let the amount of the effort you've put into making a technology work influence your decision on whether or not to continue to try to make the technology work. Decide whether to keep trying based only on the estimate of how much more resources are needed.

3.2.15.2   Make Decisions for Unidentifiable Risks

Operational risks cannot always be anticipated. Once the system or architecture is fielded and people operate it, inputs will be encountered that were not fully anticipated which will put the system or architecture in an unanticipated state. You may ask why? Certainly one should think through or test in advance all the possible inputs and all the possible combinations! Unfortunately, for most complicated systems and architectures the combinations and permutations involved are beyond human comprehension. What we are discussing here is how can we do our best given finite resources to anticipate the potential gotcha's before they get us.

For unidentifiable risks, the risk management process is as follows:

Step 1. Provide redundancy. That is, enable independent means for the system or architecture to continue to function as desired if a portion of the system or architecture fails.

Step 2. Provide self-sufficiency. That is, enable the system or architecture to continue to function if cut off from external entities normally associated with operations.

Step 3. Provide adaptability. That is, enable the system or architecture to change how it functions should a portion of the system or architecture fail or interfaces disappear.

Unfortunately, the killer risks are always the ones we don't see coming. Even though someone somewhere wrote an email that proves to be an accurate prediction, my experience is for every one of these apparently sanguine emails, there are thousands of false alarm emails that never see the light of day because their prediction never came true. Many of these apparently accurate predictions were storytelling, that is, lucky guesses. If you get enough people to guess about anything, a few of them will guess what actually happens. Still, people in general are poor at anticipating things that could have huge impacts. That means I, and probably you, being people in general, are poor at so doing as well. We have to acknowledge this about ourselves. One must consciously seek out inputs from as wide a set or sources as possible about what could go wrong and one needs to find the time to decide what to do about each. The reverse situation is also true; one can do too much trying to avoid risks. A useful heuristic, though fictional, comes from a movie in which the characters mysteriously wake up inside a cube connected on all six sides by a door to another cube. They find they can move from cube to cube, but any given cube may have in it a means to kill them quite spectacularly. Meanwhile, every once in a while they can feel the cubes move so the sides open to new cubes. The poor characters gradually diminish in number as they make their way from cube to cube, with a few survivors ending up in exactly the cube they woke up in, with now one door opening up to apparent freedom. The heuristic is, unless your actions reduce risk, there is no point taking them. Still, I'm afraid in the end, we will fail to anticipate a risk and it will get us. Recognizing this, three features useful for system and architecture survival, be it during development, or during operations, are redundancy, self-sufficiency, and adaptability. A redundant system or architecture can achieve its functionality utilizing numerous ways using some of its parts. Self-sufficiency is much harder to measure than talk about. You may grow your own food, but if you buy the seeds, or fertilizer, or water and power, or need police to guard ownership, you really are not much more self-sufficient than if buy your food. Good measures for self-sufficiency for systems and architecture are

- Percent of information needed to operate that is self-generated (meaning have absolutely no need to involve another system or architecture)
- Percent of energy needed to operate that is self-generated

Adaptability is partly measurable by the ratio of ultimate capability to nominal operation performance. For example, if your car can go 200 mph, but you normally operate at 60 mph, you have a speed adaptability of better than 3 to 1. Adaptability is also potentially measurable by the functionality one could do beyond that for nominal operation. For example, your car may normally be used to get you from point A to point B, but your car could also be used as a home. Extra functional adaptability can be measured by estimating the economic value of the alternative functions that could be performed by the system or architecture from those stipulated to create it in the first place.

When an unknown risk manifests itself, quite possibly with catastrophic implications, then the risk management process is

Step 1. Formulate a review board of people who are not associated with the design, manufacture, or operation of the failed system or architecture, but who have the fundamental education to comprehend the system or architecture.

Step 2. Invoke no penalty for the failure provided no law was broken. Any inferred penalty will impede the ability to get truthful evidence.

Step 3. Collect and preserve any available evidence of what the system or architecture was doing prior to, during, and after the risk manifested itself. This evidence is the only hope we have to determine cause and effect. The evidence must be assembled as quickly as possible and protected from contamination by accidental or purposeful means.

Step 4. Hypothesize causes.

Step 5. Select the most likely cause and if necessary secondary causes if judged credible.

Step 6. Formulate actions for the system owners and operators to take to prevent the cause in the future.

Step 7. Impose the prevention measures with a means to verify adoption.

3.3   DECISION HEURISTICS

- Making a good decision requires considering all seven of the following questions:
  1. What needs to be decided?
  2. Who needs to make the decision?
  3. When is the decision needed?
  4. What options are there to choose from?
  5. What is it about each option that distinguishes it from the other options with respect to what needs to be decided?
  6. How can appropriate and accurate information be obtained to assess each option with respect to what needs to be decided?

7. How robust is the decision should the goodness criteria assessment be flawed?

- Formal methods improve our ability to assess the robustness of the decision. Only after a type of decision was made many times, so all the seven factors have knowable finite ranges, or at least bounds, can formal methods be established to find the best decision.
- When making proxy decisions, test the candidate decision with the proxy.
- Complicated situations require a hierarchy of decisions to be made, so make sure the top decision is made correctly.
- The wrong people will make wrong decisions.
- People unaffected by the result of the decision will most likely make the best decision.
- To the extent that implementing a decision requires acceptance and cooperation of multiple people, the more the affected people participate in making the decision, the more likely the selected result will be implemented.
- The more people involved in the decision-making process, the more likely a good decision will be reached.
- Hubris comes before the fall. Don't get talked into fixing other people's problems.
- Make decisions when you have too, not before.
- Constantly test if your decision is wrong; give yourself options to alter your decision as new information becomes available.
- Do your work so you trust your work.
- What is good for a system is not necessarily good for an architecture.
- We cannot predict the truly revolutionary events, so we must react wisely when they occur.
- Use a leadership style of direction when there is one correct solution and follower acceptance to implement is low.
- Use a leadership style of consultation when there is one correct solution and follower acceptance to implement is high.
- Use a leadership style of convenience when there are many correct solutions and follower acceptance to implement is low.
- Use consensus when there are many correct solutions and follower acceptance to implement is high.
- Use collaborative decision-making method when follower expertise is high and have ample time to make decision.
- Sell the decision to followers when their expertise is low and have ample time to make decision.
- Delegate the decision when followers have high expertise and time to make the decision is limited.
- Command the decision when followers have low expertise and time to make the decision is limited.
- There are at least 14 different types of decisions that a system engineer or architect may need to help make. Table 3.9 summarizes the types, recommended measures of goodness, and formal methods to use.

TABLE 3.9

Formal Methods to Use for Each Type of Decision to Make

Type of Decision	Measure of Goodness	Formal Method to Use
Determine action based on diagnosis.	Probability of correct diagnosis	*ProbTrueGivenPosTest*
A sequence of actions to optimize a return.	The return	*probDecGateEval* *fuzzDecisionNode*
Give people a fair means to choose from a static, finite set of options.	Percent of people accepting the decision	*BordaCount*
Allocate a static, finite set of items so that the recipients feel equally satisfied.	People's willingness to pay for items	*DutchAuctionforSameItems* *DutchAuction for DifferentItems*
Evolve options to best satisfy multiple criteria.	Better than (+), equal to (0), or worse than (–)	*PughComparison*
Select from a static, finite set of options that one which best achieves multiple dissimilar criteria.	Weighted sum of criteria satisfaction	*FuzzyDecisionMaker*
Select from a static, finite set of options those which maximizes benefits while minimizing detriments.	On or very near the efficient frontier of benefit versus detriment	*EfficientFrontier*
Do the optional endeavor with the best financial return.	Whatever the endeavor seeks to achieve	*BinomialLattice* *LeaseValue* *LeaseValuewithOption*
Select a portfolio of investments which maximize the return and minimize the risk.	On or very near the efficient frontier of return versus risk	*EfficientFrontier* Or invest equally in uncorrelated risky returns and rebalance periodically.
Choose the independent variable values that optimize an index of performance, perhaps subject to constraints, when the index of performance is inexpensive to determine.	The index of performance value augmented with a penalty times the extent each constraint is not satisfied.	*FindMinimum* *NMinimize*
Choose the independent variable value that optimizes an index of performance, perhaps subject to constraints, when the index of performance is expensive to determine.	Taguchi quality function	*TaguchiResponseTable* *InteractionResponse*

TABLE 3.9

Formal Methods to Use for Each Type of Decision to Make

Type of Decision	Measure of Goodness	Formal Method to Use
Define a dynamic control law to optimize a dynamic index of performance.	The index of performance value	*ContFunctforLinConstEOM* Formulate a fuzzy controller.
Define the optimal configuration.	What the configured item is to achieve	*FindMinGA* *FindMinEs*
Determine the best strategy relative to a competitor.	What the strategy seeks to maximize or minimize	*SaddlePoint* *StrategyProbabilities* Derive from game theory.

- Prior to implementing any decision-making process, it is critical that those empowered to make the decision, or those trusted to provide information to those who will make the decision, first perceive all data associated with the decision-making process to be completely and totally wrong, until substantial evidence is presented to prove otherwise.
- To avoid bias, avoid making decisions based primarily on expert opinion, or accepting decisions made by people who will directly benefit from the decision.
- When making a decision, it is wise to know how different the inputs can be for the decision to be the same, so you know when the decision is no longer valid and you need to make a new decision.
- Do the right thing. All systems engineers and architects must realize politics always trumps analysis. This is usually appropriate, since as we have already discussed, all analysis is inherently flawed at some level, so the political considerations may override the analytical conclusion. But there is a line that must be drawn; we each have an ethical responsibility to challenge decisions that are wrong.
- Truth is very hard to know. But we must be motivated to attempt to express it, and to recognize we may be wrong about what we believe most strongly.
- Human nature tends to limit the options considered to those that worked in the past or those known to fail. To counter this human limitation, command the definition of at least six options prior to doing an assessment.
- Only use voting as a decision-making process if you are certain the voters will abide by the result.
- To optimize the return from a portfolio of investments:
  1. Presume the possibility of what could happen, not what did happen.
  2. Buy low, sell high. A periodic rebalancing strategy does this automatically.
  3. If the likely result is positive, maximize the number of opportunities since more periods provided better return statistics.
  4. Diversify; note that two independent risky investments provide a better return than one risky investment and one certain investment.
  5. Invest an amount in proportion to what can lose.

- Always check the obtained **x*** from a parameter optimization for robustness before using.
- A Taguchi method heuristic is to use approximately half of the control factors that have the most impact on the signal-to-noise ratio. A second Taguchi method heuristic is to stop selecting control factors with decreasing results when the difference in response results between it and its predecessor first increases.
- To find optimal configurations, use genetic algorithms coded in the context of the problem.
- Competition heuristics that should be heeded:
  1. When the payoff matrix shows value is in your favor, then compete, if the worse possible loss can be tolerated. Remember, it is very rare you are absolutely guaranteed to win. Your competitor is probably just as smart as you. They will recognize a situation that shows they are unlikely to win, and so most likely they will avoid competition. This is potentially to your advantage.
  2. When the payoff matrix shows you will lose, then don't compete. It is amazing how often people fail to heed this advice. Rather, seek to change the situation so the payoff matrix has at least a saddle point of 0 value or too low a value for your competitor to be interested in competing.
  3. If cooperation either minimizes your maximum loss or maximizes your minimum gain, then cooperate.
  4. You cannot beat an optimum strategy unless your intelligence is very specific. If your competitor has found an optimum strategy, intelligence is of value only if you can determine exactly what the competitor will do immediately prior to doing it.
- For identifiable risks, the risk management process is as follows:
  Step 1.   Identify potential risks.
  Step 2.   Assess the likelihood and consequence.
  Step 3.   Prepare a plan to attempt to keep the risk consequence acceptable. (Note that the plan could be null, that is, accept the risk and do nothing.)
  Step 4.   Execute the risk mitigation plan.
- The four primary ways to identify risks are as follows:
  1. Recall what bad things happened previously on similar endeavors.
  2. Utilize thought prompting questions based on prior experience.
  3. The cost, schedule, and performance for each portion of the system or architecture not previously used.
  4. Challenge the veracity of assumptions made.
- The following describe recommended ways to mitigate identifiable risks:
  1. Retain financial reserves to compensate for cost risks.
  2. Promise late and deliver early to compensate for schedule risks.
  3. Require the entity to tolerate off-nominal conditions for performance risks and predict performance acknowledging the likelihood of uncertainty.

4. Mature technology with exit gates and avoid reliance on immature technologies for key system or architecture performance, and:

- Compete the technology development.

1. Utilize cost plus award fee contracts.
2. Parcel out the money in parts, consistent with the planned activities.
3. Review the planned activities, object to and remove all unnecessary expenses other than those needed to mature the technology.
4. Proceed only if the estimated cost and time to complete are acceptable, ignoring the time and money already spent.

- People in general are poor at anticipating things that could have huge impacts.
- One must consciously seek out inputs from as wide a set or sources as possible what could go wrong and one needs to find the time to decide what to do about each.
- Unless your actions reduce risk, there is no point taking them.
- For unidentifiable risks, the risk management process is the following:

    Step 1. Provide redundancy. That is, enable independent means for the system or architecture to continue to function as desired if a portion of the system or architecture fails.

    Step 2. Provide self-sufficiency. That is, enable the system or architecture to continue to function if cut off from external entities normally associated with operations.

    Step 3. Provide adaptability. That is, enable the system or architecture to change how it functions should a portion of the system or architecture fail.

- When an unknown risk manifests itself, quite possibly with catastrophic implications, then the risk management process is the following:

    Step 1. Formulate a review board of people not associated with the design, manufacture, or operation of the failed system or architecture but with the fundamental education to comprehend the system or architecture.

    Step 2. Invoke no penalty for the failure provided no law was broken. Any inferred penalty will impede the ability to get truthful evidence.

    Step 3. Collect and preserve any available evidence of what the system or architecture was doing prior to, during, and after the risk manifested itself. This evidence is the only hope we have to determine cause and effect. The evidence must be assembled as quickly as possible and protected from contamination by accidental or purposeful means.

    Step 4. Hypothesize causes, and determine to the extent the evidence supports the cause.

    Step 5. Select the most likely cause and if necessary secondary causes if judged credible.

    Step 6. Formulate prevention measures to prevent the cause in the future.

    Step 7. Impose the prevention measures with a means to verify adoption.

APPENDIX 3A: RESULTS OF TESTING *MATHEMATICA'S* PARAMETER OPTIMIZATION ROUTINES

Reference 12 documents 35 problems consisting of 39 numerical optimization problems with reported solutions for testing parameter optimization routines. Unfortunately, upon careful review, 12 of the reported solutions do not evaluate to the reported value of the index of performance, so presumably there are typographical errors in the statement of the problem or solution. So, 27 problems are available to test *Mathematica's* built-in parameter optimization routines. The following documents the attempts to utilize the *Mathematica* to solve all 39 test cases.

PROBLEM 1

Mathematica finds the documented solution.

```
In[185]:=  J = (x1 - 2) ^2 + (x2 - 1) ^2;
           NMinimize[{J, x1 - 2 * x2 + 1 == 0, -x1^2 / 4 - x2^2 + 1 ≥ 0}, {x1, x2}]
Out[186]=  {1.39346, {x1 → 0.822876, x2 → 0.911438}}
```

PROBLEM 2

Mathematica finds the documented solution.

```
In[187]:=  J = 100 * (x2 - x1^2) ^2 + (1 - x1) ^2;
           NMinimize[J, {x1, x2}]
Out[188]=  {1.97215 × 10⁻³¹, {x1 → 1., x2 → 1.}}
```

PROBLEM 3

The reported solution is $x^* = (75, 65)$ with $J^* = 58.903$. The documented equations fail to evaluate to J^* at x^*, so a *Mathematica* solution was not attempted.

```
In[189]:=  Clear[J]; J[x1_, x2_] := 75.196 - 3.8112 * x1 + 0.12694 * x1^2 - (2.0567 * 10^-3) * x1^3 +
           (1.0345 * 10^-5) * x1^4 - 6.8306 * x2 + 0.030234 * x1 * x2 - (1.28134 * 10^-3) * x2 * x1^2 +
           (3.5256 * 10^-5) * x2 * x1^3 - (2.266 * 10^-7) * x2 * x1^4 + 0.25645 * x2^2 -
           (3.4604 * 10^-3) * x2^3 + (1.3514 * 10^-5) * x2^4 - 28.106 / (x2 + 1) -
           (5.2375 * 10^-6) * x1^2 * x2^2 - (6.3 * 10^-8) * x1^3 * x2^2 + (7 * 10^-10) * x1^3 * x2^3 +
           (3.4054 * 10^-4) * x1 * x2^2 - (1.6638 * 10^-6) * x1 * x2^3 - 2.8673 * Exp[0.0005 * x1 * x2];
           J[
           75,
           65]
Out[190]=  -42.1353
```

Problem 4

There are five reported solutions, with $\mathbf{x}^*$, depending on how well the three equality constraints are satisfied. For all constraints no larger than 10^{-6}, $\mathbf{x}^*$ is (0.0406, 0.1477, 0.7832, 0.0014, 0.4853, 0.0007, 0.0274, 0.018, 0.0375, 0.0969) and $J^* = {}^-47.761$. A solution is shown with no constraint larger than $2*10^{\wedge-11}$, with $J^* = {}^-47.751$ and $\mathbf{x}^* = $ (0.0350, 0.1142, 0.8306, 0.0012, 0.4887, 0.0005, 0.0209, 0.0157, 0.0289, 0.0751). *NMinimimum* fails to find a solution using the default settings, but *FindMinimum* does.

```
In[4]:= NMinimize[{x1*(-6.089 + Log[x1/(x1 + x10 + x2 + x3 + x4 + x5 + x6 + x7 + x8 + x9)])
    + x10*(-22.179 + Log[x10/(x1 + x10 + x2 + x3 + x4 + x5 + x6 + x7 + x8 + x9)]) +
    x2*(-17.164
      + Log[x2/(x1 + x10 + x2 + x3 + x4 + x5 + x6 + x7 + x8 + x9)]) + x3*(-34.054
      + Log[x3/(x1 + x10 + x2 + x3 + x4 + x5 + x6 + x7 + x8 + x9)]) +
    x4*(-5.914
      + Log[x4/(x1 + x10 + x2 + x3 + x4 + x5 + x6 + x7 + x8 + x9)]) + x5*(-24.721
      + Log[x5/(x1 + x10 + x2 + x3 + x4 + x5 + x6 + x7 + x8 + x9)]) +
    x6*(-14.986 + Log[x6/(x1 + x10 + x2 + x3 + x4 + x5 + x6 + x7 + x8 + x9)]) + x7*(-24.1
      + Log[x7/(x1 + x10 + x2 + x3 + x4 + x5 + x6 + x7 + x8 + x9)])
      + x8*(-10.708
      + Log[x8/(x1 + x10 + x2 + x3 + x4 + x5 + x6 + x7 + x8 + x9)])
      + x9*(-26.662 + Log[x9/(x1 + x10 + x2 + x3 + x4 + x5 + x6 + x7 + x8 + x9)]),
    x1 + 2*x2 + 2*x3 + x6 +x10-2 ==0,
    x4 + 2*x5 + x6 + x7 -1 ==0,
    x3 + x7 + x8 + 2*x9 +x10 -1 ==0,
    x1>0,
    x2>0,
    x3>0,
    x4>0,
    x5>0,
    x6>0,
    x7>0,
    x8>0,
    x9>0,
    x10>0},
      {x1,x2,x3,x4,x5,x6,x7,x8,x9,x10}]
```

NMinimize::nrmum :
   The function value −44.3752 − 0.170338 i is not a real number at {x1, x10, x2, x3, x4, x5, x6, x7, x8, x9} =
      {0.449097, 0.327184, 0.00937872, 0.572661, 0.384604, 0.19171, 0.0596398, 0.172336, 0.0362603, −
      0.0542204}. ≫

Out[4]= {−46.1229, {x1 → 0.35429, x10 → 0.253813, x2 → 0.248197, x3 → 0.37162, x4 → 0.0438535,
      x5 → 0.335432, x6 → 0.152263, x7 → 0.133019, x8 → 0.00282593, x9 → 0.119361}}

```
In[5]:= FindMinimum[{x1*(-6.089 + Log[x1/(x1 + x10 + x2 + x3 + x4 + x5 + x6 + x7 + x8 + x9)])
    + x10*(-22.179 + Log[x10/(x1 + x10 + x2 + x3 + x4 + x5 + x6 + x7 + x8 + x9)])
    + x2*(-17.164 + Log[x2/(x1 + x10 + x2 + x3 + x4 + x5 + x6 + x7 + x8 + x9)])
    + x3*(-34.054 + Log[x3/(x1 + x10 + x2 + x3 + x4 + x5 + x6 + x7 + x8 + x9)])
    + x4*(-5.914 + Log[x4/(x1 + x10 + x2 + x3 + x4 + x5 + x6 + x7 + x8 + x9)])
    + x5*(-24.721 + Log[x5/(x1 + x10 + x2 + x3 + x4 + x5 + x6 + x7 + x8 + x9)])
    + x6*(-14.986 + Log[x6/(x1 + x10 + x2 + x3 + x4 + x5 + x6 + x7 + x8 + x9)])
    + x7*(-24.1 + Log[x7/(x1 + x10 + x2 + x3 + x4 + x5 + x6 + x7 + x8 + x9)])
    + x8*(-10.708 + Log[x8/(x1 + x10 + x2 + x3 + x4 + x5 + x6 + x7 + x8 + x9)])
    + x9*(-26.662 + Log[x9/(x1 + x10 + x2 + x3 + x4 + x5 + x6 + x7 + x8 + x9)]),
    x1 + 2*x2 + 2*x3 + x6 +x10-2 ==0,
    x4 + 2*x5 + x6 + x7 -1 ==0,
    x3 + x7 + x8 + 2*x9 +x10 -1 ==0,
    x1>0,
    x2>0,
    x3>0,
    x4>0,
    x5>0,
    x6>0,
    x7>0,
    x8>0,
    x9>0,
    x10>0},
      {x1,x2,x3,x4,x5,x6,x7,x8,x9,x10}]
```

Problem 4a

This is the same problem as Problem 4, with a change in variables to **y[[i]]** = *Exp*[[**x**[[**i**]]]]. *Mathematica* finds the documented solution.

```
In[193]:= c = {-6.089, -17.164, -34.054, -5.914,
        -24.721, -14.986, -24.100, -10.708, -26.662, -22.179};
    Clear[J, x1, x2, x3, x4, x5, x6, x7, x8, x9, x10];
    J[x_, c_] := Module[{s},
        s = Log[ Sum[Exp[x[[i]]], {i, 1, 10}]];
        Sum[Exp[x[[i]]] * (c[[i]] + x[[i]] - s), {i, 1, 10}]];
    Jf = J[{y1, y2, y3, y4, y5, y6, y7, y8, y9, y10}, c];

    solution = NMinimize[{Jf,
        Exp[y1] + 2 * Exp[y2] + 2 * Exp[y3] + Exp[y6] + Exp[y10] - 2 == 0,
        Exp[y4] + 2 * Exp[y5] + Exp[y6] + Exp[y7] - 1 == 0,
        Exp[y3] + Exp[y7] + Exp[y8] + 2 * Exp[y9] + Exp[y10] - 1 == 0},
        {y1, y2, y3, y4, y5, y6, y7, y8, y9, y10}]
    y = {y1, y2, y3, y4, y5, y6, y7, y8, y9, y10};
    Exp[ y /. solution[[2]]]

Out[197]= {-47.7611, {y1 → -3.20231, y2 → -1.91237, y3 → -0.244427, y4 → -6.56118, y5 → -0.723098,
        y6 → -7.27416, y7 → -3.59724, y8 → -4.02032, y9 → -3.28838, y10 → -2.33437}}

Out[199]= {0.0406681, 0.14773, 0.783153, 0.00141422, 0.485247,
        0.000693219, 0.0273993, 0.0179473, 0.0373144, 0.0968713}
```

Problem 5

NMinimize fails to find the documented solution. *FindMinimum* does.

```
In[200]:= J = 1000 - x1^2 - 2 * x2^2 - x3^2 - x1 * x2 - x1 * x3;
    NMinimize[{J, x1^2 + x2^2 + x3^2 - 25 == 0,
        8 * x1 + 14 * x2 + 7 * x3 - 56 == 0, x1 ≥ 0, x2 ≥ 0, x3 ≥ 0}, {x1, x2, x3}]

    NMinimize::nosat :
        Obtained solution does not satisfy the following constraints within Tolerance -> 0.001`: {25 - x1² - x2² - x3² = 0}. ≫

Out[201]= {968., {x1 → 0., x2 → 4., x3 → 0.}}
```

```
In[202]:= J = 1000 - x1^2 - 2 * x2^2 - x3^2 - x1 * x2 - x1 * x3;
    FindMinimum[{J, x1^2 + x2^2 + x3^2 - 25 == 0,
        8 * x1 + 14 * x2 + 7 * x3 - 56 == 0, x1 ≥ 0, x2 ≥ 0, x3 ≥ 0}, {x1, x2, x3}]

Out[203]= {961.715, {x1 → 3.51212, x2 → 0.216988, x3 → 3.55217}}
```

Problem 6

The documentation fails to define a key parameter, so a *Mathematica* solution was not attempted.

Problem 7

The problem requires that seven functions in terms of the three independent variables be recursively determined using logical expressions. As it is not clear how *Mathematica* routines can be used to solve this problem, a solution was not attempted.

PROBLEM 8

Mathematica finds the documented solution.

```
In[204]:= J = 100 * (x2 - x1^2)^2 + (1 - x1)^2 + 90 * (x4 - x3^2)^2 +
          (1 - x3)^2 + 10.1 * ((x2 - 1)^2 + (x4 - 1)^2) + 19.8 * (x2 - 1) * (x4 - 1);
       NMinimize[J, {{x1, -10, 10}, {x2, -10, 10}, {x3, -10, 10}, {x4, -10, 10}}]

Out[205]= {0., {x1 → 1., x2 → 1., x3 → 1., x4 → 1.}}
```

PROBLEM 9

The reported **x*** = (12.277, 4.632, 0.313, 2.029) with *J** = 0.0075. *NMinimize* finds a slightly better *J** but at a different **x***; however it needs limits imposed on the independent variables to find the solution. *FindMinimum* fails to find the solution.

```
In[206]:= Clear[x1, x2, x3, x4, J];
       c = {0.1, 1, 2, 3, 4, 5, 6, 7, 8, 9, 10, 11, 12, 13, 14, 15, 16, 17, 18};
       yobs = {0.00189, 0.1038, 0.268, 0.506, 0.577, 0.604, 0.725, 0.898, 0.947,
          0.845, 0.702, 0.528, 0.385, 0.257, 0.159, 0.0869, 0.0453, 0.01509, 0.00189};
       b = x3 + (1 - x3) * x4;
       ycal = Table[x3 * b^x2 * (x2 / 6.2832)^.5 *
          (c[[i]] / 7.658)^(x2 - 1) * Exp[x2 - b * c[[i]] * x2 / (7.658)] / (1 + 1 / (12 * x2)) +
          (1 - x3) * (b / x4)^x1 * (x1 / 6.2832)^.5 * (c[[i]] / 7.658)^(x1 - 1) *
          Exp[x1 - b * c[[i]] * x1 / (7.658 * x4)] / (1 + 1 / (12 * x1)), {i, 1, 19}];
       J = Sum[(ycal[[i]] - yobs[[i]])^2, {i, 1, 19}];
       NMinimize[{J,
          x3 + (1 - x3) * x4 ≥ 0,
          x4 ≥ 0,
          x3 ≤ 1}, {{x1, .001, 5}, {x2, .001, 5}, {x3, 0, 1}, {x4, .001, 5}}]

Out[212]= {0.00749846, {x1 → 12.277, x2 → 4.63175, x3 → 0.312865, x4 → 2.02928}}
```

```
In[213]:= Clear[x1, x2, x3, x4, J];
       c = {0.1, 1, 2, 3, 4, 5, 6, 7, 8, 9, 10, 11, 12, 13, 14, 15, 16, 17, 18};
       yobs = {0.00189, 0.1038, 0.268, 0.506, 0.577, 0.604, 0.725, 0.898, 0.947,
          0.845, 0.702, 0.528, 0.385, 0.257, 0.159, 0.0869, 0.0453, 0.01509, 0.00189};
       b = x3 + (1 - x3) * x4;
       ycal = Table[x3 * b^x2 * (x2 / 6.2832)^.5 *
          (c[[i]] / 7.658)^(x2 - 1) * Exp[x2 - b * c[[i]] * x2 / (7.658)] / (1 + 1 / (12 * x2)) +
          (1 - x3) * (b / x4)^x1 * (x1 / 6.2832)^.5 * (c[[i]] / 7.658)^(x1 - 1) *
          Exp[x1 - b * c[[i]] * x1 / (7.658 * x4)] / (1 + 1 / (12 * x1)), {i, 1, 19}];
       J = Sum[(ycal[[i]] - yobs[[i]])^2, {i, 1, 19}];
       FindMinimum[{J,
          x3 + (1 - x3) * x4 ≥ 0,
          x4 ≥ 0,
          x3 ≤ 1}, {x1, x2, x3, x4}]

FindMinimum::eit :
    The algorithm does not converge to the tolerance of 4.806217383937354`*^-6 in 500 iterations. The
      best estimated solution, with feasibility residual, KKT residual, or complementary
      residual of {2.15438×10⁻⁹, 0.00218159, 4.96106×10⁻¹³}, is returned. ≫

Out[219]= {0.0510198, {x1 → 36.5487, x2 → 3.53205, x3 → 0.766879, x4 → 1.24751}}
```

PROBLEM 10

The reported solution is $\mathbf{x}^* = (0.3000, 0.3335, 0.4000, 0.4285, 0.224)$ with $J^* = -32.349$. *NMinimize* and *FindMinimum* fail to find a solution.

```
In[220]:= b = {-40, -2, -.25, -4, -4, -1, -40, -60, 5, 1};
          e = {-15, -27, -36, -18, -12};
          c = {{30, -20, -10, 32, -10},
               {-20, 39, -6, -31, 32},
               {-10, -6, 10, -6, -10},
               {32, -31, -6, 39, -20},
               {-10, 32, -10, -20, 30}};
          d = {4, 8, 10, 6, 2};
          a = {{-16, 2, 0, 1, 0},
               {0, -2, 0, .4, 2},
               {-3.5, 0, 2, 0, 0},
               {0, -2, 0, -4, -1},
               {0, -9, -2, 1, -2.8},
               {2, 0, -4, 0, 0},
               {-1, -1, -1, -1, -1},
               {-1, -2, -3, -2, -1},
               {1, 2, 3, 4, 5},
               {1, 1, 1, 1, 1}};
          Clear[J, x1, x2, x3, x4, x5]
          J[x_] := Sum[e[[i]] * x[[i]], {i, 5}] +
            Sum[Sum[c[[i, j]] * x[[i]] * x[[j]], {j, 5}] , {i, 5}] + Sum[d[[i]] * x[[i]]^3, {i, 5}]
          Jf = J[{x1, x2, x3, x4, x5}];
          constr = a.{x1, x2, x3, x4, x5};
          NMinimize[{Jf, constr[[1]] ≥ 0, constr[[2]] ≥ 0, constr[[3]] ≥ 0, constr[[4]] ≥ 0,
            constr[[5]] ≥ 0, x1 ≥ 0, x2 ≥ 0, x3 ≥ 0, x4 ≥ 0, x5 ≥ 0}, {x1, x2, x3, x4, x5}]
```

NMinimize::incst : NMinimize was unable to generate any initial points satisfying the inequality constraints
$\{3.5 x1 - 2 x3 \le 0, 16 x1 - 2 x2 - x4 \le 0, 2 x2 - 0.4 x4 - 2 x5 \le 0, 2 x2 + 4 x4 + x5 \le 0, 9 x2 + 2 x3 - x4 + 2.8 x5 \le 0\}$. The initial region specified may not contain any feasible points.
Changing the initial region or specifying explicit initial points may provide a better solution. ≫

NMinimize::cvmit : Failed to converge to the requested accuracy or precision within 100 iterations. ≫

Out[229]= $\{-1.57228 \times 10^{-7}, \{x1 \to 0., x2 \to 0., x3 \to 4.36745 \times 10^{-9}, x4 \to 0., x5 \to 0.\}\}$

```
In[230]:= b = {-40, -2, -.25, -4, -4, -1, -40, -60, 5, 1};
          e = {-15, -27, -36, -18, -12};
          c = {{30, -20, -10, 32, -10},
               {-20, 39, -6, -31, 32},
               {-10, -6, 10, -6, -10},
               {32, -31, -6, 39, -20},
               {-10, 32, -10, -20, 30}};
          d = {4, 8, 10, 6, 2};
          a = {{-16, 2, 0, 1, 0},
               {0, -2, 0, .4, 2},
               {-3.5, 0, 2, 0, 0},
               {0, -2, 0, -4, -1},
               {0, -9, -2, 1, -2.8},
               {2, 0, -4, 0, 0},
               {-1, -1, -1, -1, -1},
               {-1, -2, -3, -2, -1},
               {1, 2, 3, 4, 5},
               {1, 1, 1, 1, 1}};
          Clear[J, x1, x2, x3, x4, x5]
          J[x_] := Sum[e[[i]] * x[[i]], {i, 5}] +
            Sum[Sum[c[[i, j]] * x[[i]] * x[[j]], {j, 5}] , {i, 5}] + Sum[d[[i]] * x[[i]]^3, {i, 5}]
          Jf = J[{x1, x2, x3, x4, x5}];
          constr = a.{x1, x2, x3, x4, x5};
          FindMinimum[{Jf, constr[[1]] ≥ 0, constr[[2]] ≥ 0, constr[[3]] ≥ 0, constr[[4]] ≥ 0,
            constr[[5]] ≥ 0, x1 ≥ 0, x2 ≥ 0, x3 ≥ 0, x4 ≥ 0, x5 ≥ 0}, {x1, x2, x3, x4, x5}]
```

Out[239]= $\{-2.13615 \times 10^{-8}, \{x1 \to 5.21699 \times 10^{-11}, x2 \to 2.16936 \times 10^{-11},$
$x3 \to 4.20407 \times 10^{-10}, x4 \to 1.29734 \times 10^{-10}, x5 \to 2.10282 \times 10^{-10}\}\}$

PROBLEM 11

Mathematica finds the documented solution.

```
In[240]:=  J = 5.3578547 * x3^2 + 0.8356891 * x1 * x5 + 37.293239 * x1 - 40792.141;
           c1 = 85.334407 + 0.0056858 * x2 * x5 + 0.0006262 * x1 * x4 - 0.0022053 * x3 * x5;
           c2 = 80.51249 + 0.0071317 * x2 * x5 + 0.0029955 * x1 * x2 + 0.0021813 * x3^2;
           c3 = 9.300961 + 0.0047026 * x3 * x5 + 0.0012547 * x1 * x3 + 0.0019085 * x3 * x4;
           NMinimize[{J, c1 ≥ 0, c1 ≤ 92, c2 ≥ 90, c2 ≤ 110, c3 ≥ 20, c3 ≤ 25, x1 ≥ 78, x1 ≤ 102,
               x2 ≥ 33, x2 ≤ 45, x3 ≥ 27, x3 ≤ 45, x4 ≥ 27, x4 ≤ 45, x5 ≥ 27, x5 ≤ 45},
             {{x1, 78, 102}, {x2, 33, 45}, {x3, 27, 45}, {x4, 27, 45}, {x5, 27, 45}}]

Out[244]=  {-30665.5, {x1 → 78., x2 → 33., x3 → 29.9953, x4 → 45., x5 → 36.7758}}
```

PROBLEM 12

The reported solution is $x^* = (705.06, 68.600, 102.900, 282.341, 35.627)$ with $J^* = 1.905$. The documented equations fail to evaluate to J^* at x^*, so a *Mathematica* solution was not attempted.

```
In[245]:=  Clear[x1, x2, x3, x4, x5];
           y1 = x2 + x3 + 41.6;
           c1 = 0.024 * x4 - 4.62;
           y2 = 12.5 / c1 + 12;
           c2 = 0.0003535 * x1^2 + 0.5311 * x1 + 0.08705 * y2 * x1;
           c3 = 0.052 * x1 + 78 + .002377 * y2 * x1;
           y3 = c2 / c3;
           y4 = 19 * y3;
           c4 = 0.04782 * (x1 - y3) + 0.19556 * (x1 - y3)^2 / x2 + 0.6376 * y4 + 1.594 * y3;
           c5 = 100 * x2;
           c6 = x1 - y3 - y4;
           c7 = 0.950 - c4 / c5;
           y5 = c6 * c7;
           y6 = x1 - y5 - y4 - y3;
           c8 = (y5 + y4) * 0.995;
           y7 = c8 / y1;
           y8 = c8 / 3798;
           c9 = y7 - 0.0663 * y7 / y8 - 0.3153;
           y9 = 96.82 / c9 + 0.321 * y1;
           y10 = 1.29 * y5 + 1.258 * y4 + 2.29 * y3 + 1.71 * y6;
           y11 = 1.71 * x1 - 0.452 * y4 + 0.580 * y3;
           c10 = 12.3 / 752.3;
           c11 = (1.75 * y2) * (0.995 * x1);
           c12 = 0.995 * y10 + 1998;
           y12 = c10 * x1 + c11 / c12;
           y13 = c12 - 1.75 * y2;
           y14 = 3623 + 64.4 * x2 + 58.4 * x3 + 146312 / (y9 + x5);
           c13 = 0.995 * y10 + 60.8 * x2 + 48 * x4 - 0.1121 * y14 - 5095;
           y15 = y13 / c13;
           y16 = 148000 - 331000 * y15 + 40 * y13 - 61 * y15 * y13;
           c14 = 2324 * y10 - 28740000 * y2;
           y17 = 14130000 - 1328 * y10 - 531 * y11 + c14 / c12;
           c15 = y13 / y15 - y13 / 0.52;
           c16 = 1.104 - 0.72 * y15;
           c17 = y9 + x5;
           J = 0.0000005843 * y17 - 0.000117 * y14 - 0.1365 - 0.00002358 * y13 -
               0.000001502 * y16 - 0.0321 * y12 - 0.004324 * y5 - 0.0001 c15 / c6 - 37.48 * y2 / c12;
           x1 = 705.060;
           x2 = 68.6;
           x3 = 102.9;
           x4 = 282.341;
           x5 = 35.627;
           J

Out[286]=  2.58216
```

PROBLEM 13

The reported solution is $\mathbf{x}^* = $ (4.538, 2.400, 60.000, 9.300, 7.000) with $J^* = 5280.254$. The documented equations fail to evaluate to J^* at $\mathbf{x}^*$, so a *Mathematica* solution was not attempted.

```
In[287]:= k1 = -145 421.402;
          k2 = 2931.1506;
          k3 = -40.427932;
          k4 = 5106.192;
          k5 = 15 711.36;
          k6 = -161 622.577;
          k7 = 4176.15328;
          k8 = 2.8260078;
          k9 = 9200.476;
          k10 = 13 160.295;
          k11 = -21 686.9194;
          k12 = 123.56928;
          k13 = -21.1188894;
          k14 = 706.834;
          k15 = 2898.573;
          k16 = 28 298.388;
          k17 = 60.81096;
          k18 = 31.242116;
          k19 = 329.574;
          k20 = -2882.082;
          k21 = 74 095.3845;
          k22 = -306.262544;
          k23 = 16.243649;
          k24 = -3094.252;
          k25 = -5566.2628;
          k26 = -26 237;
          k27 = 99;
          k28 = -0.42;
          k29 = 1300;
          k30 = 2100;
          k31 = 925 548.252;
          k32 = -61 968.8432;
          k33 = 23.3088196;
          k34 = -27 097.648;
          k35 = -50 843.766;
          x6 = (k1 + k2 * x2 + k3 * x3 + k4 * x4 + k5 * x5) * x1;
          y1 = k6 + k1 * x2 + k8 * x3 + k9 * x4 + k10 * x5;
          y2 = k11 + k12 * x2 + k13 * x3 + k14 * x4 + k15 * x5;
          y3 = k16 + k17 * x2 + k18 * x3 + k19 * x4 + k20 * x5;
          y4 = k21 + k22 * x2 + k23 * x3 + k24 * x4 + k25 * x5;
          x7 = (y1 + y2 + y3) * x1;
          x8 = (k26 + k27 * x2 + k28 * x3 + k29 * x4 + k30 * x5) * x1 + x6 + x7;
          J = (50 * y1 + 9.583 * y2 + 20 * y3 + 15 * y4 - 852 960 - 38 100 * (x2 + 0.01 * x3) +
               k31 + k32 * x2 + k33 * x3 + k34 * x4 + k35 * x5) * x1 - 24 345 + 15 * x6;
          Clear[x1, x2, x3, x4, x5]
          x1 = 2.52;
          x2 = 2;
          x3 = 37.5;
          x4 = 9.25;
          x5 = 6.8;
          J

Out[336]= -2.47486 × 10^11
```

Problem 14

The reported solution is $\mathbf{x}^* = $ (11884, 3288, 20000, 4000, 114.18, −155.03) with $J^* = $ 250799.9. The documented equations fail to evaluate to J^* at $\mathbf{x}^*$, so a *Mathematica* solution was not attempted.

```
In[337]:= Clear[J, x1, x2, x3, x4, x5, x6, c]
         c[x_] := 2.7 * x + 1300;
         T1 = (0.0285 * x1 + 300) / (1 + 0.0001425 * x1);
         t1 = 500 - T1;
         a2 = -0.0001665 * x2;
         T2 = (200 - 350 * Exp[-a2]) / (1 - 1.5 * Exp[-a2]);
         t2 = 300 + (200 - T2) * Exp[a2];
         a3 = (0.085) * (9.36) * 10^-5 * x3;
         T3 = (t1 + (29.75 - t1) * Exp[-a3]) / (1 - 0.915 * Exp[-a3]);
         t3 = 350 + (t1 - T3) * Exp[a3];
         a4 = 0.00025 * x4;
         T4 = (t2 + (70 - t2) * Exp[-a4]) / (1 - 0.8 * Exp[-a4]);
         t4 = 350 + (t2 - T4) * Exp[a4];
         Tj2 = 0.8 * T3 + 0.2 * T4;
         a5 = 0.000375 * x5;
         T5 = 80 + (Tj2 - 80) * Exp[-a5];
         Tj1 = 0.7 * T1 + 0.3 * T2;
         a6 = 0.0003 * x6;
         T6 = 80 + (Tj1 - 80) * Exp[-a6];
         J = c[x1] + c[x2] + c[x3] + c[x4] + 100 * c[x5] + 100 * c[x6];
         x1 = 11 884;
         x2 = 3288;
         x3 = 20 000;
         x4 = 4000;
         x5 = 114.18;
         x6 = -155.03;
         J

Out[363]= 359 935.
```

Problem 15

Different results are reported depending on the numerical precision used. For high precision, the reported result is $\mathbf{x}^* = $ (10.7.81, 196.32, 373.83, 420.00, 21.31, 0.153) with $J^* = $ 8927.5888. The documented equations fail to evaluate to J^* at $\mathbf{x}^*$, and the documented constraints are not satisfied at $\mathbf{x}^*$, so a *Mathematica* solution was not attempted.

```
In[364]:= Clear[J, x1, x2, x3, x4, x5, x6, c1, c2, c3, c4]
          f1[x_] := If[x ≥ 0 && x < 300, 30*x, If[x ≥ 300 && x ≤ 400, 31*x]];
          f2[x_] := If[x ≥ 0 && x < 100, 28*x,
              If[x ≥ 100 && x < 200, 29*x, If[x ≥ 200 && x ≤ 1000, 30*x2]]];
          J = f1[x1] + f2[x2];
          c1 = 300 - x3*x4*Cos[1.48577 - x6] / 131.078 + 0.90798*x3^2*Cos[1.47588] - x1;
          c2 = -x3*x4*Cos[1.48477 + x6] / 131.078 + 0.90798*x4^2*Cos[1.47588] / 131.078 - x2;
          c3 = -x3*x4*Sin[1.48477 + x6] + 0.90798*x4^2*Sin[1.47588] / 131.078 - x5;
          c4 = 200 - x3*x4*Sin[1.48477 - x6] / 131.078 + 0.90798*x3^2*Sin[1.47588] / 131.078;
          x1 = 107.81;
          x2 = 196.32;
          x3 = 373.83;
          x4 = 420;
          x5 = 21.31;
          x6 = 0.153;
          J
          c1
          c2
          c3
          c4
Out[378]= 8927.58

Out[379]= 11 935.5

Out[380]= -0.350472

Out[381]= -155 461.

Out[382]= -0.0843339
```

PROBLEM 16

The reported solution is $\mathbf{x}^* = (0.9971, -0.0758, 0.5530, 0.8331, 0.9981, -0.0623, 0.5642, 0.8256, 0.0000024)$ with $J^* = 0.8660$. *Mathematica* returns the same value for J^* to 4 digits, but at a different $\mathbf{x}^*$.

```
In[383]:= Clear[x1, x2, x3, x4, x5, x6, x7, x8, x9];
          J = 0.5*(x1*x4 - x2*x3 + x3*x9 - x5*x9 + x5*x8 - x6*x7);
          NMaximize[{J,
              1 - x3^2 - x4^2 ≥ 0,
              1 - x9^2 ≥ 0,
              1 - x5^2 - x6^2 ≥ 0,
              1 - x1^2 - (x2 - x9)^2 ≥ 0,
              1 - (x1 - x5)^2 - (x2 - x6)^2 ≥ 0,
              1 - (x1 - x7)^2 - (x2 - x8)^2 ≥ 0,
              1 - (x3 - x5)^2 - (x4 - x6)^2 ≥ 0,
              1 - (x3 - x7)^2 - (x4 - x8)^2 ≥ 0,
              1 - x7^2 - (x8 - x9)^2 ≥ 0,
              x1*x4 - x2*x3 ≥ 0,
              x3*x9 ≥ 0,
              -x5*x9 ≥ 0,
              x5*x8 - x6*x7 ≥ 0,
              x9 ≥ 0}, {x1, x2, x3, x4, x5, x6, x7, x8, x9}]
Out[384]= {0.866025, {x1 → -0.0422934, x2 → -0.526077, x3 → 0.844104, x4 → -0.53618,
              x5 → -0.0422934, x6 → -0.999105, x7 → 0.844104, x8 → -0.0631518, x9 → 0.473028}}
```

PROBLEM 17

The reported solution is $\mathbf{x}^* = (9.351, 9.351, 9.351, 9.351, 9.351, 9.351, 9.351, 9.351, 9.351, 9.351)$ with $J^* = -45.778$. *Mathematica* finds a slightly different solution.

```
In[408]:= Clear[J, x1, x2, x3, x4, x5, x6, x7, x8, x9, x10]
          J[x_] :=
              Sum[Log[x[[i]] - 2]^2 + Log[(10 - x[[i]])]^2, {i, 10}] - Product[x[[i]], {i, 10}]^.2;
          Jf = J[{x1, x2, x3, x4, x5, x6, x7, x8, x9, x10}];
          NMinimize[{Jf, x1 > 2.001, x1 < 9.999, x2 > 2.001, x2 < 9.999, x3 > 2.001,
              x3 < 9.999, x4 > 2.001, x4 < 9.999, x5 > 2.001, x5 < 9.999, x6 > 2.001,
              x6 < 9.999, x7 > 2.001, x7 < 9.999, x8 > 2.001, x8 < 9.999, x9 > 2.001,
              x9 < 9.999, x10 > 2.001, x10 < 9.999}, {x1, x2, x3, x4, x5, x6, x7, x8, x9, x10}]

Out[411]= {-45.7785, {x1 → 9.35027, x2 → 9.35027, x3 → 9.35027, x4 → 9.35027, x5 → 9.35027,
              x6 → 9.35027, x7 → 9.35027, x8 → 9.35027, x9 → 9.35027, x10 → 9.35027}}
```

Problem 18

The reported $\mathbf{x}^* = (0, 0, 5.174, 0, 3.0611, 11.8395, 0, 0, 0.1039, 0, 0.3, 0.3335, 0.4, 0.4283, 0.224)$ with $J^* = -32.386$. The documented equations fail to evaluate to J^* at $\mathbf{x}^*$, so a *Mathematica* solution was not attempted.

```
In[412]:= b = {-40, -2, -.25, -4, -4, -1, -40, -60, 5, 1};
          e = {-15, -27, -36, -18, -12};
          c = {{30, -20, -10, 32, -10},
               {-20, 39, -6, -31, 32},
               {-10, -6, 10, -6, -10},
               {32, -31, -6, 39, -20},
               {-10, 32, -10, -20, 30}};
          d = {4, 8, 10, 6, 2};
          a = {{-16, 2, 0, 1, 0},
               {0, -2, 0, .4, 2},
               {-3.5, 0, 2, 0, 0},
               {0, -2, 0, -4, -1},
               {0, -9, -2, 1, -2.8},
               {2, 0, -4, 0, 0},
               {-1, -1, -1, -1, -1},
               {-1, -2, -3, -2, -1},
               {1, 2, 3, 4, 5},
               {1, 1, 1, 1, 1}};
          J[x_] :=
              Sum[b[[i]] * x[[i]], {i, 10}] + Sum[c[[i, j]] * x[[10 + i]] * x[[10 + j]], {i, 5}, {j, 5}] -
              2 * Sum[d[[i]] * x[[10 + i]]^3, {i, 5}]
          Jf = J[{x1, x2, x3, x4, x5, x6, x7, x8, x9, x10, x11, x12, x13, x14, x15}];
          x1 = 0;
          x2 = 0;
          x3 = 5.1740;
          x4 = 0;
          x5 = 3.0611;
          x6 = 11.8395;
          x7 = 0;
          x8 = 0;
          x9 = 0.1039;
          x10 = 0;
          x11 = 0.3;
          x12 = 0.3335;
          x13 = 0.4;
          x14 = 0.4283;
          x15 = 0.2240;
          Jf

Out[434]= -23.5218
```

Problem 19

Mathematica finds the reported solution.

```
In[435]:= Clear[a, b, c, J, Jf, x1, x2, x3, x4, x5, x6,
     x7, x8, x9, x10, x11, x12, x13, x14, x15, x16, Constr]
a = Table[0, {i, 16}, {j, 16}];
a[[1, 1]] = 1;
a[[1, 4]] = 1;
a[[1, 7]] = 1;
a[[1, 8]] = 1;
a[[1, 16]] = 1;
a[[2, 2]] = 1;
a[[2, 3]] = 1;
a[[2, 7]] = 1;
a[[2, 10]] = 1;
a[[3, 3]] = 1;
a[[3, 7]] = 1;
a[[3, 9]] = 1;
a[[3, 10]] = 1;
a[[3, 14]] = 1;
a[[4, 4]] = 1;
a[[4, 7]] = 1;
a[[4, 11]] = 1;
a[[4, 15]] = 1;
a[[5, 5]] = 1;
a[[5, 6]] = 1;
a[[5, 10]] = 1;
a[[5, 12]] = 1;
a[[5, 16]] = 1;
a[[6, 6]] = 1;
a[[6, 8]] = 1;
a[[6, 15]] = 1;
a[[7, 7]] = 1;
a[[7, 11]] = 1;
a[[7, 13]] = 1;
a[[8, 8]] = 1;
a[[8, 10]] = 1;
a[[8, 15]] = 1;
a[[9, 9]] = 1;
a[[9, 12]] = 1;
a[[9, 16]] = 1;
a[[10, 10]] = 1;
a[[10, 14]] = 1;
a[[11, 11]] = 1;
a[[11, 13]] = 1;
a[[12, 12]] = 1;
a[[12, 14]] = 1;
a[[13, 13]] = 1;
a[[13, 14]] = 1;
a[[14, 14]] = 1;
a[[15, 15]] = 1;
a[[16, 16]] = 1;
```

```
b = Table[0, {i, 1, 8}, {j, 1, 16}];
b[[1, 1]] = 0.22;
b[[1, 2]] = 0.2;
b[[1, 3]] = 0.19;
b[[1, 4]] = 0.25;
b[[1, 5]] = 0.15;
b[[1, 6]] = 0.11;
b[[1, 7]] = 0.12;
b[[1, 8]] = 0.13;
b[[1, 9]] = 1;
b[[2, 1]] = -1.46;
b[[2, 3]] = -1.3;
b[[2, 4]] = 1.82;
b[[2, 5]] = -1.15;
b[[2, 7]] = 0.8;
b[[2, 10]] = 1;
b[[3, 1]] = 1.29;
b[[3, 2]] = -0.89;
b[[3, 5]] = -1.16;
b[[3, 6]] = -0.96;
b[[3, 8]] = -0.49;
b[[3, 11]] = 1;
b[[4, 1]] = -1.1;
b[[4, 2]] = -1.06;
b[[4, 3]] = 0.95;
b[[4, 4]] = -0.54;
b[[4, 6]] = -1.78;
b[[4, 7]] = -0.41;
b[[4, 12]] = 1;
b[[5, 4]] = -1.43;
b[[5, 5]] = 1.51;
b[[5, 6]] = 0.59;
b[[5, 7]] = -0.33;
b[[5, 8]] = -0.43;
b[[5, 13]] = 1;
b[[6, 2]] = -1.72;
b[[6, 3]] = -0.33;
b[[6, 5]] = 1.62;
b[[6, 6]] = 1.24;
b[[6, 7]] = 0.21;
b[[6, 8]] = -0.26;
b[[6, 14]] = 1;
b[[7, 1]] = 1.12;
b[[7, 4]] = 0.31;
b[[7, 7]] = 1.12;
b[[7, 9]] = -0.36;
b[[7, 15]] = 1;
b[[8, 2]] = 0.45;
b[[8, 3]] = 0.26;
b[[8, 4]] = -1.1;
b[[8, 5]] = 0.58;
b[[8, 7]] = -1.03;
b[[8, 8]] = 0.10;
b[[8, 16]] = 1;
c = {2.5, 1.1, -3.1, -3.5, 1.3, 2.1, 2.3, -1.5};
J[x_] := - Sum[
    a[[i, j]] * (x[[i]]^2 + x[[i]] + 1) * (x[[j]]^2 + x[[j]] + 1), {j, 1, 16}, {i, 1, 16}];
Jf = J[{x1, x2, x3, x4, x5, x6, x7, x8, x9, x10, x11, x12, x13, x14, x15, x16}];
Constr = b.{x1, x2, x3, x4, x5, x6, x7, x8, x9, x10, x11, x12, x13, x14, x15, x16} - c;
NMaximize[{Jf, Constr[[1]] == 0, Constr[[2]] == 0, Constr[[3]] == 0,
    Constr[[4]] == 0, Constr[[5]] == 0, Constr[[6]] == 0, Constr[[7]] == 0,
    Constr[[8]] == 0, x1 ≥ 0, x2 ≥ 0, x3 ≥ 0, x4 ≥ 0, x5 ≥ 0, x6 ≥ 0, x7 ≥ 0,
    x8 ≥ 0, x9 ≥ 0, x10 ≥ 0, x11 ≥ 0, x12 ≥ 0, x13 ≥ 0, x14 ≥ 0, x15 ≥ 0,
    x16 ≥ 0, x1 ≤ 5, x2 ≤ 5, x3 ≤ 5, x4 ≤ 5, x5 ≤ 5, x6 ≤ 5, x7 ≤ 5, x8 ≤ 5,
    x9 ≤ 5, x10 ≤ 5, x11 ≤ 5, x12 ≤ 5, x13 ≤ 5, x14 ≤ 5, x15 ≤ 5, x16 ≤ 5},
    {x1, x2, x3, x4, x5, x6, x7, x8, x9, x10, x11, x12, x13, x14, x15, x16}]
```

Out[541]= $\{-244.9, \{x1 \to 0.0398474, x2 \to 0.791983, x3 \to 0.20287,$
$x4 \to 0.844358, x5 \to 1.26991, x6 \to 0.934739, x7 \to 1.68196, x8 \to 0.155301,$
$x9 \to 1.56787, x10 \to 1.9455 \times 10^{-11}, x11 \to 3.02317 \times 10^{-11}, x12 \to 1.6275 \times 10^{-10},$
$x13 \to 0.660204, x14 \to 7.37698 \times 10^{-11}, x15 \to 0.674256, x16 \to 1.22892 \times 10^{-11}\}\}$

PROBLEM 20

Three solutions are reported. For the one with no constraints violated, $\mathbf{x}^* = (7.804$ 1^{-3}, $1.121\ 10^{-1}$, $1.136\ 10^{-1}$, 0, 0, 0, $6.609\ 10^{-2}$, 0, 0, 0, $1.914\ 10^{-2}$, $6.009\ 10^{-3}$, 5.008 10^{-2}, $1.844\ 10^{-1}$, $2.693\ 10^{-1}$, 0, 0, 0, $1.704\ 10^{-1}$, 0, 0, 0, $8.453\ 10^{-4}$, $1.98\ 10^{-4}$), with $J^* = 0.057$. The documented equations fail to evaluate to J^* at $\mathbf{x}^*$, so a *Mathematica* solution was not attempted.

```
In[542]:= Clear[ a, J, x1, x2, x3, x4, x5, x6, x7, x8, x9, x10, x11,
         x12, x13, x14, x15, x16, x17, x18, x19, x20, x21, x22, x23, x24];
       a = {0.0693, 0.0577, 0.05, 0.2, 0.26, 0.55, 0.06, 0.10, 0.12, 0.18, 0.10, 0.09,
            0.0693, 0.0577, 0.05, 0.20, 0.26, 0.55, 0.06, 0.10, 0.12, 0.18, 0.10, 0.09};
       J = a.{x1, x2, x3, x4, x5, x6, x7, x8, x9, x10, x11, x12, x13,
            x14, x15, x16, x17, x18, x19, x20, x21, x22, x23, x14};
       x1 = 7.804^10^-3;
       x2 = 1.121 * 10^-1;
       x3 = 1.136 * 10^-1;
       x4 = 0;
       x5 = 0;
       x6 = 0;
       x7 = 6.609 * 10^-2;
       x8 = 0;
       x9 = 0;
       x10 = 0;
       x11 = 1.914 * 10^-2;
       x12 = 6.009 * 10^-3;
       x13 = 5.008 * 10^-2;
       x14 = 1.844 * 10^-1;
       x15 = 2.693 * 10^-1;;
       x16 = 0;
       x17 = 0;
       x18 = 0;
       x19 = 1.704 * 10^-1;
       x20 = 0;
       x21 = 0;
       x22 = 0;
       x23 = 8.453 * 10^-4;
       x24 = 1.98 * 10^-4;
       J

Out[569]= 0.142491
```

PROBLEM 21

The reported solution is $\mathbf{x}^* = (50, 25, 1.5)$ with $J^* = 0$. The documented equations fail to evaluate to J^* at $\mathbf{x}^*$, so a *Mathematica* solution was not attempted.

```
In[605]:= Clear[J, x1, x2, x3, x4, x5, x6, x7, x8, x9, x10, u];
       u[i_] := 25 + (-50 * Log[0.01 * i]) ^ (1 / 1.5);
       J = Sum[Exp[- (u[i] - x2) ^x3 / x1 - 0.01 * i] ^2, {i, 99}];
       x1 = 50;
       x2 = 25;
       x3 = 1.5;
       J

Out[611]= 8.01542
```

Problem 22

This problem contains four cases.

For case 1, **b** = (4.97, −1.88, −29.08, −78.02), the reported **x*** = (0, 0, 0, 0, 0, 0.00333) with J^* = 0.0156. *NMinimize* finds a different solution.

For case 2, **b** = (4.97, −1.88, −69.08, −118.02), the reported **x*** = (0, 0, 0, 0, 0. 0.00332) with J^* = 0.0156. *NMinimize* finds this solution.

For case 3, **b** = (32.97, 25.12, −29. 08, −78.02), the reported **x*** = (0, 0, 0.0633, 0, 0, 0.0134) with J^* = 4.070. *NMinimize* finds a similar solution while *FindMinimum* finds a better solution at a different point.

For case 4, **b** = (32.97, 25.12, −124.08, −173.02), the reported **x*** = (0, 0, 0.0633, 0, 0, 0.0134) with J^* = 4.070. *NMinimize* finds a better solution.

```
In[612]:= Clear[x1, x2, x3, x4, x5, x6]
J = 4.3 * x1 + 31.8 * x2 + 63.3 * x3 + 15.8 * x4 + 68.5 * x5 + 4.7 * x6;
c1 = 17.1 * x1 + 38.2 * x2 + 204.2 * x3 + 212.3 * x4 + 623.4 * x5 + 1495.5 * x6 -
     169 * x1 * x3 - 3580 * x3 * x5 - 3810 * x4 * x5 - 18 500 * x4 * x6 - 24 300 * x5 * x6;
c2 = 17.9 * x1 + 36.8 * x2 + 113.9 * x3 + 169.7 * x4 + 337.8 * x5 + 1385.2 * x6 -
     139 * x1 * x3 - 2450 * x4 * x5 - 16 600 * x4 * x6 - 17 200 * x5 * x6;
c3 = -273 * x2 - 70 * x4 - 819 * x5 + 26 000 * x4 * x5;
c4 = 159.9 * x1 - 311 * x2 + 587 * x4 + 391 * x5 + 2198 * x6 - 14 000 * x1 * x6;
```

The first case:

```
In[618]:= b = {4.97, -1.88, -29.08, -78.02};
NMinimize[{J, c1 ≥ b[[1]], c2 ≥ b[[2]], c3 ≥ b[[3]], c4 ≥ b[[4]], x1 ≥ 0,
    x1 ≤ 0.31, x2 ≥ 0, x2 ≤ 0.046, x3 ≥ 0, x3 ≤ 0.068, x4 ≥ 0, x4 ≤ 0.042,
    x5 ≥ 0, x5 ≤ 0.028, x6 ≥ 0, x6 ≤ 0.0134}, {{x1, 0, .31}, {x2, 0, 0.046},
    {x3, 0, .068}, {x4, 0, 0.042}, {x5, 0, 0.028}, {x6, 0, 0.0134}}]
```

$Out[619]=$ $\{0.0156196, \{x1 \to 1.04181 \times 10^{-9}, x2 \to 1.70474 \times 10^{-10},$
$x3 \to 7.26881 \times 10^{-11}, x4 \to 3.00529 \times 10^{-10}, x5 \to 9.61106 \times 10^{-11}, x6 \to 0.00332331\}\}$

The second case:

```
In[620]:= b = {4.97, -1.88, -69.08, -118.02};
NMinimize[{J, c1 ≥ b[[1]], c2 ≥ b[[2]], c3 ≥ b[[3]], c4 ≥ b[[4]], x1 ≥ 0,
    x1 ≤ 0.31, x2 ≥ 0, x2 ≤ 0.046, x3 ≥ 0, x3 ≤ 0.068, x4 ≥ 0, x4 ≤ 0.042,
    x5 ≥ 0, x5 ≤ 0.028, x6 ≥ 0, x6 ≤ 0.0134}, {{x1, 0, .31}, {x2, 0, 0.046},
    {x3, 0, .068}, {x4, 0, 0.042}, {x5, 0, 0.028}, {x6, 0, 0.0134}}]
```

$Out[621]=$ $\{0.0156196, \{x1 \to 2.34224 \times 10^{-9}, x2 \to 3.1394 \times 10^{-10},$
$x3 \to 1.58727 \times 10^{-10}, x4 \to 6.49753 \times 10^{-10}, x5 \to 1.49028 \times 10^{-10}, x6 \to 0.00332331\}\}$

The third case:

```
In[626]:= b = {32.97, 25.12, -29.08, -78.02};
NMinimize[{J, c1 ≥ b[[1]], c2 ≥ b[[2]], c3 ≥ b[[3]], c4 ≥ b[[4]], x1 ≥ 0,
    x1 ≤ 0.31, x2 ≥ 0, x2 ≤ 0.046, x3 ≥ 0, x3 ≤ 0.068, x4 ≥ 0, x4 ≤ 0.042,
    x5 ≥ 0, x5 ≤ 0.028, x6 ≥ 0, x6 ≤ 0.0134}, {{x1, 0, .31}, {x2, 0, 0.046},
    {x3, 0, .068}, {x4, 0, 0.042}, {x5, 0, 0.028}, {x6, 0, 0.0134}}]
```

$Out[627]=$ $\{4.07125, \{x1 \to 1.0962 \times 10^{-9}, x2 \to 3.07519 \times 10^{-10},$
$x3 \to 0.0633217, x4 \to 4.16841 \times 10^{-10}, x5 \to 1.11202 \times 10^{-10}, x6 \to 0.0134\}\}$

```
In[628]:= b = {32.97, 25.12, -29.08, -78.02};
        FindMinimum[{J, c1 ≥ b[[1]], c2 ≥ b[[2]], c3 ≥ b[[3]], c4 ≥ b[[4]], x1 ≥ 0,
            x1 ≤ 0.31, x2 ≥ 0, x2 ≤ 0.046, x3 ≥ 0, x3 ≤ 0.068, x4 ≥ 0, x4 ≤ 0.042,
            x5 ≥ 0, x5 ≤ 0.028, x6 ≥ 0, x6 ≤ 0.0134}, {x1, x2, x3, x4, x5, x6}]
```

$$\text{Out[629]= }\{3.13581, \{x1 \to 0.268565, x2 \to 1.05227 \times 10^{-9},$$
$$x3 \to 4.58734 \times 10^{-10}, x4 \to 4.24348 \times 10^{-10}, x5 \to 0.028, x6 \to 0.0134\}\}$$

The fourth case:

```
b = {32.97, 25.12, -124.08, -173.02};
NMinimize[{J, c1 ≥ b[[1]], c2 ≥ b[[2]], c3 ≥ b[[3]], c4 ≥ b[[4]], x1 ≥ 0,
    x1 ≤ 0.31, x2 ≥ 0, x2 ≤ 0.046, x3 ≥ 0, x3 ≤ 0.068, x4 ≥ 0, x4 ≤ 0.042,
    x5 ≥ 0, x5 ≤ 0.028, x6 ≥ 0, x6 ≤ 0.0134}, {{x1, 0, .31}, {x2, 0, 0.046},
    {x3, 0, .068}, {x4, 0, 0.042}, {x5, 0, 0.028}, {x6, 0, 0.0134}}]
```

$$\{3.13581, \{x1 \to 0.268565, x2 \to 8.14685 \times 10^{-11},$$
$$x3 \to 3.69363 \times 10^{-11}, x4 \to 3.53055 \times 10^{-11}, x5 \to 0.028, x6 \to 0.0134\}\}$$

Problem 23

Two different solutions are reported, each with $J^* = 1732$. The documented equations fail to evaluate to J^* at $\mathbf{x}^*$, so a *Mathematica* solution was not attempted.

```
In[630]:= Clear[J, Jf, b, u, c, a, x, x11, x12, x13, x14, x15, x21, x22, x23, x24, x25, x31, x32,
        x33, x34, x35, x41, x42, x43, x44, x45, x51, x52, x53, x54, x55, x61, x62,
        x63, x64, x65, x71, x72, x73, x74, x75, x81, x82, x83, x84, x85, x91, x92,
        x93, x94, x95, x101, x102, x103, x104, x105, x111, x112, x113, x114, x115,
        x121, x122, x123, x124, X125,
        x131, x132, x133, x134, x135,
        x141, x142, x143, x144, x145,
        x151, x152, x153, x154, X155,
        x161, x162, x163, x164, x165,
        x171, x172, x173, x174, x175,
        x181, x182, x183, x184, x185,
        x191, x192, x193, x194, x195, x201, x202, x203, x204, x205];
```

```
In[631]:= b = {30, 0, 0, 0, 0, 100, 0, 0, 0, 40, 0, 0, 0, 50, 70, 35, 0, 0, 0, 10};
          u = {60, 50, 50, 75, 40, 60, 35, 30,
             25, 150, 30, 45, 125, 200, 200, 130, 100, 100, 100, 150};
          c = {200, 100, 300, 150, 250};
          a = {{1, .84, .96, 1, .92},
               {.95, .83, .95, 1, .94},
               {1, .85, .96, 1, .92},
               {1, .84, .96, 1, .95},
               {1, .85, .96, 1, .95},
               {.85, .81, .90, 1, .98},
               {.90, .81, .92, 1, .98},
               {.85, .82, .91, 1, 1},
               {.80, .80, .92, 1, 1},
               {1, .86, .95, .96, .90},
               {1, 1, .99, .91, .95},
               {1, .98, .98, .92, .96},
               {1, 1, .99, .91, .91},
               {1, .88, .98, .92, .98},
               {1, .87, .97, .98, .99},
               {1, .88, .98, .93, .99},
               {1, .85, .95, 1, 1},
               {.95, .84, .92, 1, 1},
               {1, .85, .93, 1, 1},
               {1, .85, .92, 1, 1}};
          J[x_] := Sum[u[[j]] * Product[a[[j, i]]^x[[j, i]] - 1, {i, 5}], {j, 20}];
```

```
In[636]:= Jf = J[{{x11, x12, x13, x14, x15},
                   {x21, x22, x23, x24, x25},
                   {x31, x32, x33, x34, x35},
                   {x41, x42, x43, x44, x45},
                   {x51, x52, x53, x54, x55},
                   {x61, x62, x63, x64, x65},
                   {x71, x72, x73, x74, x75},
                   {x81, x82, x83, x84, x85},
                   {x91, x92, x93, x94, x95},
                   {x101, x102, x103, x104, x105},
                   {x111, x112, x113, x114, x115},
                   {x121, x122, x123, x124, x125},
                   {x131, x132, x133, x134, x135},
                   {x141, x142, x143, x144, x145},
                   {x151, x152, x153, x154, x155},
                   {x161, x162, x163, x164, x165},
                   {x171, x172, x173, x174, x175},
                   {x181, x182, x183, x184, x185},
                   {x191, x192, x193, x194, x195},
                   {x201, x202, x203, x204, x205}}];
          Jf
```

```
Out[637]= 0
```

```
In[638]:= x = {{x11, x12, x13, x14, x15},
           {x21, x22, x23, x24, x25},
           {x31, x32, x33, x34, x35},
           {x41, x42, x43, x44, x45},
           {x51, x52, x53, x54, x55},
           {x61, x62, x63, x64, x65},
           {x71, x72, x73, x74, x75},
           {x81, x82, x83, x84, x85},
           {x91, x92, x93, x94, x95},
           {x101, x102, x103, x104, x105},
           {x111, x112, x113, x114, x115},
           {x121, x122, x123, x124, x125},
           {x131, x132, x133, x134, x135},
           {x141, x142, x143, x144, x145},
           {x151, x152, x153, x154, x155},
           {x161, x162, x163, x164, x165},
           {x171, x172, x173, x174, x175},
           {x181, x182, x183, x184, x185},
           {x191, x192, x193, x194, x195},
           {x201, x202, x203, x204, x205}};
       x21 = 24;
       x61 = 32;
       x71 = 37;
       x81 = 28;
       x91 = 22;
       x151 = 5;
       x181 = 52;
       x12 = 1;
       x22 = 8;
       x32 = 2;
       x42 = 18;
       x52 = 11;
       x152 = 29;
       x162 = 9;
       x172 = 21;
       x23 = 9;
       x53 = 29;
       x63 = 62;
       x153 = 35;
       x173 = 17;
       x183 = 25;
       x193 = 62;
       x203 = 60;
       x114 = 9;
       x124 = 39;
       x144 = 58;
       x164 = 44;
       x15 = 47;
       x25 = 5;
       x35 = 36;
       x45 = 12;
       x65 = 6;
       x105 = 50;
       x115 = 42;
       x135 = 51;
       x155 = 1;
       Jf

Out[675]= 0
```

PROBLEM 24

Mathematica finds the documented solution.

```
In[676]:= J = 4 * (x1 - 5) ^2 + (x2 - 6) ^2;
          NMinimize[J, {x1, x2}]
Out[677]= {0., {x1 → 5., x2 → 6.}}
```

PROBLEM 25

Mathematica finds the documented solution.

```
In[678]:= J = 4 * (x1 - 5) ^2 + (x2 - 6) ^2;
          NMinimize[J, {x1, x2}]
Out[679]= {0., {x1 → 5., x2 → 6.}}
```

PROBLEM 26

Mathematica finds the documented solution.

```
In[680]:= J = (x1 + 10 * x2) ^2 + 5 * (x3 - x4) ^2 + (x2 - 2 * x3) ^4 + 10 * (x1 - x4) ^4;
          NMinimize[J, {x1, x2, x3, x4}]
Out[681]= {7.68407 × 10^{-31},
          {x1 → 2.14964 × 10^{-8}, x2 → -2.14964 × 10^{-9}, x3 → 1.28029 × 10^{-8}, x4 → 1.28029 × 10^{-8}}}
```

PROBLEM 27

The reported solutions are all $J^* = 0$ with $\mathbf{x}^* = (1, \text{unbounded})$, $(0, \text{unbounded})$ or (unbounded, 0). *NMinimize* finds a J^* near 0, but cannot determine one variable may be unbounded.

```
In[682]:= J = (x1 * x2) ^2 * (1 - x1) ^2 * (1 - x1 - x2 * (1 - x1) ^5) ^2;
          NMinimize[J, {x1, x2}]
Out[683]= {1.74967 × 10^{-68}, {x1 → 1., x2 → 3.47142 × 10^{-8}}}
```

PROBLEM 28

The reported solutions are $\mathbf{x}^* = (3, 2)$ or $(3.58443, -1.84813)$ with $J^* = 0$. *NMinimize* finds one solution, but not the other, unless force $\mathbf{x2} < 0$.

```
In[684]:= J = (x1^2 + x2 - 11) ^2 + (x1 + x2^2 - 7) ^2;
          NMinimize[J, {x1, x2}]
Out[685]= {0., {x1 → 3., x2 → 2.}}
```

Using Random search from 2000 starting points doesn't find the other solution.

In[686]:= **NMinimize[J, {x1, x2}, Method → {"RandomSearch", "SearchPoints" → 2000}]**

Out[686]= $\left\{1.97215 \times 10^{-31}, \{x1 \rightarrow -3.77931, x2 \rightarrow -3.28319\}\right\}$

Forcing **x**[[2]] to be less than 0 does find the other solution.

In[687]:= **NMinimize[{J, x2 < 0}, {x1, x2}]**

Out[687]= $\left\{4.93038 \times 10^{-32}, \{x1 \rightarrow 3.58443, x2 \rightarrow -1.84813\}\right\}$

PROBLEM 29

The reported solution is $\mathbf{x}^* = (-21.026653, -36.760090)$ with $J^* = 0$, but there is a local minimum $\mathbf{x}^* = (0.28581, 0.27936)$ with $J^* = 5.9225$. *NMinimize* with default settings finds the local minimum, even if force *NMinimize* to use the Simulated Annealing optimization method which is more likely to avoid local minimums. The global minimum is almost found only if constrain the search space for the independent variables. This can easily be done for two-dimensional problems because *Mathematica* offers elaborate three-dimensional and contour plot capabilities that enable one to visualize the likely location for the global minimum.

In[688]:= **J = (x1^2 + 12 * x2 - 1)^2 + (49 * x1^2 + 49 * x2^2 + 84 * x1 + 2324 * x2 - 681)^2;**
   NMinimize[J, {x1, x2}, Method → {"SimulatedAnnealing"}]

Out[689]= {5.92256, {x1 → 0.286328, x2 → 0.279301}}

Forcing **x**[[2]] to be less than 0 does find the other solution.

In[691]:= **NMinimize[J, {{x1, -1000, 1000}, {x2, -1000, 1000}}, Method → {"SimulatedAnnealing"}]**

Out[691]= $\left\{1.41089 \times 10^{-6}, \{x1 \rightarrow -21.0266, x2 \rightarrow -36.76\}\right\}$

PROBLEM 30

Mathematica finds the documented solution.

In[692]:= **J = 100 * (x3 - ((x1 + x2) / 2)^2)^2 + (1 - x1)^2 + (1 - x2)^2;**
   NMinimize[J, {x1, x2, x3}]

Out[693]= {0., {x1 → 1., x2 → 1., x3 → 1.}}

PROBLEM 31

The reported solution is $\mathbf{x}^* = (1.7954, 1.3779)$ with $J^* = 0.16904$ which is reported to be a local minimum. *NMinimize* finds the solution.

In[694]:= **J = (x1 - 2)^2 + (x2 - 1)^2 + 0.04 / (-x1^2 / 4 - x2^2 + 1) + (x1 - 2 * x2 + 1)^2 / .2;**
   NMinimize[J, {x1, x2}]

Out[695]= {0.169043, {x1 → 1.7954, x2 → 1.37786}}

PROBLEM 32

The reported solution is $\mathbf{x}^* = (2.714, 140.4, 1707, 31.51)$ with $J^* = 318.572$. The documented equations fail to evaluate to J^* at $\mathbf{x}^*$, so a *Mathematica* solution was not attempted.

```
In[896]:= Clear[a, b, x1, x2, x3, x4, J, Jf];
          a = {0, 0.000428, 0.001, 0.00161, 0.00209, 0.00348, 0.00525};
          b = {7.391, 11.18, 16.44, 16.2, 22.2, 24.02, 31.32};
          J[x_] := 10^4 * Sum[
              ((x[[1]]^2 + a[[i]] * x[[2]]^2 + a[[i]]^2 * x[[3]]^2) / (1 + a[[i]] * x[[4]]^2) - b[[i]]) /
              b[[i]], {i, 7}]
          J[{2.714, 140.4, 1707, 31.51}]

Out[700]= -343.575
```

PROBLEM 33

The reported solution is $\mathbf{x}^* = (0, 1)$ or $(0, {}^-1)$ with $J^* = 1.1.036$. But this cannot be correct as $Exp[{}^-3]$ is not the same as $Exp[3]$. So did not attempt a *Mathematica* solution.

PROBLEM 34

Mathematica finds the documented solution.

```
In[701]:= t = If[x1 > 0, ArcTan[x2 / x1] / (2 * Pi), .5 + ArcTan[x2 / x1] / (2 * Pi)];
          J = 100 * ((x3 - 10 * t)^2 + ((x1^2 + x2^2)^.5 - 1)^2) + x3^2;
          NMinimize[J, {x1, x2, x3}]

Out[703]= {4.98033 × 10^-39, {x1 → 1., x2 → -4.33201 × 10^-20, x3 → -6.991 × 10^-20}}
```

PROBLEM 35

Mathematica finds the documented solution.

```
In[704]:= J = (1.5 - x1 * (1 - x2))^2 + (2.25 - x1 * (1 - x2^2))^2 + (2.625 - x1 * (1 - x2^3))^2;
          NMinimize[J, {x1, x2}]

Out[705]= {4.93038 × 10^-32, {x1 → 3., x2 → 0.5}}
```

APPENDIX 3B: DATA FOR FUZZY LOGIC CONTROLLER FOR ROCKET

Our purpose is formulate a fuzzy closed-loop controller for the rocket problem, so that if the actual gravity, thrust, or thrust misalignment is not as we assumed in our model, we will get to the final desired state vector with less variance than if we fly an open-loop controller with the thrust steering constants fixed.

Our plan is to fly the original open-loop optimal **a** and **b** to the time decision gate, then based on how much our actual horizontal and vertical velocity differs from what the open-loop optimal solution predicted, use fuzzy rules to determine the new **a** and **b** to use the rest of the trajectory.

To formulate these rules we need training data. Let's try five trial values, **trialA**, **trialg**, and **trialthetabias**, for each uncertain parameter, **A**, **g**, and **thrustbias**, respectively, the nominal value, the two extremes and two intermediate values, as follows:

```
In[146]:= trialA = {.95 * A, (1 - (1 / 2) * .05) * A, A, (1 + (1 / 2) * .05) * A, 1.05 * A};
          trialg = {.99 * g, (1 - (1 / 2) * .01) * g, g, (1 + (1 / 2) * .01) * g, 1.01 * g};
          trialthetabias = {-.1 * Pi / 180, -.1 * Pi / 2 / 180, 0, .1 * Pi / 2 / 180, .1 * Pi / 180};
```

The formal requirements *FlatNoAtmConstTEOMrev* flies the nominal steering law to the decision gate time (chosen to be 3 seconds), then flies a commanded law the remainder until the **x3** = **U** stopping condition. Where the needed constants are

H = 50000
U = 5444
A = 20.82
g = 5.32

So we have at least two optional ways to implement the fuzzy controller.

Option 1 is to base the controller on fuzzy rules relating the estimated **A**, **g**, and **thetabias**, that we'll denote as: **Aest**, **gest,** and **biasest**, respectively. From the equations of motion we note, assuming the bias is small enough to be neglected:

$$\textbf{Aest} = \text{d}\textbf{x3}/\text{d}\textbf{t\_measured} / Cos[\textbf{theta}],$$

where

$$\text{d}\textbf{x3}/\text{d}\textbf{t\_measured} = \textbf{a\_measured} * Cos[\textbf{theta} + \textbf{thetabias}] + \textbf{Error}$$

and **theta** is the commanded thrust angle and **Error** is potentially a measurement or alignment bias or dither.

$$\textbf{gest} = \text{d}\textbf{x3}/\text{d}\textbf{t\_measured} * Tan[\textbf{theta}] - \text{d}\textbf{x4}/\text{d}\textbf{t\_measured,}$$

where

$$\text{d}\textbf{x4}/\text{d}\textbf{t\_measured} = \textbf{a\_measured} * Sin[\textbf{theta} + \textbf{thetabias}] + \textbf{Error}$$

Since the real acceleration

$$\textbf{a\_measured} = \textbf{A}\char`\^2 + \textbf{g}\char`\^2 - 2\textbf{A} * \textbf{g} * Sin[\textbf{theta} + \textbf{thetabias}]$$

and

$$Sin[\textbf{theta} + \textbf{thetabias}] = Sin[\textbf{theta}] + \textbf{thetabias} * Cos[\textbf{theta}]$$

For small **thetabias**, then

biasest = (**Aest**^2 + **gest**^2 – 2**Aest** * **gest** * *Sin*[**theta**] ⁻ **a_measured**) / (2**Aest** * **gest** * *Cos*[**theta**]

From t = 0 to t = **tgate**, we can obtain estimates for **Aest**, **gest**, and **biasest**, and use the mean or fuzzy value as our independent variables in fuzzy rules used to determine the new **a** and **b** needed from that point forward.

Option 2 bases the fuzzy rules for the new **a** and **b** directly on two fuzzy inputs:

$$(\textbf{x3\_measured}[[\textbf{tgate}]] - \textbf{x3nominal}[[\textbf{tgate}]])$$

and

$$(\textbf{x4\_measured}[[\textbf{tgate}]] - \textbf{x4\_nominal}[[\textbf{tgate}]]).$$

For this option, we need to integrate **a_measured** from **t** = 0 to **tgat**e and decompose the result into **x3** and **x4**. Since there will be both bias and dither errors in measuring the acceleration and the **x3** and **x4** obtained will also feel the effect of the as of yet unknown bias, the obtained difference in state vectors obtained at **tgate** will have some error.

Which approach to use is partly the determination of whether or not we prefer to integrate an accelerometer or estimate parameters based on noisy accelerometer data.

Usually, we prefer the integration, and it tends to eliminate dither errors, but not bias errors. But our choice is also dependent on whether we prefer to have fuzzy rules in terms of two independent inputs (difference in **x3** and **x4** measured from nominal values at **tgate**) or three independent variables (**Aest**, **gest**, and **biaest**). Usually, the smaller the number of independent variables, the better, provided excluding the third does not miss important information.

So at this point, one typically must get some data to see which approach is preferred. So we need to get some data combining the uncertainties to derive the fuzzy rules. Since we have three uncertainties and we can easily assume at least five trial values for each, we could evaluate 125 cases of values. But to keep the workload down we'll take advantage of the Taguchi method and use an orthogonal array to reduce the number of need optimizations. The **L25(5⁶)** array shows we can assess combinations of up to six variables at five levels with only 25 evaluations, so we'll use that array.

Table 3.10 shows which of the five value levels each independent variable should be set to, per the first three columns of the **L25(5⁶)** orthogonal array.

TABLE 3.10
Level Settings for the Independent Variables to
Obtain Training Date for the Fuzzy Controller

Case to Optimize	Level for A	Level for g	Level for Bias
1	1	1	1
2	1	2	2
3	1	3	3
4	1	4	4
5	1	5	5
6	2	1	2
7	2	2	3
8	2	3	4
9	2	4	5
10	2	5	1
11	3	1	3
12	3	2	4
13	3	3	5
14	3	4	1
15	3	5	2
16	4	1	4
17	4	2	5
18	4	3	1
19	4	4	2
20	4	5	3
21	5	1	5
22	5	2	1
23	5	3	2
24	5	4	3
25	5	5	4

Here are the results for the 25 cases. First we determine the optimal steering law for each of the 25 conditions. Then we use that law to determine what the trajectory will be at 3 seconds. The fuzzy logic controller relates the velocity components at 3 seconds to the optimal steering constants.

CASE 1: LEVELS 1, 1, 1

```
In[153]:= J[x_ ? (VectorQ[#, NumericQ] &)] := Module[{intresult, retresult},
             intresult = FlatNoAtmConstTEOMrev[x[[1]],
               x[[2]], trialA[[1]], trialg[[1]], U, trialthetabias[[1]]];
             retresult = intresult[[1]] + 10000 * (intresult[[2]] / H - 1)^2 + 1 * (intresult[[3]])^2;
             Return[retresult]
             ];
          FindMinimum[J[{z1, z2}], {z1, -.0015, -.002, -.001}, {z2, .5, .25, .75}]
```

FindMinimum::lstol :
    The line search decreased the step size to within the tolerance specified by AccuracyGoal and PrecisionGoal
        but was unable to find a sufficient decrease in the function. You may need more
        than MachinePrecision digits of working precision to meet these tolerances. ≫

```
Out[154]= {287.5, {z1 → -0.00144458, z2 → 0.491329}}
```

```
In[155]:= FlatNoAtmConstTEOMrev[-0.00144458, 0.491329,
            trialA[[1]], trialg[[1]], U, trialthetabias[[1]]]
          Print["horizontal speed at 3 sec: ", x3history[[7]]];
          Print["vertical speend at 3 sec: ", x4history[[7]]];
          Print["Velocity at 3 sec is: ", Sqrt[x3history[[7]]^2 + x4history[[7]]^2]];
          Print["Acceleration at 3 seconds is: ", Sqrt[trialA[[1]]^2 + trialg[[1]]^2 -
            2 * trialA[[1]] * trialg[[1]] * Sin[thetahistory[[7]] + trialthetabias[[1]]]]]
```

```
Out[155]= {287.5, 50003.8, 0.00773811}
```

```
          horizontal speed at 3 sec: 53.4292

          vertical speend at 3 sec: 10.0104

          Velocity at 3 sec is: 54.3589

          Acceleration at 3 seconds is: 18.1297
```

CASE 2: LEVELS 1, 2, 2

```
In[160]:= J[x_ ? (VectorQ[#, NumericQ] &)] := Module[{intresult, retresult},
             intresult = FlatNoAtmConstTEOMrev[x[[1]],
               x[[2]], trialA[[1]], trialg[[2]], U, trialthetabias[[2]]];
             retresult = intresult[[1]] + 10000 * (intresult[[2]] / H - 1)^2 + 1 * intresult[[3]]^2;
             Return[retresult]
             ];
          FindMinimum[J[{z1, z2}], {z1, -.0015, -.002, -.001}, {z2, .5, .25, .75}]
```

FindMinimum::lstol :
    The line search decreased the step size to within the tolerance specified by AccuracyGoal and PrecisionGoal
        but was unable to find a sufficient decrease in the function. You may need more
        than MachinePrecision digits of working precision to meet these tolerances. ≫

```
Out[161]= {288.003, {z1 → -0.00143856, z2 → 0.491367}}
```

```
In[162]:= FlatNoAtmConstTEOMrev[-.00143856, 0.491367,
            trialA[[1]], trialg[[2]], U, trialthetabias[[2]]]
          Print["horizontal speed at 3 sec: ", x3history[[7]]];
          Print["vertical speend at 3 sec: ", x4history[[7]]];
          Print["Velocity at 3 sec is: ", Sqrt[x3history[[7]]^2 + x4history[[7]]^2]];
          Print["Acceleration at 3 seconds is: ", Sqrt[trialA[[1]]^2 + trialg[[2]]^2 -
            2 * trialA[[1]] * trialg[[2]] * Sin[thetahistory[[7]] + trialthetabias[[2]]]]]
```

```
Out[162]= {288., 50018., -0.0385708}
```

```
          horizontal speed at 3 sec: 53.4067

          vertical speend at 3 sec: 9.97719

          Velocity at 3 sec is: 54.3306

          Acceleration at 3 seconds is: 18.1203
```

Case 3: Levels 1, 3, 3

The weight on the altitude term had to be increased from 10,000 to 20,000 to obtain an acceptable solution.

```
In[167]:= J[x_ ? (VectorQ[#, NumericQ] &)] := Module[{intresult, retresult},
            intresult = FlatNoAtmConstTEOMrev[x[[1]],
              x[[2]], trialA[[1]], trialg[[3]], U, trialthetabias[[3]]];
            retresult = intresult[[1]] + 20000 * (intresult[[2]] / H - 1)^2 + 1 * (intresult[[3]])^2;
            Return[retresult]
            ];
          FindMinimum[J[{z1, z2}], {z1, -.0015, -.002, -.001}, {z2, .5, .25, .75}]

          FindMinimum::lstol :
            The line search decreased the step size to within the tolerance specified by AccuracyGoal and PrecisionGoal
              but was unable to find a sufficient decrease in the function. You may need more
              than MachinePrecision digits of working precision to meet these tolerances. ≫

Out[168]= {288.001, {z1 → -0.00143928, z2 → 0.492063}}

In[169]:= FlatNoAtmConstTEOMrev[-.001439282, 0.4920635,
            trialA[[1]], trialg[[3]], U, trialthetabias[[3]]]
          Print["horizontal speed at 3 sec: ", x3history[[7]]];
          Print["vertical speend at 3 sec: ", x4history[[7]]];
          Print["Velocity at 3 sec is: ", Sqrt[x3history[[7]]^2 + x4history[[7]]^2];
          Print["Acceleration at 3 seconds is: ", Sqrt[trialA[[1]]^2 + trialg[[3]]^2 -
            2 * trialA[[1]] * trialg[[3]] * Sin[thetahistory[[7]] + trialthetabias[[3]]]]]

Out[169]= {288., 50004.8, -0.0325578}

          horizontal speed at 3 sec: 53.3841

          vertical speend at 3 sec: 9.94399

          Velocity at 3 sec is: 54.3023

          Acceleration at 3 seconds is: 18.1109
```

Case 4: Levels 1, 4, 4

The weight on the altitude term had to be increased from 10,000 to 20,000 to obtain an acceptable solution.

```
In[174]:= J[x_ ? (VectorQ[#, NumericQ] &)] := Module[{intresult, retresult},
            intresult = FlatNoAtmConstTEOMrev[x[[1]],
              x[[2]], trialA[[1]], trialg[[4]], U, trialthetabias[[4]]];
            retresult = intresult[[1]] + 20000 * (intresult[[2]] / H - 1)^2 + 1 * (intresult[[3]])^2;
            Return[retresult]
            ];
          FindMinimum[J[{z1, z2}], {z1, -.0015, -.002, -.001}, {z2, .5, .25, .75}]

          FindMinimum::lstol :
            The line search decreased the step size to within the tolerance specified by AccuracyGoal and PrecisionGoal
              but was unable to find a sufficient decrease in the function. You may need more
              than MachinePrecision digits of working precision to meet these tolerances. ≫

Out[175]= {288., {z1 → -0.00144029, z2 → 0.492805}}

In[176]:= FlatNoAtmConstTEOMrev[-.00144029, 0.492805,
            trialA[[1]], trialg[[4]], U, trialthetabias[[4]]]
          Print["horizontal speed at 3 sec: ", x3history[[7]]];
          Print["vertical speend at 3 sec: ", x4history[[7]]];
          Print["Velocity at 3 sec is: ", Sqrt[x3history[[7]]^2 + x4history[[7]]^2];
          Print["Acceleration at 3 seconds is: ", Sqrt[trialA[[1]]^2 + trialg[[4]]^2 -
            2 * trialA[[1]] * trialg[[4]] * Sin[thetahistory[[7]] + trialthetabias[[4]]]]]

Out[176]= {288., 50001.1, -0.0242222}

          horizontal speed at 3 sec: 53.3614

          vertical speend at 3 sec: 9.91076

          Velocity at 3 sec is: 54.274

          Acceleration at 3 seconds is: 18.1015
```

Case 5: Levels 1, 5, 5

The weight on the altitude term had to be increased to 30,000 from 10,000 to obtain an acceptable result.

```
In[181]:= J[x_ ? (VectorQ[#, NumericQ] &)] := Module[{intresult, retresult},
            intresult = FlatNoAtmConstTEOMrev[x[[1]],
              x[[2]], trialA[[1]], trialg[[5]], trialthetabias[[5]]];
            retresult = intresult[[1]] + 30000 * (intresult[[2]] / H - 1)^2 + 1 * (intresult[[3]])^2;
            Return[retresult]
            ];
        FindMinimum[J[{z1, z2}], {z1, -.0015, -.002, -.001}, {z2, .5, .25, .75}]
```

FindMinimum::lstol :
    The line search decreased the step size to within the tolerance specified by AccuracyGoal and PrecisionGoal
        but was unable to find a sufficient decrease in the function. You may need more
        than MachinePrecision digits of working precision to meet these tolerances. ≫

Out[182]= {288., {z1 → -0.00144138, z2 → 0.49356}}

```
In[183]:= FlatNoAtmConstTEOMrev[-.00144138, 0.49356,
            trialA[[1]], trialg[[5]], U, trialthetabias[[5]]]
        Print["horizontal speed at 3 sec: ", x3history[[7]]];
        Print["vertical speend at 3 sec: ", x4history[[7]]];
        Print["Velocity at 3 sec is: ", Sqrt[x3history[[7]]^2 + x4history[[7]]^2]];
        Print["Acceleration at 3 seconds is: ", Sqrt[trialA[[1]]^2 + trialg[[5]]^2 -
            2 * trialA[[1]] * trialg[[5]] * Sin[thetahistory[[7]] + trialthetabias[[5]]]]]
```

Out[183]= {288., 50000.1, -0.0151963}

horizontal speed at 3 sec: 53.3388

vertical speend at 3 sec: 9.87752

Velocity at 3 sec is: 54.2456

Acceleration at 3 seconds is: 18.0921

Case 6: Levels 2, 1, 2

The constraints are better satisfied using a weight of 20,000 on the altitude term rather than 10,000.

```
In[188]:= J[x_ ? (VectorQ[#, NumericQ] &)] := Module[{intresult, retresult},
            intresult = FlatNoAtmConstTEOMrev[x[[1]],
              x[[2]], trialA[[2]], trialg[[1]], U, trialthetabias[[2]]];
            retresult = intresult[[1]] + 20000 * (intresult[[2]] / H - 1)^2 + 1 * (intresult[[3]])^2;
            Return[retresult]
            ];
        FindMinimum[J[{z1, z2}], {z1, -.0015, -.002, -.001}, {z2, .5, .25, .75}]
```

Out[189]= {280.003, {z1 → -0.00151469, z2 → 0.487227}}

```
In[190]:= FlatNoAtmConstTEOMrev[-.00151469, 0.487227,
            trialA[[2]], trialg[[1]], U, trialthetabias[[2]]]
        Print["horizontal speed at 3 sec: ", x3history[[7]]];
        Print["vertical speend at 3 sec: ", x4history[[7]]];
        Print["Velocity at 3 sec is: ", Sqrt[x3history[[7]]^2 + x4history[[7]]^2]];
        Print["Acceleration at 3 seconds is: ", Sqrt[trialA[[2]]^2 + trialg[[1]]^2 -
            2 * trialA[[2]] * trialg[[1]] * Sin[thetahistory[[7]] + trialthetabias[[2]]]]]
```

Out[190]= {280., 50017.1, -0.0194367}

horizontal speed at 3 sec: 54.8121

vertical speend at 3 sec: 10.7374

Velocity at 3 sec is: 55.8539

Acceleration at 3 seconds is: 18.628

Case 7: Levels 2, 2, 3

```
In[195]:= J[x_ ? (VectorQ[#, NumericQ] &)] := Module[{intresult, retresult},
            intresult = FlatNoAtmConstTEOMrev[x[[1]],
              x[[2]], trialA[[2]], trialg[[2]], U, trialthetabias[[3]]];
            retresult = intresult[[1]] + 20000 * (intresult[[2]] / H - 1)^2 + 1 * (intresult[[3]])^2;
            Return[retresult]
          ];
          FindMinimum[J[{z1, z2}], {z1, -.0015, -.002, -.001}, {z2, .5, .25, .75}]

Out[196]= {280.003, {z1 → -0.0015158, z2 → 0.487933}}

In[197]:= FlatNoAtmConstTEOMrev[-.0015158, 0.487933,
            trialA[[2]], trialg[[2]], U, trialthetabias[[3]]]
          Print["horizontal speed at 3 sec: ", x3history[[7]]];
          Print["vertical speend at 3 sec: ", x4history[[7]]];
          Print["Velocity at 3 sec is: ", Sqrt[x3history[[7]]^2 + x4history[[7]]^2]];
          Print["Acceleration at 3 seconds is: ", Sqrt[trialA[[2]]^2 + trialg[[2]]^2 -
            2 * trialA[[2]] * trialg[[2]] * Sin[thetahistory[[7]] + trialthetabias[[3]]]]]

Out[197]= {280., 50017.4, -0.0159686}

          horizontal speed at 3 sec: 54.7889

          vertical speend at 3 sec: 10.7055

          Velocity at 3 sec is: 55.825

          Acceleration at 3 seconds is: 18.6184
```

Case 8: Levels 2, 3, 4

The weight on the altitude term had to be increased to 30,000 from 10,000 to obtain an acceptable result.

```
In[202]:= J[x_ ? (VectorQ[#, NumericQ] &)] := Module[{intresult, retresult},
            intresult = FlatNoAtmConstTEOMrev[x[[1]],
              x[[2]], trialA[[2]], trialg[[3]], U, trialthetabias[[4]]];
            retresult = intresult[[1]] + 20000 * (intresult[[2]] / H - 1)^2 + 1 * (intresult[[3]])^2;
            Return[retresult]
          ];
          FindMinimum[J[{z1, z2}], {z1, -.0015, -.002, -.001}, {z2, .5, .25, .75}]

Out[203]= {280.003, {z1 → -0.00151693, z2 → 0.488641}}

In[204]:= FlatNoAtmConstTEOMrev[-.00151693, 0.488641,
            trialA[[2]], trialg[[3]], U, trialthetabias[[4]]]
          Print["horizontal speed at 3 sec: ", x3history[[7]]];
          Print["vertical speend at 3 sec: ", x4history[[7]]];
          Print["Velocity at 3 sec is: ", Sqrt[x3history[[7]]^2 + x4history[[7]]^2]];
          Print["Acceleration at 3 seconds is: ", Sqrt[trialA[[2]]^2 + trialg[[3]]^2 -
            2 * trialA[[2]] * trialg[[3]] * Sin[thetahistory[[7]] + trialthetabias[[4]]]]]

Out[204]= {280., 50017., -0.0209969}

          horizontal speed at 3 sec: 54.7657

          vertical speend at 3 sec: 10.6735

          Velocity at 3 sec is: 55.7961

          Acceleration at 3 seconds is: 18.6088
```

CASE 9: LEVELS 2, 4, 5

The weight on the altitude term was increased to 20,000 from 10,000 to obtain an acceptable result.

```
In[209]:= J[x_? (VectorQ[#, NumericQ] &)] := Module[{intresult, retresult},
         intresult = FlatNoAtmConstTEOMrev[x[[1]],
             x[[2]], trialA[[2]], trialg[[4]], U, trialthetabias[[5]]];
         retresult = intresult[[1]] + 20000 * (intresult[[2]] / H - 1)^2 + 1 * (intresult[[3]])^2;
         Return[retresult]
         ];
     FindMinimum[J[{z1, z2}], {z1, -.0015, -.002, -.001}, {z2, .5, .25, .75}]
```

FindMinimum::lstol :
    The line search decreased the step size to within the tolerance specified by AccuracyGoal and PrecisionGoal
        but was unable to find a sufficient decrease in the function. You may need more
        than MachinePrecision digits of working precision to meet these tolerances. ≫

```
Out[210]= {280.002, {z1 → -0.00151796, z2 → 0.489337}}
```

```
In[211]:= FlatNoAtmConstTEOMrev[-.00151796, 0.489337,
         trialA[[2]], trialg[[4]], U, trialthetabias[[5]]]
     Print["horizontal speed at 3 sec: ", x3history[[7]]];
     Print["vertical speend at 3 sec: ", x4history[[7]]];
     Print["Velocity at 3 sec is: ", Sqrt[x3history[[7]]^2 + x4history[[7]]^2]];
     Print["Acceleration at 3 seconds is: ", Sqrt[trialA[[2]]^2 + trialg[[4]]^2 -
         2 * trialA[[2]] * trialg[[4]] * Sin[thetahistory[[7]] + trialthetabias[[5]]]]]
```

```
Out[211]= {280., 50014.7, -0.0154313}
```

    horizontal speed at 3 sec: 54.7424

    vertical speend at 3 sec: 10.6415

    Velocity at 3 sec is: 55.7671

    Acceleration at 3 seconds is: 18.5992

CASE 10: LEVELS 2, 5, 1

The weight on the altitude term was increased to 20,000 from 10,000 to get an acceptable solution.

```
In[216]:= J[x_? (VectorQ[#, NumericQ] &)] := Module[{intresult, retresult},
         intresult = FlatNoAtmConstTEOMrev[x[[1]],
             x[[2]], trialA[[2]], trialg[[5]], U, trialthetabias[[1]]];
         retresult = intresult[[1]] + 20000 * (intresult[[2]] / H - 1)^2 + 1 * (intresult[[3]])^2;
         Return[retresult]
         ];
     FindMinimum[J[{z1, z2}], {z1, -.0015, -.002, -.001}, {z2, .5, .25, .75}]
```

FindMinimum::lstol :
    The line search decreased the step size to within the tolerance specified by AccuracyGoal and PrecisionGoal
        but was unable to find a sufficient decrease in the function. You may need more
        than MachinePrecision digits of working precision to meet these tolerances. ≫

```
Out[217]= {280.502, {z1 → -0.00151534, z2 → 0.494674}}
```

```
In[218]:= FlatNoAtmConstTEOMrev[-.00151534, 0.494674,
         trialA[[2]], trialg[[5]], U, trialthetabias[[1]]]
     Print["horizontal speed at 3 sec: ", x3history[[7]]];
     Print["vertical speend at 3 sec: ", x4history[[7]]];
     Print["Velocity at 3 sec is: ", Sqrt[x3history[[7]]^2 + x4history[[7]]^2]];
     Print["Acceleration at 3 seconds is: ", Sqrt[trialA[[2]]^2 + trialg[[5]]^2 -
         2 * trialA[[2]] * trialg[[5]] * Sin[thetahistory[[7]] + trialthetabias[[1]]]]]
```

```
Out[218]= {280.5, 50010.7, -0.0285211}
```

    horizontal speed at 3 sec: 54.8352

    vertical speend at 3 sec: 10.3704

    Velocity at 3 sec is: 55.8072

    Acceleration at 3 seconds is: 18.6127

Case 11: Levels 3, 1, 3

The weight on the altitude term was increased to 20,000 from 10,000 to obtain an acceptable solution.

```
In[223]:= J[x_ ? (VectorQ[#, NumericQ] &)] := Module[{intresult, retresult},
          intresult = FlatNoAtmConstTEOMrev[x[[1]],
            x[[2]], trialA[[3]], trialg[[1]], U, trialthetabias[[3]]];
          retresult = intresult[[1]] + 20000 * (intresult[[2]] / H - 1)^2 + 1 * (intresult[[3]])^2;
          Return[retresult]
          ];
       FindMinimum[J[{z1, z2}], {z1, -.0015, -.002, -.001}, {z2, .5, .25, .75}]

Out[224]= {272.503, {z1 → -0.00159324, z2 → 0.484222}}

In[225]:= FlatNoAtmConstTEOMrev[-.00159324, 0.484222,
          trialA[[3]], trialg[[1]], U, trialthetabias[[3]]]
       Print["horizontal speed at 3 sec: ", x3history[[7]]];
       Print["vertical speend at 3 sec: ", x4history[[7]]];
       Print["Velocity at 3 sec is: ", Sqrt[x3history[[7]]^2 + x4history[[7]]^2]];
       Print["Acceleration at 3 seconds is: ", Sqrt[trialA[[3]]^2 + trialg[[1]]^2 -
          2 * trialA[[3]] * trialg[[1]] * Sin[thetahistory[[7]] + trialthetabias[[3]]]]]

Out[225]= {272.5, 50017.3, -0.0162941}

       horizontal speed at 3 sec: 56.1938

       vertical speend at 3 sec: 11.467

       Velocity at 3 sec is: 57.3518

       Acceleration at 3 seconds is: 19.1273
```

Case 12: Levels 3, 2, 4

The weight on the altitude term was increased to 20,000 from 10,000 to obtain an acceptable solution.

```
In[230]:= J[x_ ? (VectorQ[#, NumericQ] &)] := Module[{intresult, retresult},
          intresult = FlatNoAtmConstTEOMrev[x[[1]],
            x[[2]], trialA[[3]], trialg[[2]], U, trialthetabias[[4]]];
          retresult = intresult[[1]] + 20000 * (intresult[[2]] / H - 1)^2 + 1 * (intresult[[3]])^2;
          Return[retresult]
          ];
       FindMinimum[J[{z1, z2}], {z1, -.0015, -.002, -.001}, {z2, .5, .25, .75}]

       FindMinimum::lstol :
          The line search decreased the step size to within the tolerance specified by AccuracyGoal and PrecisionGoal
            but was unable to find a sufficient decrease in the function. You may need more
            than MachinePrecision digits of working precision to meet these tolerances. ≫

Out[231]= {272.503, {z1 → -0.00159433, z2 → 0.484879}}

In[232]:= FlatNoAtmConstTEOMrev[-.001594334, 0.4848789,
          trialA[[3]], trialg[[2]], U, trialthetabias[[4]]]
       Print["horizontal speed at 3 sec: ", x3history[[7]]];
       Print["vertical speend at 3 sec: ", x4history[[7]]];
       Print["Velocity at 3 sec is: ", Sqrt[x3history[[7]]^2 + x4history[[7]]^2]];
       Print["Acceleration at 3 seconds is: ", Sqrt[trialA[[3]]^2 + trialg[[2]]^2 -
          2 * trialA[[3]] * trialg[[2]] * Sin[thetahistory[[7]] + trialthetabias[[4]]]]]

Out[232]= {272.5, 50017., -0.0196059}

       horizontal speed at 3 sec: 56.1699

       vertical speend at 3 sec: 11.4362

       Velocity at 3 sec is: 57.3223

       Acceleration at 3 seconds is: 19.1175
```

CASE 13: LEVELS 3, 3, 5

The weight on the altitude term was increased to 20,000 from 10,000 to obtain an acceptable solution.

```
In[237]:= J[x_ ? (VectorQ[#, NumericQ] &)] := Module[{intresult, retresult},
          intresult = FlatNoAtmConstTEOMrev[x[[1]],
              x[[2]], trialA[[3]], trialg[[3]], U, trialthetabias[[5]]];
          retresult = intresult[[1]] + 20000 * (intresult[[2]] / H - 1)^2 + 1 * (intresult[[3]])^2;
          Return[retresult]
          ];
      FindMinimum[J[{z1, z2}], {z1, -.0015, -.002, -.001}, {z2, .5, .25, .75}]

Out[238]= {272.503, {z1 → -0.00159543, z2 → 0.485538}}

In[239]:= FlatNoAtmConstTEOMrev[-.00159543, 0.485538,
          trialA[[3]], trialg[[3]], U, trialthetabias[[5]]]
      Print["horizontal speed at 3 sec: ", x3history[[7]]];
      Print["vertical speend at 3 sec: ", x4history[[7]]];
      Print["Velocity at 3 sec is: ", Sqrt[x3history[[7]]^2 + x4history[[7]]^2]];
      Print["Acceleration at 3 seconds is: ", Sqrt[trialA[[3]]^2 + trialg[[3]]^2 -
          2 * trialA[[3]] * trialg[[3]] * Sin[thetahistory[[7]] + trialthetabias[[5]]]]]]

Out[239]= {272.5, 50017.4, -0.0166869}

      horizontal speed at 3 sec: 56.1461

      vertical speend at 3 sec: 11.4054

      Velocity at 3 sec is: 57.2928

      Acceleration at 3 seconds is: 19.1077
```

CASE 14: LEVELS 3, 4, 1

The weight on the altitude term was increased to 20,000 from 10,000 to obtain an acceptable solution.

```
In[244]:= J[x_ ? (VectorQ[#, NumericQ] &)] := Module[{intresult, retresult},
          intresult = FlatNoAtmConstTEOMrev[x[[1]],
              x[[2]], trialA[[3]], trialg[[4]], U, trialthetabias[[1]]];
          retresult = intresult[[1]] + 20000 * (intresult[[2]] / H - 1)^2 + 1 * (intresult[[3]])^2;
          Return[retresult]
          ];
      FindMinimum[J[{z1, z2}], {z1, -.0015, -.002, -.001}, {z2, .5, .25, .75}]

      FindMinimum::lstol :
          The line search decreased the step size to within the tolerance specified by AccuracyGoal and PrecisionGoal
              but was unable to find a sufficient decrease in the function. You may need more
              than MachinePrecision digits of working precision to meet these tolerances. ≫

Out[245]= {273.001, {z1 → -0.00159198, z2 → 0.490727}}

In[246]:= FlatNoAtmConstTEOMrev[-.00159198, 0.490727,
          trialA[[3]], trialg[[4]], U, trialthetabias[[1]]]
      Print["horizontal speed at 3 sec: ", x3history[[7]]];
      Print["vertical speend at 3 sec: ", x4history[[7]]];
      Print["Velocity at 3 sec is: ", Sqrt[x3history[[7]]^2 + x4history[[7]]^2]];
      Print["Acceleration at 3 seconds is: ", Sqrt[trialA[[3]]^2 + trialg[[4]]^2 -
          2 * trialA[[3]] * trialg[[4]] * Sin[thetahistory[[7]] + trialthetabias[[1]]]]]]

Out[246]= {273., 50006.6, -0.0306739}

      horizontal speed at 3 sec: 56.2413

      vertical speend at 3 sec: 11.1294

      Velocity at 3 sec is: 57.3319

      Acceleration at 3 seconds is: 19.1208
```

CASE 15: LEVELS 3, 5, 2

The weight on the altitude term was increased to 20,000 from 10,000 to obtain an acceptable solution.

```
In[251]:= J[x_? (VectorQ[#, NumericQ] &)] := Module[{intresult, retresult},
            intresult = FlatNoAtmConstTEOMrev[x[[1]],
              x[[2]], trialA[[3]], trialg[[5]], U, trialthetabias[[2]]];
            retresult = intresult[[1]] + 20000 * (intresult[[2]] / H - 1)^2 + 1 * (intresult[[3]])^2;
            Return[retresult]
            ];
          FindMinimum[J[{z1, z2}], {z1, -.0015, -.002, -.001}, {z2, .5, .25, .75}]
```

FindMinimum::lstol :
   The line search decreased the step size to within the tolerance specified by AccuracyGoal and PrecisionGoal
      but was unable to find a sufficient decrease in the function. You may need more
      than MachinePrecision digits of working precision to meet these tolerances. ≫

```
Out[252]= {273.001, {z1 → -0.00159309, z2 → 0.491391}}
```

```
In[253]:= FlatNoAtmConstTEOMrev[-.00159309, 0.49139,
            trialA[[3]], trialg[[5]], U, trialthetabias[[2]]]
          Print["horizontal speed at 3 sec: ", x3history[[7]]];
          Print["vertical speend at 3 sec: ", x4history[[7]]];
          Print["Velocity at 3 sec is: ", Sqrt[x3history[[7]]^2 + x4history[[7]]^2]];
          Print["Acceleration at 3 seconds is: ", Sqrt[trialA[[3]]^2 + trialg[[5]]^2 -
            2 * trialA[[3]] * trialg[[5]] * Sin[thetahistory[[7]] + trialthetabias[[2]]]]]
```

```
Out[253]= {273., 50006.4, -0.0319694}
```

```
horizontal speed at 3 sec: 56.2175

vertical speend at 3 sec: 11.0987

Velocity at 3 sec is: 57.3026

Acceleration at 3 seconds is: 19.1111
```

CASE 16: LEVELS 4, 1, 4

The weight on the altitude term was increased to 30,000 from 10,000 to obtain an acceptable solution.

```
In[258]:= J[x_? (VectorQ[#, NumericQ] &)] := Module[{intresult, retresult},
            intresult = FlatNoAtmConstTEOMrev[x[[1]],
              x[[2]], trialA[[4]], trialg[[1]], U, trialthetabias[[4]]];
            retresult = intresult[[1]] + 30000 * (intresult[[2]] / H - 1)^2 + 1 * (intresult[[3]])^2;
            Return[retresult]
            ];
          FindMinimum[J[{z1, z2}], {z1, -.0015, -.002, -.001}, {z2, .5, .25, .75}]
```

FindMinimum::lstol :
   The line search decreased the step size to within the tolerance specified by AccuracyGoal and PrecisionGoal
      but was unable to find a sufficient decrease in the function. You may need more
      than MachinePrecision digits of working precision to meet these tolerances. ≫

```
Out[259]= {265.502, {z1 → -0.00167183, z2 → 0.481433}}
```

```
In[260]:= FlatNoAtmConstTEOMrev[-.00167183, 0.481433,
            trialA[[4]], trialg[[1]], U, trialthetabias[[4]]]
          Print["horizontal speed at 3 sec: ", x3history[[7]]];
          Print["vertical speend at 3 sec: ", x4history[[7]]];
          Print["Velocity at 3 sec is: ", Sqrt[x3history[[7]]^2 + x4history[[7]]^2]];
          Print["Acceleration at 3 seconds is: ", Sqrt[trialA[[4]]^2 + trialg[[1]]^2 -
            2 * trialA[[4]] * trialg[[1]] * Sin[thetahistory[[7]] + trialthetabias[[4]]]]]
```

```
Out[260]= {265.5, 50011.9, -0.0161129}
```

```
horizontal speed at 3 sec: 57.5742

vertical speend at 3 sec: 12.1989

Velocity at 3 sec is: 58.8524

Acceleration at 3 seconds is: 19.6275
```

CASE 17: LEVELS 4, 2, 5

The weight on the altitude term was increased to 30,000 from 10,000 to obtain an acceptable solution.

```
In[265]:= J[x_ ? (VectorQ[#, NumericQ] &)] := Module[{intresult, retresult},
          intresult = FlatNoAtmConstTEOMrev[x[[1]],
              x[[2]], trialA[[4]], trialg[[2]], U, trialthetabias[[5]]];
          retresult = intresult[[1]] + 30000 * (intresult[[2]] / H - 1) ^2 + 1 * (intresult[[3]]) ^2;
          Return[retresult]
          ];
      FindMinimum[J[{z1, z2}], {z1, -.0015, -.002, -.001}, {z2, .5, .25, .75}]
```

FindMinimum::lstol :
   The line search decreased the step size to within the tolerance specified by AccuracyGoal and PrecisionGoal
       but was unable to find a sufficient decrease in the function. You may need more
       than MachinePrecision digits of working precision to meet these tolerances. »

Out[266]= {265.502, {z1 → -0.0016729, z2 → 0.482044}}

```
      FlatNoAtmConstTEOMrev[-.0016729, 0.482044,
       trialA[[4]], trialg[[2]], U, trialthetabias[[5]]]
      Print["horizontal speed at 3 sec: ", x3history[[7]]];
      Print["vertical speend at 3 sec: ", x4history[[7]]];
      Print["Velocity at 3 sec is: ", Sqrt[x3history[[7]]^2 + x4history[[7]]^2]];
      Print["Acceleration at 3 seconds is: ", Sqrt[trialA[[4]]^2 + trialg[[2]]^2 -
          2 * trialA[[4]] * trialg[[2]] * Sin[thetahistory[[7]] + trialthetabias[[5]]]]]
```

{265.5, 50 011.7, -0.0190495}

horizontal speed at 3 sec: 57.5497

vertical speend at 3 sec: 12.1693

Velocity at 3 sec is: 58.8223

Acceleration at 3 seconds is: 19.6175

CASE 18: LEVELS 4, 3, 1

The weight on the altitude term was increased to 30,000 from 10,000 to obtain an acceptable solution.

```
In[267]:= J[x_ ? (VectorQ[#, NumericQ] &)] := Module[{intresult, retresult},
          intresult = FlatNoAtmConstTEOMrev[x[[1]],
              x[[2]], trialA[[4]], trialg[[3]], U, trialthetabias[[1]]];
          retresult = intresult[[1]] + 30000 * (intresult[[2]] / H - 1) ^2 + 1 * (intresult[[3]]) ^2;
          Return[retresult]
          ];
      FindMinimum[J[{z1, z2}], {z1, -.0015, -.002, -.001}, {z2, .5, .25, .75}]
```

FindMinimum::lstol :
   The line search decreased the step size to within the tolerance specified by AccuracyGoal and PrecisionGoal
       but was unable to find a sufficient decrease in the function. You may need more
       than MachinePrecision digits of working precision to meet these tolerances. »

Out[268]= {265.502, {z1 → -0.00167895, z2 → 0.488112}}

```
In[269]:= FlatNoAtmConstTEOMrev[-.00167895, 0.488112,
       trialA[[4]], trialg[[3]], U, trialthetabias[[1]]]
      Print["horizontal speed at 3 sec: ", x3history[[7]]];
      Print["vertical speend at 3 sec: ", x4history[[7]]];
      Print["Velocity at 3 sec is: ", Sqrt[x3history[[7]]^2 + x4history[[7]]^2]];
      Print["Acceleration at 3 seconds is: ", Sqrt[trialA[[4]]^2 + trialg[[3]]^2 -
          2 * trialA[[4]] * trialg[[3]] * Sin[thetahistory[[7]] + trialthetabias[[1]]]]]
```

Out[269]= {265.5, 50 011.9, -0.0165188}

horizontal speed at 3 sec: 57.6473

vertical speend at 3 sec: 11.8885

Velocity at 3 sec is: 58.8604

Acceleration at 3 seconds is: 19.6303

Case 19: Levels 4, 4, 2

The weight on the altitude term was increased to 30,000 from 10,000 to obtain an acceptable solution.

```
In[274]:= J[x_ ? (VectorQ[#, NumericQ] &)] := Module[{intresult, retresult},
          intresult = FlatNoAtmConstTEOMrev[x[[1]],
            x[[2]], trialA[[4]], trialg[[4]], U, trialthetabias[[2]]];
          retresult = intresult[[1]] + 30000 * (intresult[[2]] / H - 1)^2 + 1 * (intresult[[3]])^2;
          Return[retresult]
          ];
       FindMinimum[J[{z1, z2}], {z1, -.0015, -.002, -.001}, {z2, .5, .25, .75}]

       FindMinimum::lstol :
          The line search decreased the step size to within the tolerance specified by AccuracyGoal and PrecisionGoal
             but was unable to find a sufficient decrease in the function. You may need more
             than MachinePrecision digits of working precision to meet these tolerances. ≫

Out[275]= {266.001, {z1 → -0.00167011, z2 → 0.487771}}

       FlatNoAtmConstTEOMrev[-.00167011, 0.487771,
        trialA[[4]], trialg[[4]], U, trialthetabias[[2]]]
       Print["horizontal speed at 3 sec: ", x3history[[7]]];
       Print["vertical speend at 3 sec: ", x4history[[7]]];
       Print["Velocity at 3 sec is: ", Sqrt[x3history[[7]]^2 + x4history[[7]]^2]];
       Print["Acceleration at 3 seconds is: ", Sqrt[trialA[[4]]^2 + trialg[[4]]^2 -
          2 * trialA[[4]] * trialg[[4]] * Sin[thetahistory[[7]] + trialthetabias[[2]]]]]

       {266., 50008.9, -0.0201411}

       horizontal speed at 3 sec: 57.623

       vertical speend at 3 sec: 11.859

       Velocity at 3 sec is: 58.8306

       Acceleration at 3 seconds is: 19.6204
```

Case 20: Level 4, 5, 3

The weight on the altitude term was increased to 30,000 from 10,000 to obtain an acceptable solution.

```
In[276]:= J[x_ ? (VectorQ[#, NumericQ] &)] := Module[{intresult, retresult},
          intresult = FlatNoAtmConstTEOMrev[x[[1]],
            x[[2]], trialA[[4]], trialg[[5]], U, trialthetabias[[3]]];
          retresult = intresult[[1]] + 30000 * (intresult[[2]] / H - 1)^2 + 1 * (intresult[[3]])^2;
          Return[retresult]
          ];
       FindMinimum[J[{z1, z2}], {z1, -.0015, -.002, -.001}, {z2, .5, .25, .75}]

       FindMinimum::lstol :
          The line search decreased the step size to within the tolerance specified by AccuracyGoal and PrecisionGoal
             but was unable to find a sufficient decrease in the function. You may need more
             than MachinePrecision digits of working precision to meet these tolerances. ≫

Out[277]= {266.002, {z1 → -0.00167128, z2 → 0.488401}}

In[278]:= FlatNoAtmConstTEOMrev[-.00167128, 0.488401,
        trialA[[4]], trialg[[5]], U, trialthetabias[[3]]]
       Print["horizontal speed at 3 sec: ", x3history[[7]]];
       Print["vertical speend at 3 sec: ", x4history[[7]]];
       Print["Velocity at 3 sec is: ", Sqrt[x3history[[7]]^2 + x4history[[7]]^2]];
       Print["Acceleration at 3 seconds is: ", Sqrt[trialA[[4]]^2 + trialg[[5]]^2 -
          2 * trialA[[4]] * trialg[[5]] * Sin[thetahistory[[7]] + trialthetabias[[3]]]]]

Out[278]= {266., 50012.2, -0.0140457}

       horizontal speed at 3 sec: 57.5986

       vertical speend at 3 sec: 11.8294

       Velocity at 3 sec is: 58.8008

       Acceleration at 3 seconds is: 19.6105
```

Case 21: Levels 5, 1, 5

The weight on the altitude term was increased to 30,000 from 10,000 to obtain an acceptable solution.

```
In[283]:= J[x_ ? (VectorQ[#, NumericQ] &)] := Module[{intresult, retresult},
            intresult = FlatNoAtmConstTEOMrev[x[[1]],
              x[[2]], trialA[[5]], trialg[[1]], U, trialthetabias[[5]]];
            retresult = intresult[[1]] + 30000 * (intresult[[2]] / H - 1)^2 + 1 * (intresult[[3]])^2;
            Return[retresult]
            ];
          FindMinimum[J[{z1, z2}], {z1, -.0015, -.002, -.001}, {z2, .5, .25, .75}]
```

FindMinimum::lstol :
    The line search decreased the step size to within the tolerance specified by AccuracyGoal and PrecisionGoal
        but was unable to find a sufficient decrease in the function. You may need more
        than MachinePrecision digits of working precision to meet these tolerances. ≫

```
Out[284]= {259., {z1 → -0.00174927, z2 → 0.478731}}
```

```
In[285]:= FlatNoAtmConstTEOMrev[-.00174927, 0.478731,
            trialA[[5]], trialg[[1]], U, trialthetabias[[5]]]
          Print["horizontal speed at 3 sec: ", x3history[[7]]];
          Print["vertical speend at 3 sec: ", x4history[[7]]];
          Print["Velocity at 3 sec is: ", Sqrt[x3history[[7]]^2 + x4history[[7]]^2]];
          Print["Acceleration at 3 seconds is: ", Sqrt[trialA[[5]]^2 + trialg[[1]]^2 -
            2 * trialA[[5]] * trialg[[1]] * Sin[thetahistory[[7]] + trialthetabias[[5]]]]]
```

```
Out[285]= {259., 50 002.2, 0.00243007}
```

    horizontal speed at 3 sec: 58.9534

    vertical speend at 3 sec: 12.9333

    Velocity at 3 sec is: 60.3554

    Acceleration at 3 seconds is: 20.1285

Case 22: Levels 5, 2, 1

The weight on the altitude term was increased to 30,000 from 10,000 to obtain an acceptable solution.

```
In[290]:= J[x_ ? (VectorQ[#, NumericQ] &)] := Module[{intresult, retresult},
            intresult = FlatNoAtmConstTEOMrev[x[[1]],
              x[[2]], trialA[[5]], trialg[[2]], U, trialthetabias[[1]]];
            retresult = intresult[[1]] + 30000 * (intresult[[2]] / H - 1)^2 + 1 * (intresult[[3]])^2;
            Return[retresult]
            ];
          FindMinimum[J[{z1, z2}], {z1, -.0015, -.002, -.001}, {z2, .5, .25, .75}]
```

FindMinimum::lstol :
    The line search decreased the step size to within the tolerance specified by AccuracyGoal and PrecisionGoal
        but was unable to find a sufficient decrease in the function. You may need more
        than MachinePrecision digits of working precision to meet these tolerances. ≫

```
Out[291]= {259.002, {z1 → -0.00175584, z2 → 0.484794}}
```

```
In[292]:= FlatNoAtmConstTEOMrev[-.00175585, 0.484796,
            trialA[[5]], trialg[[2]], U, trialthetabias[[1]]]
          Print["horizontal speed at 3 sec: ", x3history[[7]]];
          Print["vertical speend at 3 sec: ", x4history[[7]]];
          Print["Velocity at 3 sec is: ", Sqrt[x3history[[7]]^2 + x4history[[7]]^2]];
          Print["Acceleration at 3 seconds is: ", Sqrt[trialA[[5]]^2 + trialg[[2]]^2 -
            2 * trialA[[5]] * trialg[[2]] * Sin[thetahistory[[7]] + trialthetabias[[1]]]]]
```

```
Out[292]= {259., 50 011.8, -0.0177422}
```

    horizontal speed at 3 sec: 59.0533

    vertical speend at 3 sec: 12.6475

    Velocity at 3 sec is: 60.3925

    Acceleration at 3 seconds is: 20.1409

CASE 23: LEVELS 5, 3, 2

The weight on the altitude term was increased to 30,000 from 10,000 to obtain an acceptable solution.

```
In[297]:= J[x_? (VectorQ[#, NumericQ] &)] := Module[{intresult, retresult},
            intresult = FlatNoAtmConstTEOMrev[x[[1]],
                x[[2]], trialA[[5]], trialg[[3]], U, trialthetabias[[2]]];
            retresult = intresult[[1]] + 30000 * (intresult[[2]] / H - 1) ^ 2 + 1 * (intresult[[3]]) ^ 2;
            Return[retresult]
            ];
        FindMinimum[J[{z1, z2}], {z1, -.0015, -.002, -.001}, {z2, .5, .25, .75}]
```

FindMinimum::lstol :
    The line search decreased the step size to within the tolerance specified by AccuracyGoal and PrecisionGoal
        but was unable to find a sufficient decrease in the function. You may need more
        than MachinePrecision digits of working precision to meet these tolerances. ≫

Out[298]= $\{259.002, \{z1 \rightarrow -0.00175686, z2 \rightarrow 0.485362\}\}$

```
In[299]:= FlatNoAtmConstTEOMrev[-.00175686, 0.485362,
            trialA[[5]], trialg[[3]], U, trialthetabias[[2]]]
        Print["horizontal speed at 3 sec: ", x3history[[7]]];
        Print["vertical speend at 3 sec: ", x4history[[7]]];
        Print["Velocity at 3 sec is: ", Sqrt[x3history[[7]]^2 + x4history[[7]]^2]];
        Print["Acceleration at 3 seconds is: ", Sqrt[trialA[[5]]^2 + trialg[[3]]^2 -
            2 * trialA[[5]] * trialg[[3]] * Sin[thetahistory[[7]] + trialthetabias[[2]]]]]
```

Out[299]= $\{259., 50011.4, -0.0122695\}$

```
horizontal speed at 3 sec: 59.0284

vertical speend at 3 sec: 12.6192

Velocity at 3 sec is: 60.3622

Acceleration at 3 seconds is: 20.1308
```

CASE 24: LEVELS 5, 4, 3

The weight on the altitude term was increased to 30,000 from 10,000 to obtain an acceptable solution.

```
In[304]:= J[x_? (VectorQ[#, NumericQ] &)] := Module[{intresult, retresult},
            intresult = FlatNoAtmConstTEOMrev[x[[1]],
                x[[2]], trialA[[5]], trialg[[4]], U, trialthetabias[[3]]];
            retresult = intresult[[1]] + 30000 * (intresult[[2]] / H - 1) ^ 2 + 1 * (intresult[[3]]) ^ 2;
            Return[retresult]
            ];
        FindMinimum[J[{z1, z2}], {z1, -.0015, -.002, -.001}, {z2, .5, .25, .75}]
```

FindMinimum::lstol :
    The line search decreased the step size to within the tolerance specified by AccuracyGoal and PrecisionGoal
        but was unable to find a sufficient decrease in the function. You may need more
        than MachinePrecision digits of working precision to meet these tolerances. ≫

Out[305]= $\{259.001, \{z1 \rightarrow -0.00175769, z2 \rightarrow 0.485902\}\}$

```
In[306]:= FlatNoAtmConstTEOMrev[-.00175769, 0.485902,
            trialA[[5]], trialg[[4]], U, trialthetabias[[3]]]
        Print["horizontal speed at 3 sec: ", x3history[[7]]];
        Print["vertical speend at 3 sec: ", x4history[[7]]];
        Print["Velocity at 3 sec is: ", Sqrt[x3history[[7]]^2 + x4history[[7]]^2]];
        Print["Acceleration at 3 seconds is: ", Sqrt[trialA[[5]]^2 + trialg[[4]]^2 -
            2 * trialA[[5]] * trialg[[4]] * Sin[thetahistory[[7]] + trialthetabias[[3]]]]]
```

Out[306]= $\{259., 50005., -0.0157101\}$

```
horizontal speed at 3 sec: 59.0034

vertical speend at 3 sec: 12.5909

Velocity at 3 sec is: 60.3319

Acceleration at 3 seconds is: 20.1208
```

CASE 25, LEVELS 5, 5, 4

The weight on the altitude term was increased to 30,000 from 10,000 to obtain an acceptable solution.

```
In[311]:= J[x_? (VectorQ[#, NumericQ] &)] := Module[{intresult, retresult},
      intresult = FlatNoAtmConstTEOMrev[x[[1]],
        x[[2]], trialA[[5]], trialg[[5]], U, trialthetabias[[4]]];
      retresult = intresult[[1]] + 30000 * (intresult[[2]] / H - 1)^2 + 1 * (intresult[[3]])^2;
      Return[retresult]
      ];
FindMinimum[J[{z1, z2}], {z1, -.0015, -.002, -.001}, {z2, .5, .25, .75}]
```

FindMinimum::lstol :
   The line search decreased the step size to within the tolerance specified by AccuracyGoal and PrecisionGoal
      but was unable to find a sufficient decrease in the function. You may need more
      than MachinePrecision digits of working precision to meet these tolerances. ≫

Out[312]= {259.002, {z1 → -0.00175901, z2 → 0.486511}}

```
FlatNoAtmConstTEOMrev[-.00175901, 0.486511,
   trialA[[5]], trialg[[5]], U, trialthetabias[[4]]]
Print["horizontal speed at 3 sec: ", x3history[[7]]];
Print["vertical speend at 3 sec: ", x4history[[7]]];
Print["Velocity at 3 sec is: ", Sqrt[x3history[[7]]^2 + x4history[[7]]^2]];
Print["Acceleration at 3 seconds is: ", Sqrt[trialA[[5]]^2 + trialg[[5]]^2 -
   2 * trialA[[5]] * trialg[[5]] * Sin[thetahistory[[7]] + trialthetabias[[4]]]]]
```

{259., 50012.1, -0.0140788}

horizontal speed at 3 sec: 58.9784

vertical speend at 3 sec: 12.5626

Velocity at 3 sec is: 60.3015

Acceleration at 3 seconds is: 20.1107

REFERENCES

1. Mlodinow, Leonard, *The Drunkard's Walk: How Randomness Rules Our Lives*, New York: Pantheon Books, 2008.
2. Surowiecki, James, *The Wisdom of Crowds: Why the Many Are Smarter Than the Few and How Collective Wisdom Shapes Business, Economies, Societies and Nations*, New York: Doubleday, 2004.
3. Taleb, Nassim Nicholas, *The Black Swan: The Impact of the Highly Improbable*, New York: Random House, 2007
4. Pugh, Stuar, Clausing, Don, and Andrade, Ron, *Creating Innovative Products Using Total Design*, Reading, MA: Addison Wesley Longman, 1996.
5. Peace, Glen Stuart, *Taguchi Methods*, Reading, MA: Addison-Wesley, 1993.
6. Bennis, Warren, and Biederman, Patricia Ward, *Organizing Genius: The Secrets of Creative Collaboration*, New York: Perseus, 1997.
7. Zander, Rosamund Stone, and Zander, Benjamin, *The Art of the Possible*, New York: Penguin, 2000.
8. Luenberger, David G., *Investment Science*, New York: Oxford University Press, 1998.
9. Citron, S. J., *Elements of Optimal Control*, New York: Holt, Rinehart and Winston, 1969.
10. Michalewicz, Zbigniew, *Genetic Algorithms + Data Structures = Evolving Programs*, New York: Springer, 1992.
11. U.S. Air Force, *USAF Risk Identification: Integration & Ilities Guidebook*, DTD, December 2008, https://acc.dau.mil/CommunityBrowser.aspx?id=318289&lang=en-US. The corresponding calculator is available at http://www.afit.edu/cse/docs/RI3_Calculator_Ver%201.8.4%20beta_2009.xlsm.
12. Himmelblau, David M., *Applied Nonlinear Programming*, New York: McGraw-Hill, 1972.

4 Establish Natural Language Requirements

A natural language requirement is a statement in a language such as English defining a feature a system or architecture must possess. Later in this chapter I'll suggest which "features" need be addressed. To help people comprehend the requirements, attributes for documenting such things as the requirement's source, rationale, and verification method are associated with the requirement statement.

Since all natural languages are inherently ambiguous, many people believe that only a model, ideally an executing model, of the system or architecture can clearly communicate the requirements. In "model-based systems engineering," what the system is to achieve is physically or mathematically modeled to predict, or at least mimic, the subsequent behavior of the system or architecture. In order to make realistic models, it is usually necessary to mathematically model physical behavior, which means we must first assume how the system or architecture requirements will be achieved. So model-based systems engineering may presume a solution preference before it should be presumed. As we discussed in the prior chapter, the way to keep solution selection options open is to make different models for the different implementation options, and compare suitability based on one or a few technical parameters and cost. Proponents of model-based systems engineering believe that if the customers and users perceive the model behavior as what they want, then one can be reasonably assured the "requirements" are correct. As discussed in Chapter 2, I share the opinion models are critical to accomplishing good systems engineering and architecting. Who would buy a house or car if offered only a collection of requirement statements, rather than at least pictures, or a three-dimensional model? However, though model-based systems engineering can enhance requirement perception, natural language requirements are still needed for the following reasons:

1. Though one can try to translate directly from a customer verbal statements to a model, inevitably for anything even remotely complex, you first have to write down what the customer wants so you can figure out what to model. So, inevitably, some set of natural language requirements must be penned if for no other purpose than to guide the model to be created.
2. To trust the model output, it needs to be verified. The verification effort needs to compare a result from the model to something expected of it. The documentation of what is expected of the model can often be documented as a mathematical expression, but not always, so once again we need natural language requirements to attempt to record the expectations.

3. Even if it is reasonable to presume a physical behavior, such as a rocket will be required to achieve orbit, and we verify we derived the necessary equations and solve them correctly, we remain uncertain that something can actually be built that will perform as the model predicts. So, though the customer may want what the model predicts, achieving the model behavior may be extremely difficult or impossible.

In a phrase, though models are less directly ambiguous, they are inherently indirectly ambiguous. I believe the system engineering and architect's job is to define achievable requirements—or to otherwise inform everyone of the impracticality of achieving their desire. Again, this infers the need to compare what the model achieves to prior documented requirements.

The following case study describes some of the real-world difficulties in establishing requirements.

Case Study 4.1: Repeated Requirements

BACKGROUND

In the early 2000s the USAF, for the third time in its history, sought to develop a space-based laser (SBL). Though the SBL could execute many missions, SBL's primary job would be to keep a laser beam on an enemy's rocket stage long enough to cause a rupture and the destruction of the rocket, presumably keeping the warhead from causing any damage, at least at the intended delivery point. (That the warhead might still cause substantial damage somewhere else seems to rarely be of concern to antimissile technology developments.) At the time, chemical lasers were the only practical means to achieve the laser power levels needed to rupture the rocket's pressure vessel. A high power chemical laser is essentially a rocket itself. A significant chemical combustion occurs that produces photons, which after application of almost magical optics, can generate substantial power. Roughly speaking, the amount of laser beam power that results is very much dependent on the corresponding volume and extent of the chemical reaction. This particular project was seeking to produce a demonstration article, not an operational system. The plan was to launch the SBL into orbit and test its capacity to destroy missiles in flight, in a very controlled test environment. Ultimately, since these tests would encompass the full functionality of an operational system, the demonstrator would go a long way to both discovery and hopefully retire the key technical risks. The effort was by far the most technically challenging I witnessed in my entire career.

To make things really interesting, rather than compete and award the contract to one company, or perhaps a few companies and ultimately neck down to a provider through a series of competitions, the government elected to formulate a national team composed of Boeing, Lockheed Martin, and then TRW, now Northrop Grumman. The contract was set up so it was mandatory that each organization receive one third of the sales and award fee.

One of the early products was the System Specification, which came together relatively painlessly. Then it was necessary to produce the segment specifications.

The SBL system was deemed to consist of the following segments: space vehicle, ground segment (for mission planning and control), and an integrated test laboratory (for developmental testing on the ground prior to testing in space). The space vehicle consisted of the following elements: beam director, beam controller, spacecraft bus, and laser. The beam director was the big mirror used to focus the beam onto the rocket, and the beam controller was the source of the command and control to both aim, focus, and maintain the beam as a coherent set of light. Boeing was the system integrator and had primary responsibility for the beam controller. Lockheed Martin had the primary responsibility for the spacecraft and beam director. TRW had primary responsibility for the laser. The functionality of the on-orbit segments was, unfortunately, unclear, due to unclear roles and responsibilities of the performing organizations, and tradition. By tradition, in the space business, the spacecraft provider is usually the space vehicle integrator, where what is integrated onto the spacecraft are the payloads. This tradition had inherent logic, as the payloads were usually much "smaller" than the spacecraft, that is, each individually needs a fraction of the spacecraft's provided attach area, electrical power, thermal control, and data processing. So Lockheed Martin, as spacecraft provider, maintained they were also responsible for the space vehicle. For SBL, the traditional situation of a "larger" spacecraft bus with many "smaller" payloads was literally the opposite. The laser payload overwhelmed the spacecraft in every conceivable physical dimension. TRW had a long history of providing various communications and surveillance payloads to Boeing and Lockheed Martin. The head of TRW at the time bitterly resented this subordinate role as payload provider, and longed for the day when the "payload is prime." If ever there was a situation this might be true, it was for SBL, so he directed under no circumstance would his laser payload group submit to the Lockheed Martin leadership for the space vehicle specification. Yet, for SBL, there was a third legitimate claim for space vehicle leadership. For any laser, making a powerful beam is all well and good, but unless the beam is kept on target very precisely, it is all but useless. Boeing, as the beam controller provider, had responsibility for this element. To the Boeing beam control guys, the situation was analogous to the old World War II bombers. The pilot (spacecraft) got the plane to and back from the target, but it is the bombardier (beam controller) that flies the plane for the last critical maneuvers to get the bombs on target.

An additional issue was how to produce the specifications for the end items. There are basically two ways to view a specification for an end item: as a compilation of all known requirements for that end item, including subelements of the item, or as a communication of just what the providers of the end item need to do. Traditionally, when requirements were documents, system engineers preferred the first approach. They liked to document all their hard work, so they produced a system specification with sections for requirements for the entire system as well as sections to record requirements for subordinate parts of the system. This approach leads to requirement written as follows:

The [insert end item name] shall [insert feature].

Which results in many "repeated" requirements. Repeated requirements state the same feature, but imposed on different items. The number of repeated requirements can skyrocket quickly because the typical specification tree has seven levels: system, segment, element, subsystem, assembly, component, and part. That means the same feature could end up being documented seven times for the main body of each specification and an untold number of times for all subordinate items. Alternatively, today, thanks to the convenience of database software, a different approach is found to be much more efficient. In this approach, a feature need be defined once. Requirement statements have the form:

The [insert function] shall [insert feature].

If function in the statement applies to multiple end items, then it is allocated to multiple items. In this second approach, the specification for an item contains only the subset of all the requirements for that item, not requirements for its subordinate items. If you want to know the requirements for the subordinate items, look at the specification for those subordinates. In this second approach, no requirement is "repeated," though many requirements may be allocated to many different items. On SBL the traditional approach was used.

What Happened

The overall program manager, who was from Boeing, knew that if he clarified the space vehicle roles and responsibilities, he would upset at least two-thirds of his team. So he purposefully never made the call. As a consequence the requirement process never did determine the key requirements for the entire program.

Though Boeing was the overall system integrator, they concurred with Lockheed Martin recommendation to follow the traditional specification production approach, where each specification had sections devoted to the item as a whole, and sections to record any application requirements for the items immediate subordinate items. The system specification had about 125 requirements naming the total system. The system specification also contained about 120 space vehicle requirements, 80 ground segment requirements, and 50 test facility requirements. A large fraction of these 250 odd segment requirements documented in the SBL system specification were exactly the same statements as in the 125 system requirements, except rather than stating,

"The SBL system shall X;"

it stated,

"The [insert SBL segment name] shall X."

Similarly, the space vehicle specification had more than 300 requirements, with more than half repeated requirements. The payload specification had nearly 400 requirements, again, more than half repeated requirements. One by one, for days, each statement was reviewed and commented upon by dozens of people to the extent that the requirement generation process consumed more budget than the technology maturation process. But, despite all this effort, two extremely important jobs were not accomplished. First, the requirement statements were recorded without rationales and without verification methods. Thanks to the long and laborious dialogues, the rationale was almost always explained, but never documented for the hundreds and even thousands of future people who would be needed to implement the requirement, increasing the odds these implementers would misinterpret the requirement. Perhaps even more importantly, a verification method for each requirement was not documented. A requirement without a verification method cannot be validated. So requirements were being allocated to end items for which end item compliance was unspecified.

Shortly after the payload specification was approved, the program was canceled despite achieving near perfect award fee scores. The predominant reason for the cancellation was the cost of the system could not be justified in relation to the relatively inexpensive countermeasures that an adversary could have utilized. Even if the project wasn't cancelled, the team was on the verge of complete self-destruction. With responsibility for key requirements unclear, and the next job being to produce all the subsystem requirements, the number of repeated requirements would bloom into the thousands—swamping all effort to review, verify, and maintain status check on all those sentences that said the same thing.

Lessons Learned

Requirement allocation requires roles and responsibility clarity. Though the SBL program was done in by a sane assessment of its cost versus benefit, it would ultimately have been done in by the byzantine politics of the teaming arrangement. Comprehensible and verifiable requirements can only result from clear roles and responsibilities.

Put only the requirements needed for the end item in the specification for the end item. Providing a system that works is hard enough. We don't need to impose on people a mountain of words that repeats requirements. Look at the situation from the end item provider's point of view. They only need to know their requirements. It is very common that a requirement may be applicable exactly as written to many end items. Think of a specification not as a stand-alone document, but as the instantiation of the subset of the requirement applicable to that end item. There is no need to allocate the requirements applicable to the lower level items in the specification for the higher level item, because they will be allocated to the lower level item. If the same requirement is applicable to many of the subitems or even all of the subitems, then the exact same requirement statement should be allocated to each. A requirement has a statement, and a bunch of attributes. One of those attributes is what the requirement is allocated to. All one needs to do is to document the allocation. Other useful attributes may address how the requirement verification is to be accomplished. If the same requirement is allocated to multiple items, it needs to be verified as stipulated by all those items.

Validate requirements first, not last. Many practitioners of systems engineering make a distinction between "verification" and "validation." They define "verification" as showing the items built comply with their requirements, while "validation" shows the item complies with the user desires. So the chronological order is verification followed by validation. This is because the provider first "verifies," that is, conducts an effort to obtain evidence the item meets requirements prior to the buyer being willing to take ownership of the item. Then, subsequently, the buyer "validates" by a separate effort to see how well what they bought or are about to make final payment on really works. So "validation" is the stronger proof of compliance than "verification." But think about it a moment, what possible good does it do to discern an item isn't validated even though verified? True, contractually one may be off the hook, but certainly the net result will be an unhappy customer and damage to the brand. I strongly recommend we follow the process described in reference 1, which proceeds through a sequence of requirement status states defined as follows:

1. *Defined.* The requirement is written in an unambiguous manner.
2. *Validated.* Have proof the requirement is needed, has an accepted rationale, is correct, comprehended, and verifiable.
3. *Verification method determined.* A means is defined to obtain the evidence that the implementing item complies with the requirement.
4. *Approved.* The requirement is to be achieved.
5. *Allocated.* The requirement is imposed on the entity to achieve the requirement.
6. *Traced to verification document.* There exists a documented procedure that explains how the verification method is to be achieved.
7. *Designed.* Instructions exist explaining how to create something to achieve the requirement.
8. *Implemented.* The entity to fulfill the requirement exists.
9. *Tested.* The procedure described in the verification document was performed to obtain the evidence the real world entity achieves the requirement.
10. *Verified.* The obtained evidence confirms the real world entity fulfills the requirement.

The next case study illustrates another common issue that exists between program phases and system or architecture functions.

Case Study 4.2: Requirement Generator

BACKGROUND

I served on an integrated product team (IPT) at Rockwell International's Space System Division, asked to formulate ways to reduce by a factor of 10 the time and cost for design and production. On the team was a middle level executive in the production organization. During the course of our investigations, he kept a steady dialogue pushing for a requirements determination tool. My perception was he was asking for something like a database tool to help keep track of requirements, but on multiple occasions when I would describe such things to him, it was clear that wasn't what he wanted. After a week or so of misinterpreting each other, we arranged to sit alone in a conference room with a white board to try to describe to each other our ideas. I showed him a demo of System Architect, a tool then in its infancy that we were using as a core component in our systems engineering tool set. He seemed impressed, but clearly it was not what he had in mind. Then he showed me a picture of his idea. The picture had hundreds of little boxes on it, each labeled "Unit n" with n running from 1 to 100, as examples. His requirement generator sketch showed a person in front of a computer screen. What was on the screen was an inflow of shall statements and as outflow a list of units. What my colleague was envisioning was a "designer" that could automatically interpret a customer's spec and identify what existing parts would be used to produce the item. My friend was asking for a revolutionary tool that completely automated the design process and ensured the resulting design could be implemented with items that were already produced. To him, design was a necessary evil, just ordering items from a catalog. I recall feeling completely dumbfounded. On the one hand, I was impressed by the belief what seemed an art to me could be turned into a process. I remember thinking if we could pull it off we'd achieve our cost and schedule reduction goal. But, on the other hand, we were basically in the business of providing the government things that accomplished ends for which there was no suitable existing item. Sooner or later it seemed to me we had to design new items, and what we were looking for was how to get the production considerations into the decision-making process, and this idea didn't help with that at all. My friend had interpreted the offer to get productions requirements into the design as being satisfied by the designer only using parts the production people already knew how to make or buy. My friend didn't see being on an IPT as a means to get his needs for new items known, he saw it as a one-time deal to set up a system for designers to reuse parts. At the end of that day I realized the whole idea of "integrated" product teams had a great deal of inertia and misconception to overcome.

What Happened

I explained to my boss what production was looking for. He basically asked, are you sure? I said yes. The proposed requirements development project was never funded. I believe my friend, to his dying day, remained convinced in the rationality of his idea, and saw the refusal to support the project as just another attempt by the pesky designers to have job security at the expense of production.

Lessons Learned

Regardless of the program implementation construct used, all systems have eight primary functions for which requirements may need to be specified. Recall in Chapter 1 that our profession has settled on eight primary functions: develop, design, manufacture, deploy, train, operate, maintain, and dispose. The "design phase" is not the "design function." The design phase is the period predominantly devoted to determining what the end items should be and how to make them. The design functions are the tasks and activities associated with coming up with the instructions to make the end items, for each of which performance requirements may be imposed. Functions and program phases are not the same, even if the same words are used for both. Program phases are a management construct used to organize the effort to produce a system. A common, but certainly not universal, set of program phases is research, design, development, test, and evaluate. Another set may be develop, produce, and service, where here develop has as its subphases determine requirements, design, implement, integrate, and accept. In this construct the develop phase produces a first or few items, while the production phase produces larger quantities of the item. Different industries have adopted relatively stable program phases, but every once in a while a revolution occurs that radically alters how an industry fields its systems. Functions are tasks and activities that the system is to perform.

Requirements can and should lead directly to design and production. Many system engineering purists insist performance and functional requirements be written as implementation neutral as possible, the rationale being this enables the best design and production approach to be found by rational decision making. There certainly is tremendous merit to this opinion. However, for many situations, where there is a wealth of perfectly good end items available to fulfill a mission; all that is needed is to translate customer and user requirements directly into the instructions as to what to build and produce. A typical example is a car or computer. The customer's requirements can literally be shown as component options to select from. Though this approach is not conducive to fostering innovative new solutions, it is a tremendously cost effective requirement determination process.

In this chapter I define three types of natural language requirements, and summarize ways to determine what requirements are needed. Formal requirements in *Mathematica*® are presented to

1. help write less ambiguous statements;
2. maintain a requirement database; and
3. prepare specifications containing subsets of the requirements applicable for those who will design and produce an end item, and retain verification evidence.

Then various processes to determine requirements are presented, along with strengths and weaknesses and recommendations when best to use that particular requirement determining process. Then processes and heuristics are presented to verify each step of the process from requirements determination to delivering the complying systems and architecture. Finally, formal requirements are presented for a means to predict when the requirements development process is mature enough to stop and formally initiate design.

4.1 DEFINE THREE TYPES OF NATURAL LANGUAGE REQUIREMENTS: FUNCTIONAL, PERFORMANCE, AND SOUGHT

Each requirement type addresses a different type of feature of a system or architecture and each has a different set of attributes. Three convenient requirement types are functional, performance, and sought. Any and all requirements for a system or architecture can be one of these three types.

As mentioned in Chapter 1, there are eight primary functions: develop, design, produce, deploy, train, operate, support, and dispose. Recall that a functional requirement declares an activity or task that the system or architecture is to do, not how well the activity or task is to be performed, nor how the function is to be implemented. Develop functions name the activities to establish what the system or architecture should be. Develop functional requirements stipulate the decision gates for the program to proceed. Develop performance requirements may state activity time or cost limits, or dictate external interfaces to honor, or stipulate a particular implementation approach to use or avoid, as well as the incremental prototypes or technology developments that should be performed. Design functions name the activities to create the instructions to build the system or architecture. For software end items, design requirements are often formally defined pictorial representations of the software constructs, or pseudo-code, or the coding language or standards to follow, with the key performance requirements indicating execution time limits, input and out throughput, or memory limits. For hardware end items, design function performance requirements are margins, factors of safety, tolerances, reliability, dimension limits, mass limits, as well as standards to adhere to, materials to use or avoid, access provisions, operating and storage environments, or specific entities to procure. Production functions name the activities to create the end items. For software, production could be manual or automated translation of the design requirements into a specific high-level language, that itself is ultimately translated into machine instructions. For hardware, production performance requirements are procedures to follow, raw material stock needs and reserves, standards, tolerances, facility cleanliness levels, throughput, yield, and safety precautions. Deploy functions name activities to get the system to its operational environment. Times, paths, and environmental factors to tolerate are typical performance requirements for deploy functional requirements. Train functions name activities to prepare and ensure users know how to operate or support the system or architecture. Highest education level for operators to accommodate, proficiency categories and levels, the nature and content of the training material (manuals, books with exercises, videos, simulations, and proficiency examinations) are typical performance requirements for train functional requirements. Operate functions name the activities the system must conduct. Typical operate performance requirements stipulate the environment to be tolerated, key performance to be achieved, lifetimes, or other measures for how often the system is to be used, as well as overall reliability and availability. Other performance requirements allocated to operate activities may address the impact the system or architecture is allowed to have on its environment. Support functions name the activities to keep the system operating. Typical support performance requirements address how to avoid damage to the system, inspection procedures, maintenance schedules, maintenance procedures, and how to ensure a safe environment for those doing the maintenance. Dispose functions name activities for the system when no longer needed. Performance requirements for dispose functions may address where to put the system and how to get it there, or how long residual material needs to be stored, or perhaps to what extent the system or architecture is to be disassembled and what to do with the resulting parts.

For many systems or architecture, large numbers of requirements may be reused from previous systems or architecture, because the new item is a derivative of an existing item. When prior requirements can be used, do so it because it reduces costs and there is a good chance these are well-crafted requirements because they have already been translated into well-working real things. Clearly, then, do not reuse requirements that failed to produce well-working real things! But new systems and architecture will need new requirements, and the generation process can be time consuming and messy. Many system engineers draft requirements with "to be determined" (TBD) or "to be reviewed" (TBR) marks next to the numbers contained in the statements. Indeed, a common practice is to count the "TBx's" and utilize the reduction to zero as evidence the requirement generation job is nearing its end.

A requirement with TBD communicates no useful information, so why create? A requirement with a TBR is a draft requirement; it doesn't matter whether it contains a TBR or not. An accepted requirement with a TBR mark is begging for a cost increase for any value other than that marked TBR, and even if the value is not changed, the system probably costs more than needed to because the implementation will proceed to the stipulated amount which usually is set conservatively to try to avoid a cost increase if changes, but thereby causes a cost increase from what could have been accomplished with an easier requirement to achieve.

We need a means to note if a requirement is active or draft, and that is easily handled by an attribute associated with the statement. We traditionally adopted the TBx markings because we see our job to be to write requirements, so we wrote requirements, even if incomplete or uncertain, so we can keep track of what we have done and still need to do. A more direct approach is to introduce a third type of requirement, the sought requirement. A sought requirement is a statement from someone stipulating what they need to know. The statement can be written how the requester hopes the requirement will read. The statement may simply ask what needs to be known, for example:

"Need to know mass limits for end item X."

For derivative systems or architectures, at the start of the project, a very large number of requirements can immediately be stipulated as "sought" because similar requirements were needed for the predecessor system or architecture. The sought requirements can be counted, and as the count goes to zero, the requirement generation job is coming to an end.

4.1.1 Stipulate Attributes for Functional, Performance, and Sought Requirements

Convenient attributes for functional requirements are as follows:

1. Identification: a unique number.
2. Status: "draft," "approved," or "discontinued."
3. Parent(s): the identification of the requirement or requirements from which this requirement is derived. Only another functional requirement can be a parent for a functional requirement. Functional requirements form a hierarchy where the lower level requirements add detail to the higher level requirements. The lower level functions should be a complete and independent set of activities associated with the next higher level functional requirement.
4. End item allocation: name of the end item that is to perform the function. This presumes a hierarchical decomposition of the end items exists. A function may be allocated to more than one end item, but do so only with extreme care or the implementation may contain unnecessary redundancy. To achieve safe or reliable systems, rather than allocating the same function to multiple end items, recommend instead to create different functions, a prime function allocated to the prime item, and one or more redundant or back-up functions, to be allocated to other end items. Or better still, use performance requirements to stipulate the fail-safe level or reliability required and let the design experts find the best implementation. An end item may have more than one function allocated to it. This is often desirable as it will tend to minimize parts count, but again, care is needed to avoid end items from becoming overly complex.
5. Acceptance record: date and name of organization (or person with contact information) who accepted responsibility to implement the function and all allocated performance requirements. In essence, this is the person the requirements were written for, who is acknowledging need, receipt, comprehension, and commitment to implement. This attribute is useful to create pressure to make sure only needed requirements are defined. If this functional requirement is to be achieved by multiple end items, then representatives of those multiple organizations must accept the requirement. This presumes an organizational hierarchy exists naming the persons or organizations responsible for all end items.
6. Rationale: a brief explanation in plain language stating why the requirement is needed and why presumed correctly stated. Rationales are very difficult and time consuming to write. However, there is often more useful information in the rationale than the requirement statement, and it is my experience that while writing the rationale, you figure out the real requirement.
7. Author: who wrote the requirement with contact information.

   Notice there are no attributes regarding verification. This is because the functional requirement just says "do something," not how well to do it, so verification is addressed with respect to the performance requirements allocated to the functional requirement. Obviously, we do need to ensure all end items have at least one functional requirement and we need to ensure all functional requirements are allocated and accepted by at least one end item, but we check for this without stipulating anymore attributes.

8. Compliance method: document how to implement this requirement. This is an optional attribute, as allocation may be all that is needed to trace the function to and end item. Compliance is more general than allocation, or documenting the basic method that will be used to implement each function—such as simply stating such things as "commercial off the shelf," "reuse from Program x," "modification of part Y from program Z," "new development," and so on. If add this attribute, can easily construct compliance reports that provide an indication of the basic methods that will be used to implement each function.

Performance requirements specify how well a function needs to be performed. More than one performance requirement may be allocated to a functional requirement. Useful attributes for performance requirements are as follows:

1. Identification: a unique number.
2. Status: "draft," "approved," or "discontinued" to denote as stated.
3. Parent(s): the identification of the requirement or requirements from which this requirement is derived. Only another performance requirement can be a parent for a performance requirement.
4. Function allocation: the identification number of the functional requirement to achieve this performance. A performance requirement may be allocated to more than one functional requirement. To enable this to happen we need to agree to a rule on how to do this. If the functions are allocated to multiple items, then clearly each of those end items must independently verify compliance with the same performance requirement. What must be clarified is if the performance requirement is allocated to a function, is it also allocated to all descendants of that function? We can build an approach that allows for either a yes or no answer. I recommend using a process for when the answer is no. I'll call this *explicit allocation*. At first, this may seem to be illogical. Surely, if a requirement is allocated a performance requirement, then for any more detailed subdivision of the function the requirement must still apply for each subfunction? Explicit allocation is both logical and vastly simplifies the verification process. To show the rule is logical, let's revisit our very first simple functional requirement example: "Lift the chair." Lifting a chair has subfunctions of accelerate chair to a velocity, decelerate chair to 0 velocity, and apply force to keep the chair at specified height. A performance requirement for this function might be "so the point of the chair closest to the floor is 0.5 meter give or take 0.05 meter." This performance requirement applies to the first and second subfunctions, but not the third, which only needs to know the mass of the chair. Explicit allocation vastly simplifies the verification planning effort because we need to only verify the performance requirements for the functions to which they have been explicitly allocated, rather than have to decipher if performance requirements are truly applicable to all subfunctions implicitly allocated. For example, if performance requirement **R** is allocated to function **F**, which in turn is allocated to end item **E**, then we must find a way to show **E** does complies with **R**. We do not need to find ways to show every subelement of **E** also complies with **R**. However, if we explicitly allocate **R** to **F**, and **F** to

E, and **R** is also explicitly allocated to a subfunction **Fsub**, which is allocated to a different end item **Esub**, then we must verify both **E** and **Esub** comply with **R**. If we presumed a performance requirement allocated to a function is ALSO allocated to all subfunctions, then when planning verification efforts, we must carefully ascertain to what extent the implicit allocation needs to be included in the verification of the end item. For the example above, once **R** was allocated to **F** which was allocated to **E**, when building the verification plan, we must now ascertain to what extent **R** applies to all sub-functions of **F** that may be allocated to many items and determine for each case whether or not and how to show the end items comply with **R**. This can be done, but is a much more difficult task and more prone to missing a verification that can lead to a failed or unaccepted systems.

5. Verification method: how it is to be shown that the performance is achieved by the end item to which allocated. Options are as follows:

   5.1. Record: Rely on a certification that was previously proven and believed to still be true. For example, a circuit board is presumed to achieve its performance requirements because it is routinely manufactured and the supplier has previously demonstrated it achieves a very low defect rate and continues to test samples of its production run to ensure product remains within tolerance.

   5.2. Inspection: Exercise a defined procedure to examine specified features for conformance. For example, measure the circuit board dimensions and mass to ensure it meets stipulated requirements.

   5.3. Analysis: Predict the performance based on a mathematical model. For example, the circuit board is verified to meet its requirement if a mathematical model of how it translates inputs into outputs shows it complies.

   5.4. Test all, or test samples: To test means to exercise a defined procedure on physical representations of the actual system or architecture, in a defined environment, usually with independent witnesses. For example, the circuit board is subjected to planned inputs while housed in a compartment in which worst case operating temperature, humidity, and electromagnetic environment are maintained, for a specified time, while the real outputs are witnessed to see if they are as required. If multiple end items are to be produced, an important question is, does every end item have to be tested, or just some? If the system is such that no failures can be tolerated, exhaustive testing of all produced end items may be desired, but this will be costly and possibly very difficult if the test itself affects, or even destroys, the end item. We often rely on the assumption that if items are manufactured by the same process, then testing a sample of items may be acceptable to believe all the produced items will perform as required. If we do not test all the items, then the design must be able to tolerate the expected fraction of items likely to be unacceptable. So, the "test all" entry clarifies that every produced end item associated with this requirement must be tested to ensure it meets requirement. The "test samples" entry enables an appropriate sample of end items to be tested, and infers the production process needs to be monitored to ensure the chosen sampling process is trustworthy.

5.5. Demonstration: Utilize in operational environment, usually with independent witnesses. For example, inputs and outputs to the circuit board are monitored in the completed system during real-time operations.

6. Verification acceptance: date and name of the organization (or person with contact information) responsible to conduct the verification effort. The organization that designs the end item that will achieve the performance requirement may also be responsible to verify that it does, or perhaps a different organization, the one producing the end item, is responsible to verify the requirement is met. This attribute again provides a means to help prevent the generation of unneeded or unverifiable requirements.

7. Verification status: one of the following: not verified, in work, or verified.

8. Verification evidence: either explicitly the evidence that shows the requirement is met, or a pointer to or name of a file that contains the evidence.

9. Rationale: an explanation in plain language explaining why the requirement is needed and why presumed correctly stated.

10. Author: who wrote the requirement with contact information.

Note there is no "level" of verification attribute. Common system level names, from highest to lowest, are *architecture, system, segment, subsystem, component, assembly,* and *part.* In this text, verification is conducted at the level of end item allocated the functional requirement to which the performance requirement is allocated. For example, if circuit board performance is to be verified, it would be at the circuit board level. If that circuit board also needs to be verified that it works with additional circuit boards for an end item at a higher level in the system, then a function and associated performance requirements need to be allocated to that end item, for which verification attributes are documented. That is because the performance requirement is allocated to the functional requirement, which is allocated to an organization to provide. When the providing organization "accepts" the functional requirement, they are accepting the allocated performance requirements as well. It's possible the providing organization is quite willing to accept all but one performance requirement; by insisting they then reject the entire functional requirement, we are ensuring the issue gets resolved.

11. Compliance: if the performance requirement stipulates a numerical level to achieve, this attribute enables recording the current predicted or demonstrated value, or the percentile exceeding the requirement. If the performance requirement stipulates adhering to a standard or, for example, materials to avoid, then can record level of current compliance along with method used to comply.

Useful attributes for sought requirements are as follows:

1. Identification: a unique number.

2. Status: "draft," "approved," or "discontinued."

3. Requesting entity: name and contact information for the person who is requesting the requirement.

4. Acceptance record: date and name of systems engineering or architecting organization or person that agreed to be responsible to provide the requirement. This provides a counterpressure to make sure only needed

requirements are generated, not something that might be an optional design detail, or even already specified, but not recognized yet by the requester.

5. Rationale: a brief statement why this requirement is sought.
6. Author: name of person or organization that needs the requirement determined.

Note that neither parent nor verification attributes are listed. The predominant reason to note parent functional and performance requirements is to enable tracing a network relationship amongst the requirements, so just in case a requirement at some level in the network is changed, requirements related to the changed requirement can be found to check if they need to change too. Though the sought requirements may be related, we need not worry about relational changes until they are formal functional or performance requirements, and to be such, we stipulate their parents. Similarly, we don't worry about verifying sought performance requirements until we turn them into approved requirements.

The above are the minimal attributes for each requirement type. In my experience, if you elect to exclude any, you eventually have problems with the verification process. Any attribute the organization finds useful to efficiently do their job can certainly be added. For example, it may be useful to know who has authority to approve to accept or change each requirement. Approval authority, or even an indication of whether or not the customer must approve the requirement or its change, is a simple attribute that can be added. An approval authority attribute was explicitly not included in the recommended set because, in my experience, all attributes are costly to maintain. At some times, the customer may not care about any requirement; at other times the customer may care passionately about any requirement. Also, it is very easy for the approval authority attribute to be used to create information walls between groups, and this usually results in many requirements that should have had much wider review and approval being created and hidden for those who might be impacted by them. Obviously, extreme care is needed when deciding what attributes to add. One should inherently resist adding attributes because it is always significantly easier for the attribute definer to add than it is for the potentially hundreds of people who will need to provide the information for thousands of requirements. A good question to ask yourself before adding an attribute is, "If I had to provide the input for every requirement, how long would it take me including every time it needs to be modified, and will having this attribute really save at least that much time?" Only if the answer is yes, consider adding the attribute. Compliance attributes are suggested as optional for this reason. Having them makes the system engineer's life easier, but for larger systems and architectures, there is a tremendous amount of work required of other people to keep up to date the true current state of compliance, and unless the system engineers and architects create an audit system, they will be uncertain as to the veracity of the attribute values.

4.1.2 Allocate Interface Requirements to an End Item as Any Other Functional Requirement

Because the very nature of an interface presumes at least a two-sided relationship, a tradition has developed to collect interface requirements separately in documents usually with "interface" in the title, that apply uniquely to the parties of the interface. So, an end item has several requirements documents to comply with, its own collection of requirements (often called a specification) plus one or more interface requirement documents. In the days of manually produced paper specifications this was efficient as the interface specifications paperwork need only be distributed amongst the parties of that interface. Yet, every possible interface is a function to be performed with associated performance desired. Clearly both parties need to concur to the requirement statements, as well how the interface will be implemented, but in the end, the "interface" requirements can and should be allocated to those end items just as any other functional requirements that end item is to implement. This can easily be recognized by using the acceptance record attribute to record all parties to the interface requirement.

4.1.3 Treat Compliance Requirements as Any Other Requirement

Inevitably, customers will care very much about some requirements, and presume as "details" many more requirements they trust the provider to determine and comply with, until of course, one of those supposed details becomes a cost, schedule, or performance concern. So, a tradition has developed to package separately requirements to be formally controlled by the customer, from a larger number of requirements that a customer may review or change if desired, for which the providing organization is responsible for "compliance" but controls without explicitly bothering the customer. Compliance requirements typically provide the detailed definition of the operating environments or explicit design details. In many companies this material represents the collective lessons learned by the implementing organization that are passed on to the program performers to help ensure program success by not making a prior mistake again. In the days of manually produced paper specifications, these compliance requirements were often packaged into documents for limited audiences, usually defined by engineering discipline, such as structures, guidance and control, or propulsion. Yet, every compliance requirement is either a function with allocated performance, or a performance requirement that needs to be allocated to a function. Given today's powerful database programs, each and every compliance requirement should be treated no differently than any other requirement. Treating the compliance material as any other requirement means each needs a parent and rationale. This is very important, because otherwise, given the esoteric nature of the material, the implementing organizations often falsely presume the need for the compliance and end up making the "wrong" thing, which isn't discovered until major reviews, or even in integration and test, causing cost and schedule delays to fix the situation.

4.2   WRITE GOOD NATURAL LANGUAGE REQUIREMENTS

Writing good natural language requirements is an art, for which we cannot yet stipulate formal requirements, but can offer heuristics for accomplishment.

The requirement statement may be codified with words such as "shall," "should," or "will," to denote mandatory, preferred, or deferred compliance, respectively. Use of these particular words is a tradition; any words or phrases that communicate the level of desirability for the feature are appropriate, as long as understood by the customer, users, and the rest of the implementation team.

Good functional requirements minimize the reader's uncertainty as to what activities are required to be implemented. Good functional requirements state what is to be done, avoiding stating things not to do, because "Not A" is usually a much bigger set than "A," and some of the items in "Not A" may be perfectly acceptable. Performing a good functional decomposition remains mostly an art. To extent intellectual property rights allow, good functional decomposition can be copied, rather than invented. Nature should be an inspiration. Inevitably, a functional decomposition proceeds to a level at which implementation decisions must be performed, because any additional decomposition is implementation dependent. When this happens, provided good decision-making process are used to choose the implementation option, smile, because you're doing your job well. A good functional decomposition will tend to minimize the unique end items needed, while simultaneously enabling low-cost and reliable end items be used. A good functional decomposition goes only deep enough to uniquely assign each needed activity to an end item. A failed functional decomposition has functions that overlap or exclude an activity that is needed. Try as we might, functional requirements will remain ambiguous since they are general statements of activities sans performance. Functional requirements are routinely read with implementation expectations in mind, and different implementations will infer different meanings. The predominant errors regarding functional requirements are as follows:

1. Missing activities that need to be accomplished
2. Imposing an implementation method prematurely
3. Decomposing functions too much, producing a lot activity names that can be implementation options rather than end item requirements

Good performance requirements minimize the readers' uncertainty as to how well a function needs to be accomplished, and are verifiable. Good performance requirements address what is critical the function accomplish, allowing the designer as much implementation freedom as possible. The thought foremost in mind is to stipulate performance, not implementation. But, as soon as the functional decomposition needs to address implementation decisions to proceed, the performance parameters required will become implementation specific. Again, provided good decision-making processes are used, you get to smile a second time when this starts to happen. Performance requirements are often extremely useful if specifying a permissible range, as this increases the possibility of using existing end items. If performance requirements are to express a minimally acceptable level (threshold) and a desired level (objective), then one must also provide a means to assess the value of providing more performance than threshold level, so rational decisions can be made as to how much resources should be expended to do so. The biggest value system engineering and architecting can provide is to find the total set of performance requirements which collectively balance any conflicting customer and user needs, for which a solution can be provided at lowest cost or quickest. Because performance requirements impact cost and schedule, with the precise consequences uncertain at time of stipulation, then painfully obvious eventually, the biggest error regarding performance requirements is to ask for more than can be provided with resources available, or more than needs to be provided.

Good sought requirements make it clear what information is needed and why. Ideally the statement is written as the requirement needed, but a question is perfectly acceptable.

Sought requirements should be declared as soon as need is known, even if it may take some time before an answer can be provided. The least valuable sought requirement is one that is asked for after the activity dependent on the answer is already underway.

4.3   REDUCE AMBIGUITIES IN THE NATURAL LANGUAGE REQUIREMENT STATEMENT

Natural language requirements are inherently ambiguous due to their very nature. The plethora of possible interpretations of all words used will lead various people to read requirements differently than intended, assuming they actually bother to read them rather than presume they know what they say, or presume they know what they should say in their opinion. A typical English word has 10 synonyms, so even a 4 word requirement statement has 10,000 possible interpretations.

You can minimize these issues by careful use of words and phrases. One method to reduce natural language ambiguity is evident in legal documents where a writing style is enforced by declaring rules concerning the words, phrases, and grammar to use. The use of the style is perfected by extensive education, and ultimately by specialization, as those who are best at it get paid by others to do it for them. New provisions often build upon existing provisions. The new is understood in part by meanings inferred from common experience and extensive dialogue. The resulting prose can be very precise to those schooled in the style rules, but incomprehensible to others.

Herein, formal rules are defined to check for the following types of ambiguities:

1. Compliance level
2. Completeness
3. Precision
4. Comprehension
5. Referencing
6. Vague words
7. Functional requirement
8. Acronyms
9. English unit usage
10. Word emphasis

Also provided are suggestions to minimize the ambiguities.

4.3.1   COMPLIANCE-LEVEL AMBIGUITY

Compliance-level ambiguity is caused by confusion regarding the extent the requirement is to be adhered to. If a requirement is a constraint, then it bounds the acceptable solution space and compliance is mandatory. If a requirement is a preference, then it determines that the goodness of the solution within the constraint space and compliance is optional, but very important to achieving customer satisfaction. Expectations are features the customer wants without an explicit declaration. Not all expectations are practical given cost and schedule constraints. If a requirement is an expectation, the requirement must be written to clearly communicate whether to comply, offer as an option, or defer incorporating the stated feature. Deferring does not mean eliminating;

in the ideal case, *defer* means leaving the possibility of future inclusion easy to do. Any feature that is not required is simply not written. The words *shall*, *should*, and *will* are used here to denote mandatory, optional, and deferred compliance, respectively.

ComplianceLevelChecker performs the compliance level check by implementing these rules.

Check for the following:

Type 1. also, anticipate, apply, applies, are to, aspire, can, could, crave, demand, desire, expect, force, forcing, ideally, goal, got to, has to, is to, might, must, necessary, necessitates, need, needed, needs, obligate, obligation, require, prefer, preference, should, stipulate, want, wants, will, would

Type 2. can't, don't, mustn't, needn't, not, shouldn't, won't

Issue warnings:

Using <check for text> could cause compliance level ambiguity.

Provide suggestions:

For type 1:

If feature **M** is mandatory for entity **E**, use: The **E** shall **M**.
If feature **O** is optional for entity **E**, use: The **E** should **O**.
If feature **D** is deferred for entity **E**, use: The **E** will **D**.

If entity **E** may exhibit feature **V** with values between **V1** and **V2**, use: The **E** shall **V** between **V1** and **V2** with equal preference over the range.

If entity **E** must exhibit feature **V** at **Vt**, but is desired to be **Vo**, use: The **E** shall **V** with threshold **Vt** and objective **Vo** with preference **Vo** = **n** **Vt** (where **n** >= 1 indicating approximately how much more valuable **Vo** is than **Vt**).

For type 2:

Phrase as a statement of inclusion rather than exclusion.

4.3.2 COMPLETENESS AMBIGUITY

Completeness ambiguity is caused when missing information is included in the requirement statement. The rules for checking completeness are as follows:

Check for the following:

not known, tbd, tbr, tbs, tbx, to be determined, to be provided, to be reviewed, to be specified, to be supplied, unknown, ?

Issue warnings:

Using <check for text> could increase cost or delay receipt.

Provide suggestions:

Delete the requirement until complete, or provide the incomplete information.

CompletenessChecker implements the completeness ambiguity check.

4.3.3 Precision Ambiguity

Precision ambiguity is uncertainty in how to interpret numerical information. The rules for precision ambiguity check are as follows.

Check for the following:

Word is type 1, 2, 3, or 4 defined below, and next word is a number using digits or letters, where:

Type 1 is above, at least, minimum of, no less than, not less than, not to be less than, exceed.

Type 2 is at most, below, maximum, no greater than, not greater than, not to be greater than, under, up to, within.

Type 3 is about, almost, approximately, at, between, close to, exactly, give or take, more or less, near, of, or so, plus or minus, roughly, tolerance, use, +/–.

Type 4 is average.

Issue warnings:

Using <check for text> could cause precision ambiguity.

Provide suggestions:

For type 1:

For a feature **F** to be >(=) **L**, an unambiguous statement to use is as follows:

The probability **F** is greater than (or equal to) **L** shall be **p**.

An equally unambiguous statement to use is as follows:

The degree of belief for **F** shall be (insert fuzzy **L**).

The most ambiguous statement to use is as follows:

The **F** shall be greater than (or equal to) **L**.

For type 2:

For a feature **F** to be <(=) **U**, an unambiguous statement to use is as follows:

The probability **F** is less than (or equal to) **U**, shall be **p**.

An equally unambiguous statement to use is as follows:

The degree of belief for **F** shall be (insert a fuzzy **U**).

The most ambiguous statement to use is as follows:

The **F** shall be less than (or equal to) **U**.

For type 3:

For feature **F** to be countable, use the following:

The **F** shall be **C**.

For feature **F** to be >(=) **L** and <(=) **U**, an unambiguous statement to use is as follows:

The probability **F** is greater than (or equal to) **L** and less than (or equal to) **U** shall be **p**.

An equally unambiguous statement to use is as follows:

The degree of belief for **F** shall be (insert fuzzy number that spans **L** and **U**).

The most ambiguous statement to use is as follows:

The **F** shall be greater than (or equal to) **L** and less than (or equal to) **U**.

For type 4:

For average: use mean, median, mode, and arithmetic or geometric, or explain how average is to be determined.

PrecisionChecker implements the precision ambiguity check.

4.3.4   COMPREHENSION AMBIGUITY

Comprehension ambiguity is caused by complex syntax that potentially hinders understanding the requirement. The rules for comprehension ambiguity check are as follows.
   Check for the following:

Type 1. Using more than two instances of *shall*, a semicolon, or a colon.
Type 2. Using more than two instances of *and* or *or*.
Type 3. Using more than one *but*.

   Issue warnings:

Type 1: >= <shall or colon limit> *shall* statements or clauses separated by colons could hinder comprehension.
Type 2: >= <and or limit> *and* or *or* could hinder comprehension.
Type 3: Using *but* could hinder comprehension.

   Provide suggestions:

Replace statement with multiple statements of the form:

(If **C1** then) the **E1** (shall, should, will) be **A1** ((else if **C2** then) the **E2** (shall, should, will) be **A2**) ...), where **Ci** are conditions, **Ei** are entities, and **Ai** are attributes.

ComprehensionChecker implements the comprehension ambiguity check.

4.3.5   REFERENCING AMBIGUITY

Referencing ambiguity occurs when citing other information sources that might not be known or followed. The rules for referencing ambiguity check are as follows:

Check for the following:

as defined, as shown, as specified, comply with, per, refer to, referenced, see, under

Issue warnings:

<check for text> information could be misinterpreted.

Provide suggestions:

Include information in external reference to comply with as additional requirements.

SiteReferenceChecker implements the check for referencing ambiguity.

4.3.6   VAGUE WORDS AMBIGUITY

Vague words are adjectives or adverbs that may not be interpreted as intended. The rules for vague words ambiguity check are as follows:
Check for the following:

All adjectives and adverbs in Lutz,[2] excluding those already listed in other style check requirements, with next word not a number.

Issue warnings:

<check for text> could be vague.

Provide suggestions:

Ask questions of those responsible to comply with requirement to determine if they interpret as intended. If not, replace with requirement(s) that provide correct answers to the question.

VaguenessChecker implements the check for vague words ambiguity. *Vagueness Checker* needs two files, **vaguewordswithbreaksfile**, and **vaguewordsfile**, holding potentially vague phrases and individual words. As each file contains a large number of entries they are not presented here, but are available online with the formal requirements.

4.3.7   FUNCTIONAL REQUIREMENT

The ambiguity checker attempts to identify likely functional requirements and suggest a specific sentence format. The rules for functional requirement check are as follows:
Check for the following:

able to, achieve, act, accomplish, be capable of, bring about, capability to, carry out, code, collect, command, complete, design, develop, disseminate, dispose, do, establish, execute, fix, fulfill, instruct, maintain, manufacture, move, operate, perform, program, provide, pull, push, realize, record, repair, ride, roll, support, teach, train, transport, watch, witness

Issue warnings:

<check for text> suggests this statement is a functional requirement.

Provide suggestions:

For entity **E** to have function **F**, use "A function of **E** shall be to **F**."
For entity *E* to have a subfunction **S** of **F**, use "A subfunction of F for E shall
be to S."

FunctionalRequirementChecker implements the check for functional requirement ambiguity.

4.3.8   ACRONYM AMBIGUITY

Acronym ambiguity occurs whenever acronyms are used but not defined. The list of
potential acronyms is very large. What can readily be checked is if a used acronym is
on an approved list with definitions. Unfortunately, a text matching program cannot
determine if an acronym that is identical to one on the approved list is used properly.
The rules for acronym ambiguity check are

Check for the following:

Approved acronyms in an array with each element the string representation of
the approved and defined acronym.

Issue warnings:

<check for text> could be an acronym that may not be understood.

Provide suggestions:

Spell out the acronym, or make sure definition is known by reader.

AcronymChecker implements the check for acronym ambiguity.

4.3.9   ENGLISH UNIT AMBIGUITY

Presuming a program has standardized the use of the metric system, English unit
ambiguity occurs when English units of measure are used. The English unit ambiguity check rules are as follows.

Check for the following:

mil, inch, inches, in, foot, feet, ft, yard, yards, yd, yds, mile, miles, mi, nm, fathom,
fathoms, league, parsec, astronomical unit, IAU, AU, light year, pica, acre,
knot, fps, ft/sec, mph, g, acre foot, barrel, board foot, gallon, gallons, gal, pint,
pints, pt, quart, quarts, qt, cup, cups, tablespoon, tablespoons, tbs, teaspoon,
teaspoons, tsp, ounce, ounces, oz, btu, calorie, cal, kilocalorie, kcal, therm,

quad, eV, kilowatt hour, kilowatt-hour, kw-hr, kw hr, kwh, kwhr, foot pound, foot-pound, foot poundal, foot-poundal, ft lb, ftlb, ft-lbf, ft lbf, explosive energy of TNT, ton equivalent of TNT, horsepower, hp, ft-lbf/sec, ft lbf/sec, ton of refrigeration, pound, lb, lbs, poundal, kip, lbf, lbm, grain, slug, ton, atm, torr, bar, psf, psi, centimeter of mercury, centimeter of water, foot of water, fahrenheit, F, celsius, C, rankine, R, deg, revolutions, rev, RPM, Faraday, Gilbert, statampere, statvolt, statohm, curie, footcandle, footlambert, lambert, roentgen

Issue warnings:

Using: <check for text> could denote an English unit of measure.

Provide suggestions:

Convert to <metric equivalent for check for text> as the unit of measure.

UnitChecker implements the English unit ambiguity check.

4.3.10   Word Emphasis Ambiguity

Nouns and verbs have many potential interpretations. A way to determine a requirement statement is being interpreted as intended is to substitute synonyms for each noun or verb and determine if any resulting modified requirement statement is preferable. The rules for synonym ambiguity are as follows.

Check for the following:

All the synonyms for all the nouns and verbs found in Lutz.[2]

Issue warnings:

None.

Provide suggestions:

If the synonyms for word **W** are **S1, S2, S3**, and so on, for **W**, consider **S1, S2, S3**, and so on.

WordEmphasisChecker implements word emphasis check an ambiguity check. *WordEmphasisChecker* imports files of nouns and verbs along with alternative meanings. Each of these files can be found online with the formal requirements definitions, to avoid having to essentially reproduce a dictionary in this text.

4.3.11   Requirement Statements for Ambiguities

ReqCheck executes each of the above ambiguity checks to identify draft requirement statement ambiguities. *ReqCheck* takes as input **text** which is the requirement statement in quotes, and a logical array string **docheck** for which if **docheck[[i]]** is **True**, then ambiguity check **i** is performed, if **False**, ambiguity check **i** is not performed.

Here is an example application:

ReqCheck['Ideally the system should have a signal to noise ratio of 12 dB (TBR).",
(True, True, True, True, True, True, True)]

The requirement is:

Ideally the system should have a signal to noise ratio of 12 dB (TBR).

POSSIBLE COMPLIANCE AMBIGUITY

Ideally the system should have a signal to noise ratio of 12 dB (TBR).

Using: 'ideally' could cause compliance level ambiguity

SUGGESTION

If feature M is mandatory for entity E, use: 'The E shall M.'
If feature O is optional for entity E, use: 'The E should O.'
If feature D is deferred for entity E, use 'The E will D.'
If entity E must exhibit feature V between V1 and V2, use: 'The E shall V from V1 to V2 with equal preference.'
If entity E must exhibit feature V at Vt but is desired to be Vo, use: 'The E shall V with threshold Vt and objective Vo, with preference Vo = n Vt.'
(n is a number ≥1 indicating approximately how much more valuable Vo is than Vt)

Ideally the system **should** have a signal to noise ratio of 12 dB (TBR).

Using: 'should' could cause compliance level ambiguity

SUGGESTION

If feature M is mandatory for entity E, use: 'The E shall M.'
If feature O is optional for entity E, use: 'The E should O.'
If feature D is deferred for entity E, use 'The E will D.'
If entity E must exhibit feature V between V1 and V2, use: 'The E shall V from V1 to V2 with equal preference.'
If entity E must exhibit feature V at Vt but is desired to be Vo, use: 'The E shall V with threshold Vt and objective Vo, with preference Vo = n Vt.'
(n is a number ≥1 indicating approximately how much more valuable Vo is than Vt)

POSSIBLE COMPLETENESS AMBIGUITY

Ideally the system should have a signal to noise ratio of 12 dB (**TBR**).

Using: 'tbr' could increase cost or delay receipt

SUGGESTION

Delete requirement until complete, or provide incomplete information.

POSSIBLE PRECISION AMBIGUITY

Ideally the system should have a signal to noise ratio **of** 12 dB (TBR).

Using: ' of ' could cause precision ambiguity

SUGGESTION:

For feature F to be countable, use 'The F shall equal C.' For feature F to be $>(=)$ L
and $<(=)$ U, an unambiguous statement to use is: 'The probability F is greater
than (or equal to) L and less than (or equal to) U shall be P.' An equally
unambiguous statement to use is: 'The degree of belief for F shall be (insert
fuzzy number that spans L and U).' The most ambiguous statement to use is:
'The F shall be greater than (or equal to) L and less than (or equal to) U.'

POSSIBLE VAGUE WORDS

Ideally the system should have a signal to noise ratio of 12 dB (TBR).

'Ideally' could be vague

SUGGESTION

Ask questions of those responsible to comply with requirement to determine if
they interpret as intended.
If not, replace with requirements(s) that provide the correct answers to the questions.

SYNONYMS FOR KEY WORDS

Ideally the **system** should have a signal to noise ratio of 12 dB (TBR).

For 'system' consider:
  1: (a set of principles) order, regularity, rules, structure, worldview,
     Weltanschauung; 2: (a way of operating) method, mode, scheme,
     way, custom, practice, modus operandi, plan, procedure, routine,
     technique; 3: (a way of organizing) organization, structure,
     method, protocol, framework, hierarchy, construct, edifice, taxonomy

Ideally the system should **have** a signal to noise ratio of 12 dB (TBR).

For 'have' consider:
  1: possess, hold, get, procure, acquire, obtain, secure, own, accept; comprise,
     contain, include; 2: bear, beget, bring forth, deliver; 3: deceive,
     cheat, dupe, fool, outwit, swindle, trick; 4: (have at) strike, thrust,
     attack; 5: (have done with) cease, desist, finish with, give up, leave,
     quit, stop, throw over, wash one's hands of; 6: (have to be) obliged,
     must, ought, should, be compelled, be forced, be obliged, be required

Ideally the system should have a **signal** to noise ratio of 12 dB (TBR).

For 'signal' consider:
  noun - sign, beacon, flag, flare, watchword, alarm, omen; verb - beckon,
     motion, gesture, nod, wave, alert

Ideally the system should have a signal to **noise** ratio of 12 dB (TBR).

For 'noise' consider:
  sound, clamor, tumult, racket, fracas, din, pandemonium, turbulence, caterwauling,
     blare, clatter, commotion, babble, stridency, uproar, ballyhoo, hubbub

Ideally the system should have a signal to noise **ratio** of 12 dB (TBR).

For 'ratio' consider: proportion, share, percentage,
     balance, correlation, rate, correspondence, quotient, relationship

Here is a second example application of *ReqCheck*:

```
ReqCheck["The OTA primary mirror diameter shall be exactly 8 feet.",
 {True, True, True, True, True, True, True, True, True, True}]
```

The requirement is:

The OTA primary mirror diameter shall be exactly 8 feet.

POSSIBLE PRECISION AMBIGUITY

The OTA primary mirror diameter shall be **exactly** 8 feet.

Using: 'exactly' could cause precision ambiguity

SUGGESTION:

For feature F to be countable, use 'The F shall equal C.' For feature F to be >(=) L and <(=) U, an unambiguous statement to use is: 'The probability F is greater than (or equal to) L and less than (or equal to) U shall be P.' An equally unambiguous statement to use is: 'The degree of belief for F shall be (insert fuzzy number that spans L and U).' The most ambiguous statement to use is: 'The F shall be greater than (or equal to) L and less than (or equal to) U.'

POSSIBLE ACRONYM AMBIGUITY

The **OTA** primary mirror diameter shall be exactly 8 feet.

'OTA' could be an acronym that may not be understood

SUGGESTION

Spell out the acronym or make sure definition is known by reader

POSSIBLE UNITS AMBIGUITY

The OTA primary mirror diameter shall be exactly 8 **feet**.

Using: 'feet' could denote an English unit of measure

SUGGESTION: Convert to 'meter' as the unit of measure.

POSSIBLE VAGUE WORDS

The OTA **primary** mirror diameter shall be exactly 8 feet.

'primary' could be vague

SUGGESTION

Ask questions of those responsible to comply with requirement to determine if they interpret as intended.
If not, replace with requirements(s) that provide the correct answers to the questions.

SYNONYMS FOR KEY WORDS

The OTA primary **mirror** diameter shall be exactly 8 feet.

For 'mirror' consider:
 noun - copy, double, replica, spitting image, likeness, representation, twin; verb - copy, emulate, imitate, mimic, reflect, represent

The OTA primary mirror diameter shall **be** exactly 8 feet.

For 'be' consider:
 1: live, exist, endure, persist, subsist, breathe, abide, survive; 2: happen, occur, transpire, take place, come about; 3: equal, comprise, amount to, consist of 4: mean, signify, denote, explain

A third example of *ReqCheck*:

```
ReqCheck[
 "The system mustn't operate when the temperature is greater than 200F, or the pressure
  exceeds 100lbf, or the average humidity its too high; under these
  conditions comply with requirements recorded elsewhere.",
 {True, True, True, True, True, True, True, True, True, True}]
```

The requirement is:

*The system mustn't operate when the temperature is greater than
  200F, or the pressure exceeds 100lbf, or the average humidity its too
  high; under these conditions comply with requirements recorded elsewhere.*

POSSIBLE FUNCTIONAL REQUIREMENT

The system mustn't **operate** when the temperature is greater than
  200F, or the pressure exceeds 100lbf, or the average humidity its too
  high; under these conditions comply with requirements recorded elsewhere.

Using: 'operate' suggests this statement is a functional requirement

The system mustn't operate when the temperature is greater than
  200F, or the pressure exceeds 100lbf, or the average humidity its too
  high; under these conditions comply with requirements **record**ed elsewhere.

Using: 'record' suggests this statement is a functional requirement

SUGGESTION

For entity E to have a function F, use:
'A function of E shall be to F.'
For entity E to have a sub-function S of function F, use:
'A sub-function of F for E shall be to S.'

POSSIBLE COMPLIANCE AMBIGUITY

The system **mustn't** operate when the temperature is greater than
  200F, or the pressure exceeds 100lbf, or the average humidity its too
  high; under these conditions comply with requirements recorded elsewhere.

Using: 'mustn't' could cause compliance level ambiguity

SUGGESTION

Phrase as a statement of inclusions rather than exclusion.

POSSIBLE PRECISION AMBIGUITY

The system mustn't operate when the temperature is greater than
  200F, or the pressure **exceed**s 100lbf, or the average humidity its too
  high; under these conditions comply with requirements recorded elsewhere.

Using: 'exceed' could cause precision ambiguity

SUGGESTION:

For feature F to be >(=) L, an unambiguous statement to use is: 'The probability F is
  greater than (or equal to) L shall be P.' An equally unambiguous statement
  to use is: 'The degree of belief for F shall be (insert fuzzy L)'. The most
  ambiguous statement to use is: 'The X shall be greater than (or equal to) L.'

POSSIBLE PRECISION AMBIGUITY

The system mustn't operate when the temperature is greater than
  200F, or the pressure exceeds 100lbf, or the **average** humidity its too
  high; under these conditions comply with requirements recorded elsewhere.

Using: 'average' could cause precision ambiguity

SUGGESTION:

For 'average', use: 'mean', 'median' or 'mode', and
  'arithmetic' or 'geometric', or otherwise state how to determine value.

```
POSSIBLE COMPREHENSION AMBIGUITY
```

The system mustn't operate when the temperature is greater than
   200F, **or** the pressure exceeds 100lbf, **or** the average humidity its too
   high; under these conditions comply with requirements recorded elsewhere.

≥2'and' or'or' could hinder comprehension

```
SUGGESTION
```

Replace statement with statement(s) of the form:
'(If C1 then) the E1 (shall, should, will) be
   A1 ((else if C2 then) the E2 (shall, should, will) be A2)...)'
where Ci are conditions, Ei are entities and Ai are attributes.

```
POSSIBLE EXTERNAL INFORMATION AMBIGUITY
```

The system mustn't operate when the temperature is greater than
   200F, or the pressure exceeds 100lbf, or the average humidity its too
   high; under these conditions **comply with** requirements recorded elsewhere.

'comply with' information could be misinterpreted

The system mustn't operate when the temperature is greater than
   200F, or the pressure exceeds 100lbf, or the average humidity its too
   high; **under** these conditions comply with requirements recorded elsewhere.

'under ' information could be misinterpreted

```
SUGGESTION
```

Include information in external reference to comply with as additional requirements.

```
POSSIBLE UNITS AMBIGUITY
```

The system mustn't operate when the temperature is greater than
   200F, or the pressure exceeds 100lbf, or the average humidity its too
   high; under these conditions comply with requirements recorded elsewhere.

Using: 'F' could denote an English unit of measure

SUGGESTION: Convert to 'degrees Kelvin' as the unit of measure.

```
POSSIBLE UNITS AMBIGUITY
```

The system mustn't operate when the temperature is greater than
   200F, or the pressure exceeds **100lbf,** or the average humidity its too
   high; under these conditions comply with requirements recorded elsewhere.

Using: 'lbf' could denote an English unit of measure

SUGGESTION: Convert to 'Newton' as the unit of measure.

```
POSSIBLE VAGUE WORDS
```

The system mustn't operate when the temperature is greater than
   200F, or the pressure exceeds 100lbf, or the average humidity **its** too
   high; under these conditions comply with requirements recorded elsewhere.

'its' could be vague

The system mustn't operate when the temperature is greater than
   200F, or the pressure exceeds 100lbf, or the average humidity its **too**
   high; under these conditions comply with requirements recorded elsewhere.

'too' could be vague

The system mustn't operate when the temperature is greater than
   200F, or the pressure exceeds 100lbf, or the average humidity its too
   high; **under** these conditions comply with requirements recorded elsewhere.

'under' could be vague

```
SUGGESTION
```

Ask questions of those responsible to comply with requirement to determine if
they interpret as intended.
If not, replace with requirements(s) that provide the correct answers to the questions.

```
SYNONYMS FOR KEY WORDS

The system mustn't operate when the temperature is greater than
   200F, or the pressure exceeds 100lbf, or the average humidity its too
   high; under these conditions comply with requirements recorded elsewhere.

For 'system' consider:
   1: (a set of principles) order, regularity, rules, structure, worldview,
      Weltanschauung; 2: (a way of operating) method, mode, scheme,
      way, custom, practice, modus operandi, plan, procedure, routine,
      technique; 3: (a way of organizing) organization, structure,
      method, protocol, framework, hierarchy, construct, edifice, taxonomy

The system mustn't operate when the temperature is greater than
   200F, or the pressure exceeds 100lbf, or the average humidity its too
   high; under these conditions comply with requirements recorded elsewhere.

For 'operate' consider:
   1: manage, handle, maneuver, manipulate, engage, use, utilize; perform, run, function,
      act, move, work; 2: (perform surgery) treat, administer, serve, excise, cut, explore

The system mustn't operate when the temperature is greater than
   200F, or the pressure exceeds 100lbf, or the average humidity its too
   high; under these conditions comply with requirements recorded elsewhere.

For 'temperature' consider: thermal reading, heat, warmth, cold, body heat

The system mustn't operate when the temperature is greater than
   200F, or the pressure exceeds 100lbf, or the average humidity its too
   high; under these conditions comply with requirements recorded elsewhere.

For 'average' consider: norm, mean, standard, medium, par, rule, mediocrity, midpoint

The system mustn't operate when the temperature is greater than
   200F, or the pressure exceeds 100lbf, or the average humidity its too
   high; under these conditions comply with requirements recorded elsewhere.

For 'comply' consider:
   observe, respect, oblige, satisfy; accommodate, conform, toe the line,
      acquiesce, defer, accede, submit, yield
```

4.4   DETERMINE THE NATURAL LANGUAGE REQUIREMENTS

There are many candidate processes to determine natural language requirements. The following sections summarize each process, strengths and weaknesses, and when best to use.

4.4.1   Natural Language Requirements by Reusing from Prior Programs

Prior systems or architecture may provide a wealth of established requirements. Prior requirements are beneficial in several ways. First, if the end items created for prior programs are acceptable for use in a new program, then the entire requirement establishment process for those elements is radically simplified. If the instructions to build the end items were retained and can simply be used again, avoiding all that work and time delay. Some caution is warranted. For some systems the time lag between requirement generation and end item production can be so long, that the material or processes to produce the end items may no longer be available. Also, how the end items are produced may now be different than how the original requirements stipulate. So even when existing end items could do the needed job, it is necessary to confirm the end items can still be produced and identify the appropriate requirements to use for their verification. If an end item for a prior program was new, but is now routinely produced, then it can simply be bought. But care should still be taken to update the verification attributes associated with the requirements. Second, end items created for prior programs may be usable with modification. Again, if the original requirements were not lost, they can be reviewed and edited, which is almost always much more efficient than starting with a clean sheet of paper. Third, even when the new program needs substantially different end items, the program is likely to need requirements addressing the same topics as those for a prior related program. At a minimum then, reviewing prior program requirement data can help identify the requirements that need to be established. If the new program and prior program share the same environments, then requirements defining the environment may be reused with little or no change, even though the end items sought to operate in those environments may be different. Fourth, requirements established to ensure end items are well produced and easy to support are almost always reusable, checking only that they apply to the reality of the end items to be produced.

4.4.1.1 Process to Determine Natural Language Requirements by Reusing from Prior System or Architecture

Step 1. Prepare the structure of requirements that the new system or architecture needs by identifying primary and subordinate functions, and associated performance criteria, acknowledging likely implementation elements.

Step 2. Gather existing requirement definitions and verification evidence from prior system or architecture and sort by new program functions and implementation elements. Note that the process of relating the prior requirements to the new program needs may suggest changes to the result of Step 1; if so, make those changes.

Step 3. For each group of prior requirements sorted by new functions, make decision whether can use as is, can use with modification, and cannot use.

Step 4. For "can use" prior requirements, confirm applicability to new program.

Step 5. For "can use with modification" prior requirements, decide how to modify.

Step 6. Complete attribute data for each requirement retained.

Step 7. Formally review and approve additions to program requirement database.

Step 8. Identify the remaining requirements to establish.

4.4.1.2 Strengths

The requirements are determined very quickly with few errors.

4.4.1.3 Weaknesses

Requirements may not have "aged" well, invoking features that may no longer be doable. Or, the requirement documentation available may not have kept up with the reality of how the end items are now produced, so if used without checking, the requirement set will not be relevant to the end items. Or, if the requirements are sufficiently old, much better approaches may now be available. Too forceful an insistence of reuse may stifle innovation, but this is mitigated somewhat if the requirements reused are implementation neutral.

4.4.1.4   Best Use

Whenever possible.

4.4.2   Natural Language Requirements by Interpreting Customer Provided Documents

A customer may provide one or more documents with statements and diagrams indicating requirements. If so, the documentation is parsed into the requirements statements, either by finding the items the customer was kind enough to formally identify as requirements, or by reading each sentence and examining each diagram and creating a requirement statement judged to be an accurate representation of what the customer seems to want. If attributes are not provided, they need to be hypothesized. Although the provided requirements may be quite specific, it is unusual that the customer will provide every requirement needed, so additional requirements need to be derived.

4.4.2.1   Process to Determine Requirements by Interpreting Customer Provided Documents

Step 1. Itemize every statement that may be construed as a candidate customer requirement into a functional, performance, or sought requirement. This may be very straightforward, as the customer specifically denotes statements as requirements, or may take considerable interpretation.

Step 2. Make sure to comprehend each requirement, both as it stands alone and in context with other relevant requirements. If the customer did not provide attribute information, hypothesize the necessary attributes.

Step 3. To extent practical, get confirmation to understand each requirement as customer intended by discussing any uncertainties with customer.

Step 4. As the customer is human, it is possible alternate requirements may better express what they really want, so to extent customer allows, recommend less ambiguous requirements for customer to adopt, or alternative requirements that achieve their apparent desires sooner, or at lower cost or risk.

Step 5. Determine for each comprehended requirement whether it is a constraint, preference, or expectation. If constraint, impose upon program. If preference, make decision as to how much of the optional feature to offer. If expectation, make decision whether constraint, option (and therefore how much to offer), or to defer, and if defer, how will defer so can most easily be added in the future.

Step 6. Formally review and approve additions to program requirement database.

4.4.2.2   Strengths

A great deal of what the customer may want is explicitly declared.

4.4.2.3   Weaknesses

Though the requirements are declared, that does not mean you comprehend them. The effort to comprehend the requirements may be quite considerable. Often when the customer goes to such extremes, they are soliciting competitive bids on the system or architecture they seek. To be fair to all bidders, they will be reluctant to explain the requirements to one bidder without the other bidders present, or to avoid the hassle, the customer may simply not want to discuss the requirements in depth. The customer is human too, and writing good requirements is difficult, so it is quite possible some requirements fail to actually communicate what the customer wants. Most customers recognize these potential problems, so they often will hire the potential providers to help develop the requirements. In theory, this should provide near-perfect comprehension of what the customer wants. But to achieve this clarity, you have to be very careful to listen to the customer reaction to what you say you can offer, rather than push the solution you feel you can provide better than your competitors. True, it's possible the customer will concur and stipulate requirements that are best fulfilled by your offering, but more likely they will know what they like best from all the potential providers and will ask for a combination of the features. In your zeal to try to convince your customer to slant the requirements in your favor, you may lose lock on what the customer wants, as you are blinded by what you want the customer to want. Sometimes the requirements do not state what the customer really wants, but indicate what their superiors want communicated, and the real requirements go unstated. For example, who approves the release of the requirement document may want a very mature, proven solution, while the direct customer may want to push the state of the art. The official requirement document may reflect the wishes of the authority with the power to approve, but the selection process may reflect the wishes of the direct acquirer. Or, the customer may perceive all available providers as roughly equally good at providing the desired solution, and what they really want is to work with the people they are most comfortable with. This overwhelming requirement may go completely unstated. For complicated systems, you will often find one or more subsets of the customer community responsible for an aspect of the system or architecture will appear to overspecify requirements. That is, they will provide so many detailed requirements they have essentially locked in the design. If the design works, give it to them. If the requirements stipulate a design that will not work, you must try to communicate this to the customer, with the best unemotional evidence you can muster. If the evidence is compelling, the customer will change their mind and usually be very grateful. If you cannot convince the customer to change unachievable requirements, then you have to decide if circumstances may change and the customer's opinion will change before it becomes critical, or to not offer a solution. Finally, it is unlikely the customer will stipulate every requirement needed to implement a solution, so the job will remain to derive the remaining necessary requirements.

4.4.2.4   Best Use

Whenever possible.

4.4.3   NATURAL LANGUAGE REQUIREMENT BY SURVEYS

A questionnaire is prepared and given to representative customers or users with their answers used to determine requirements.

4.4.3.1   Process to Determine Natural Language Requirements by Surveys

Depending on the circumstances, people asked to complete questionnaires may or may not want to. A general recommendation then is to keep any questionnaire as easy and quick to complete as possible, and therefore of limited focus.

Step 1. Establish what needs to be known.
Step 2. Determine whom to provide questionnaire to that will reliably provide the needed information.
Step 3. Determine how many people need to participate to be a valid input and by what means to solicit participation.
Step 4. For the target information providers, determine the best way to deliver the questionnaire, for example, by mail or email, by website, by calling, or by visiting in person.
Step 5. For the target information providers, determine the format of the input, that is, what to ask and how answers are to be obtained. Options include statements to mark *true*, *false*, or *don't know/not applicable*; statements to show a level of agreement or compliance with; reaction to an image; or free-form input.
Step 6. Prepare a draft survey and conduct at least one test data-gathering exercise. Modify survey based on results of test.
Step 7. Distribute the survey to target information providers.
Step 8. Tabulate the results and formulate corresponding requirements.
Step 9. Formally review and approve additions to program requirement data base.

4.4.3.2   Strengths

One can obtain direct customer or user input regarding requirements trying to ascertain. Surveys can be particularly useful to ascertain the relative importance customers and users place on conflicting requirements.

4.4.3.3   Weaknesses

Searching for requirements via surveys almost always yields conflicting requirements due to variation in what people want. So all such methods need a means to resolve these conflicts. One can attempt to appease the majority, or each cluster of similar interest by offering options, or perhaps an innovative solution can be found to appease the apparent contradictory requirements. Extreme care must be taken in selecting the survey respondents. Should the survey be given to a random sample of the all potential customers and users? Should the survey be given to those believed to be the most important customer and users? How do you really ascertain relative customer importance? Surveys themselves are notoriously easy to produce but often prove worthless. I wish I had a dime for every survey I completed that was so poorly designed that it is obvious the originator never bothered to attempt to complete it. Always test the survey on a sample population to be sure your questions are interpreted as you desire and you collect the type of input you need. We now live in a survey crazy society—you cannot rent something, stay somewhere, or browse the Internet without being asked to complete a survey. You must design the survey to be as easy and quick to complete, or the people you ask to participate may simply decline to complete the survey. The survey will not provide requirement information beyond that you design into the survey, so you could miss asking about things critical to the customer. Multiple surveys may be required, some exploring for topics of importance, not just details. Also, surveys rarely enable you to find out the customer wants something they do not have. Survey results can be dangerous as they can legitimize wrong requirements. For example, if there are 100 people affected by the system or architecture, and you get survey response from 25, how do you know that the conclusions based on those 25 responses will be agreeable to the other 75 people? It is quite possible the 25 have strong biases not representative of the majority. Finally, few customers take surveys completely seriously, so it's a good idea to find independent ways to collaborate survey conclusions.

4.4.3.4   Best Use

To establish details for narrow topics for which random sampling of users is likely to represent majority opinion or detailed sampling of most affected users is possible.

4.4.4   NATURAL LANGUAGE REQUIREMENTS BY WITNESS

The requirements authors watch the customer or user in action, so the authors can comprehend features that could enhance results or ease the work load for the users.

4.4.4.1   Process to Determine Natural Language Requirements by Witness

Step 1. Establish what needs to be known.
Step 2. Determine whom to witness doing what to get the information needed. Determine to what extent you need to protect those witnessed from having their identity become known. Often you will witness errors, and few people want their errors to be public.
Step 3. Obtain permission to witness. Sometimes the customer or user is very happy and willing to be witnessed; if so, seek their active input to accomplish Steps 1, 2, and 4.
Step 4. Determine how to capture the observations; examples are video recording, audio recordings, sketches, or notes. Be very clear in advance what you are looking for: duration of activities, range of activities, error rates, delays, and so on.
Step 5. Witness the activity, and then collect data. Much more may be going on than you expected, or much faster than you can accurately record, or even much slower than you anticipated. So repeat from Step 4 as needed and can.
Step 6. To fullest extent possible, have those whom you witnessed endorse at least the accuracy of the record. You may not have witnessed what you think you witnessed.
Step 7. Prepare draft findings into requirements. Again, to the fullest extent possible, share with those you witnessed to get their concurrence.
Step 8. Tabulate the results and formulate corresponding requirements
Step 9. Formally review and approve additions to program requirement database.

4.4.4.2 Strengths

One can get deep insight into user needs.

4.4.4.3 Weaknesses

Witnessing is time consuming and expensive. The performers need deep domain knowledge to comprehend what they are witnessing and to be able to visualize alternative functions. Not all users will welcome being witnessed, and there is always the possibility user behavior will be different when witnessed than when not. Though witnessing is quite powerful to determine what could benefit the user, that customer may not seek the system or architecture implied.

4.4.4.4 Best Use

To find requirements for potential enhancements to existing system or architectures or to get ideas for features for new systems or architecture.

4.4.5 Natural Language Requirements by Focus Groups

A small number of people are invited to review and discuss a potential offering, with their reactions recorded to help determine what features they favor and which they do not.

4.4.5.1 Process to Determine Natural Language Requirements by Focus Groups

Step 1. Establish what needs to be known.
Step 2. Determine how many people need to participate to obtain a valid input, by what means to solicit participation, and what will provide to people in exchange for participating.
Step 3. Determine the best way to present the material to be reviewed to the target participants and the best means to record their reactions. Basically, you must prepare a script for the presentation while also putting in place a means to record reaction, be it video recording, audio recording, asking the participants to complete prepared surveys, or having witnesses take notes.
Step 4. Determine whom to invite to be a focus group member and how much of their time will be need.
Step 5. Find or create an appropriate place to host the meeting and record the happenings.
Step 6. Perform at least one trial focus group meeting, and modify output of Steps 1 to 5 above based on what is learned.
Step 7. Conduct one or more focus group meetings.
Step 8. Tabulate the results and formulate corresponding requirements.
Step 9. Formally review and approve additions to program requirement database.

4.4.5.2 Strengths

You get direct input from potential customers or users. If the topic is narrow and well understood and important to the participants, you can get clear and valuable insight into what they want or need.

4.4.5.3   Weaknesses

All the same weaknesses as for surveys, only more so. Plus, you face the likelihood that the personality of one or a few of the participants will drown out inputs from many of the other participants, who may be more representative of the total customer or user community. Focus group sessions are very difficult to facilitate, and it takes considerable skill to keep the discussion on topic while not also influencing the participants. Since the focus group members really have nothing at stake, their inputs may lack conviction, or be first reactions, therefore not representative of their longer term. more stable opinions. So, just as with surveys, data obtained from focus groups need independent confirmation.

4.4.5.4   Best Use

Use focus groups only when truly representative of the customer or user community. Focus groups are best used to determine the range of possible requirements for narrow issues.

4.4.6   NATURAL LANGUAGE REQUIREMENTS BY ASSESSING PRODUCT DEFECTS

Review records of user complaints or maintenance records of a system or architecture for systematic issues.

4.4.6.1   Process to Determine Natural Language Requirements by Assessing Product Defects

Step 1. Gather complaints or error reports. To fullest extent possible, ascertain method of collection. If an existing process to capture does not exist, invent a process.

Step 2. Itemize the defects by type, frequency, and impact on customer or users. Look for those defects that are most important to prevent.

Step 3. Conduct detailed assessment to identify likely causes for defects.

Step 4. Hypothesize ways to eliminate defect causes.

Step 5. Conduct trials to determine validity of hypotheses.

Step 6. Tabulate the results and formulate corresponding requirements.

Step 7. Formally review and approve additions to program requirement database.

4.4.6.2   Strengths

Directly address known deficiencies with known impacts

4.4.6.3   Weaknesses

This method can only be applied if there is an existing, reliable record of defects, in terms of failures, user complaints, or service outages. Even if the record exists, to accurately infer requirements from fault data requires deep domain knowledge.

4.4.6.4   Best Use

Implement enhancements to existing systems or architectures and get ideas for features for new systems or architectures.

4.4.7   Determine Natural Language Requirements by Deriving Concepts of Operation

A concept of operation describes how the system or architecture is to behave from the operator's perspective. Here the word *operation* includes the functions I called training, deployment, operations, support, and disposal. The description could be a single picture, a long narrative, a movie, or an equation of how the system or architecture is to be operated. The concept of operations foreshadows the implementation approach chosen, but as much as practical remains aloof from the exact end items that will be used, instead describing what activities the user expects to perform to achieve what ends. For very complicated systems, it may be pointless to try to describe every single operation, so the concept of operations may concentrate on describing a few key activities, sometimes called *threads*.

4.4.7.1   Process to Determine Natural Language Requirements by Concepts of Operation

Step 1. Describe quantitatively what is to be accomplished by the system or architecture for whom by whom. This is called the *mission*. For complicated systems or architecture, more than one such description may be needed. However, if the number of missions exceeds three or so, then probably need to subdivide the concept of operations effort into multiple volumes of three or less missions each.

Step 2. Name the performers, both internal to the system or architecture, and those external to the system or architecture, that need to participate for the mission to be a success, and the relationships between them.

Step 3. Name what information or material that needs to move between the internal and external performers to enable the mission to be accomplished.

Step 4. Name and determine the sequence of the operations functions needed to get and move the information and material to perform the mission. This can be accomplished by examining the operations activities from both the perspective of those with highest authority to execute the mission and those with lowest authority or most limited roles.

Step 5. Indicate, chronologically, the order and duration of the activities that move the stipulated information or material to enable the mission.

Step 6. Define constraints on the activities that move information or material flow between performers to enable the mission.

Step 7. Define measures of effectiveness that quantitatively denote how well the operations activities are proceeding.

Step 8. Mathematically model the behavior to accomplish the missions, in such a manner that system or architecture feature options can be explored for effectiveness.

Step 9. To the fullest extent practical, confirm that your perception of the concept of operations is what the customer and user want.

Step 10. Utilizing the mathematical models, decide what are the best performance requirements to allocate to each operational function associated with the mission.

Step 11. Tabulate the results and formulate corresponding requirements.

Step 12. Formally review and approve additions to program requirement database.

4.4.7.2 Strengths

By describing what is desired in terms of activities and results, achieve a high level of clarity as to the operations functional requirements and desired associated performance requirements. As the key operations may involve multiple aspects of the system or architecture, this approach helps understand which interfaces are critical to success and how they should be implemented.

4.4.7.3 Weaknesses

Concept of operations development is time consuming and expensive. If the product is a long written narrative, many people won't read it, so there will be an unknown level of uncertainty if correctly interpreting customer and user desires. Sometimes the customer provides at least the top level concept of operations to put their provided requirements in some perspective. Otherwise you may produce the concept of operations you think the customer wants rather than the one they do want. Pictures or movies work better than narratives to get the basic top-level information, but details may remain elusive. Care needs to be taken to not confuse customer or user wishes with constraints. As the focus is operations, deployment and disposal can easily be addressed, training and support less so, while and design and production functions are not considered. This can easily lead to a wonderful description for the desired operations with little or no hope the result can be implemented due.

4.4.7.4 Best Use

This is for new or never-accomplished portions of systems or architectures to flesh out the basic types of requirements needed. Other methods are usually then necessary to get the details correct.

4.4.8 NATURAL LANGUAGE REQUIREMENTS BY FORMAL DIAGRAMMING TECHNIQUES

Though by definition, natural language requirements are sentences, diagrams are very useful to show to customer, users, designers, producers, and supporters to grasp what the requirements statements are trying to communicate. Seven types of diagrams are needed:

1. Mission diagram. An illustration that evokes the fundamental job the system or architecture is to achieve.
2. Hierarchy diagram. This can be shown using an indentured outline or a construct similar to an organization chart. Hierarchy diagrams are used to depict

customer and user relationships, work breakdown structures, the performing organization relationships, the system end items, and functional decomposition.

3. Activity flow. Boxes or ovals are used to name things to accomplish, ordered left to right, with arrows used to denote precedence. Unlabeled arrows denote only sequencing, but an arrow could be labeled to describe explicitly what is flowing from activity to activity. Any named activity can be decomposed into subordinate functions in a second diagram. The horizontal dimension of each box or oval could be in proportion to the duration of the activity per a time scale denoted on the diagram. The controls for each activity are denoted by named arrows flowing vertically downward into the top of each box, or oval. Or, material and items needed to implement the activity can be noted as named arrows flowing vertically upward into the bottom of each box or oval.

4. Relationship diagram. Labeled rectangles, ovals, or stick figures denote entities, with arrowhead lines used to show what information or material flows from one entity to another. An alternative is a relationship table. This is shown usually with the entities listed in the main diagonal of the table. The remaining boxes are used to record the input and output between each entity using the following rules: each box left or right of an entity is to record what is output from that entity as input to the corresponding entity on the diagonal above or below.

5. Timeline diagram. This is shown by using left to right, or top to bottom an axis depicting increasing time, and noting when activities start and stop relative to this axis.

6. Feature table. The columns of the table name what is recorded in each row. More than one table may be defined to show more than one feature.

7. Data graph. Shows dependent variables as a function of independent variables.

All formal diagramming methods utilize the seven above constructs to define a finite list of specific diagrams that the methodology judges adequate to express any aspect of interest for a system or architecture. That there are so many diagramming methodology variants is partly a result of the process maturing, but also a result of disagreements as to what constitutes useful information. The primary reason for the disagreement is that different types of systems and architecture, meaning they attempt to achieve different missions by different end items, need different depictions. Still, tremendous progress is being made and one should explore the current status of these efforts and utilize to the fullest extent.

In addition to the structural difference, diagram methodologies also differ by what aspect of the system to perceive. There are at least two potential ways to view a system or architecture—functionally or physically. But there are many other potentially valid views:

1. Maturity of items used to implement
2. What it costs or will cost
3. How much longer it will take to be in operation, or will last
4. How well it performs a mission
5. Impacts to environment

Diagrams can certainly also be used regardless of the number of views deemed adequate.

4.4.8.1 Process to Determine Requirements by Formal Diagramming Techniques

Step 1. Sketch a draft diagram, using less than seven items in a single diagram.

Step 2. Review it with customer, users and designers, producers and supporters, record edits directly on draft. Repeat as needed to settle on final answer.

Step 3. Translate into natural language requirements. Note extent the diagram itself can be used as a requirement record, use as such, particularly if tool enables the diagram to be translated into the end item.

Step 4. Formally review and approve additions to program requirement database.

4.4.8.2 Strengths

Diagrams capture a lot of information succinctly. By sketching and modifying in real time with a customer, users and designers, producers, and supporters, one can often quickly establish what basic requirements are. Diagrams can be reused, saving a lot of time and avoiding errors. Tools enable an activity diagram to be translated automatically into a hierarchy diagram or even a timeline, making it relatively easy to make different diagrams from the same source inputs.

4.4.8.3 Weaknesses

For complicated systems, huge numbers of diagrams are needed, each adding detail for an object on a higher level diagram, just as a huge number of drawings may be needed to depict the hardware end items to produce. Retaining comprehension of many diagrams is equally difficult as retaining comprehension of large amounts of text. Denoting whether something is a constraint or a preference is difficult on a diagram.

4.4.8.4 Best Use

Software intensive systems or whenever tools enable diagrams to be automatically converted into end items.

4.4.9 NATURAL LANGUAGE REQUIREMENTS BY QUALITY MATRICES

Quality matrices are a series of tables that indicate the relationship between what (typically labeled rows) and how (typically labeled columns).[3] The basic methodology has four phases: product planning, part deployment, process planning, and production planning. Each phase utilizes at least one matrix to relate what is desired to how it will be done. For product planning, the what's are the customer desires and the how's what we call the functions and performance requirements. The part deployment matrix takes the relevant functions and performance requirements from the product planning metric as rows and introduces the derived design requirements as columns. The process and production planning matrices take the relevant design requirements as rows and introduce process and production functions and performance requirements as columns, respectively. To establish what to put in each matrix requires application of the other techniques discussed above.

4.4.9.1   Process to Determine Natural Language Requirements by Quality Matrices

Step 1. Produce the product-planning matrix.
- Determine customer wants.
- Estimate importance of each want.
- Establish functions and associated performance parameters relevant to achieving each want.
- Denote the correlation of which functions will fulfill which wants.
- Determine performance importance. (That is, if **what_importance** is a vector such that **what_importance**[[i]] records the importance of the ith what relative to the others, and **C** is a matrix, such that **C**[[i,j]] is relative importance of the ith what to achieve the jth function, then the importance of each function is given by **C** . **what_importance**.)
- To extent practical, ascertain competitor's performance levels.
- Conduct trades to establish target performance levels which offer superior value to customer.

Step 2. Produce part deployment matrix.
- Identify relevant functions and performance for the part.
- Conduct a trade to select an implementing concept.
- Identify as columns the critical design features and parameters for the concept chosen.
- Establish correlation of design features to function and performance requirements.
- Calculate design feature and parameters importance (the dot product of the functional and performance importance with the correlation values).
- Conduct trades to determine best design feature parameters.

Step 3. Process the planning matrix.
- Identify the relevant design features and parameters with importance.
- Identify processes that will be used to provide each design feature.
- If no process exists, develop one.
- Establish the correlation between processes and features.
- Calculate the importance of each process.
- Establish the production values to achieve.

Step 4. Support material.
- Prepare production quality control procedures and monitor.
- Prepare maintenance schedule material.
- Prepare operating instruction material.

4.4.9.2   Strengths

Enables easy visualization of the allocation of performance requirements to functions as well as functional allocation to end items. Enables high levels of reuse from system to system or architecture to architecture. Can easily display threshold or objective performance levels as well as competitor performance levels so can visualize how might offer a more attractive product. Well suited for hardware end items with describable, static features.

4.4.9.3 Weaknesses

Even relatively simple systems need huge numbers of functions and performance requirements which rapidly can create a diagram needing a very large wall area and therefore hard to comprehend. Since the diagram is a static table, it does not show flow, change, or time.

4.4.9.4 Best Use

Concise documentation for subelements of the system or architecture for which the fundamental functions, performance requirements, and end items can all fit in readable rectangles that fit on a standard size piece of paper or computer screen, or as a means to show how well the requirements provide a more competitive offering.

4.4.10 Natural Language Requirements by Models or Prototypes

A representative version of the system or architecture is produced and displayed and possibly used by user and customer to solicit feedback regarding what to keep the same and what to change.

4.4.10.1 Process to Determine Natural Language Requirements by Models or Prototypes

Step 1. Build a model of the proposed offering with a clear idea what to demonstrate to customer to solicit reaction.
Step 2. Demonstrate model to customer, note positive and negative reactions.
Step 3. Adjust model to better satisfy customer.
Step 4. Repeat from Step 2 until customer satisfaction is maximized with resources available.

4.4.10.2 Strengths

Since the model enables the customer and user to comprehend the system less abstractly, this should increase the likelihood of comprehending what they like or dislike. When the system or architecture is relying on new technologies, incremental prototypes offer the means to ascertain if the needed technologies are maturing to the level necessary for the system or architecture to work with incremental funding.

4.4.10.3 Weaknesses

Many people perceive this method as the best way to get requirements. Yet use is fraught with issues. First, all the potential concerns associated with the reliability of the input when using surveys or focus groups still apply. Second, preliminary prototypes may simulate behavior that is desired, but very difficult to achieve in real life, or may underestimate the true manufacturing difficulties to make the real item. One may then get a "true" indication of the desirability of the feature from a user or customer perspective, but fail to comprehend how difficult the feature is to provide. Third, model production can be time consuming and expensive, though the art of the process is to mitigate these negative. Fourth, by definition, since the prototype is a representation of the end item, not the requirements for the end item, it can only be created by choosing a specific implementation. So there is some danger the use of prototypes will limit the implementation solution space. This can be mitigated considerably by insisting at least prior to agreeing to the final implementation, prototypes representing numerous potential implementations be produced. A good heuristic is to insist at least six different implementations be found. Most people can easily think of three ways to do anything—the way it is done now, a way it shouldn't be done, and the way they first think it can be done. Forcing an effort to find at least three more options has a good chance of finding a better solution.

4.4.10.4   Best Use

For simple systems or architecture for which the primary job is to determine how to mix and match features to offer product superior to those currently available.

4.4.11   Natural Language Requirements by Incremental Build

This could also be called trial and error, or simply art. For some systems, or at least some subsets of systems, there may be no practical means to obtain requirements from users or customers. In those instances, one essentially must build a product and see what customer thinks of it after built, and incorporate their changes into the next build of the product.

4.4.11.1   Process to Determine Natural Language Requirements by Self-determination

Step 1. Determine what features the customer and users want to the fullest level they can describe.

Step 2. To the extent you need additional information to fully know how to implement a customer's stated feature, use your own judgment to determine the additional needed information.

Step 3. Produce the end items.

Step 4. Show the end item to customer and users, and learn what changes or new features they want.

Step 5. Repeat from Step 2.

4.4.11.2   Strengths

This method is particularly suited for very innovative end items. When many options can be developed and proposed in a relatively short time, this can be a very efficient method, as a continuous stream of hypothetically useful end items is produced, the survivors of which are those customers want the most. Over time, this method will eventually provide more improved systems. This method is effective for many software intensive systems, particularly those with extensive graphical user interfaces. This method all but guarantees hitting cost and schedule targets since the due date for the increment is fixed and the effort can be predicted with high accuracy. What is at risk is the extent the desired features are all fully implemented by the stipulated date for the stipulated amount.

4.4.11.3   Weaknesses

This method relies on a great deal of customer interaction, which may not be sustainable. Though progress toward meeting customer or user objectives is likely, there still is no absolute guarantee that the customer and users won't change their minds so often the process will spiral out of control.

4.4.11.4   Best Use

When the requirements are very uncertain, or the effort to determine them exceeds the effort to produce a specification for the customer to react to. Also an effective way to introduce new elements to an existing system that combine the latest state of the art technologies in novel or innovative ways to satisfy well recognized shortfalls.

4.5   MAINTAIN A NATURAL LANGUAGE REQUIREMENT DATABASE

A collection of natural language requirements are best maintained using a database program. Formal requirements are presented to create and maintain functional, performance, and sought requirements, manage the database of requirements, as well as prepare specifications which are the subsets of the requirements allocated to an end item.

4.5.1   CREATE A REQUIREMENT TYPE

CreateReqType creates a requirement type. *CreateReqType* defines and stores in a file called **typename**, an array, the first element of which defines each attribute associated with the requirement type. The second element is a unique identifier for that type of requirement.

CreateReqType is first used to create a functional requirement type. For this example, the type was arbitrarily given the name **P** (to, for example, represent a program name) followed by the type name.

```
In[146]:= CreateReqType["Pfunctional",
        {{"Identification", "unique number"},
        {"Status", "1 of: draft, approved, discontinued"},
        {"End item allocation", "array of end item name(s) which will implement"},
        {"Rationale", "brief explanation of why need and correct"},
        {"Accepted by",
         "name of person/organizational entity accepting responsibility to implement"},
        {"Author", "name and contact information of definer"},
        {"Parent", "array of identifications of same
          type, that if change, could require this requirement change"},
        {"Compliance method", "state how will implement"}}]
```

To see the state of the **Pfunctional** database, use the *Mathematica* routine *Get* or <<.

```
In[147]:= << Pfunctional

Out[147]= {{{Identification, unique number}, {Status, 1 of: draft, approved, discontinued},
         {End item allocation, array of end item name(s) which will implement},
         {Rationale, brief explanation of why need and correct}, {Accepted by,
          name of person/organizational entity accepting responsibility to implement},
         {Author, name and contact information of definer},
         {Parent, array of identifications of same type, that if change, could require this
           requirement change}, {Compliance method, state how will implement}}, 1}
```

The file is an array with two elements: the first element is also an array, with two elements, the first an attribute name, and the second the attribute definition. This enables such information be utilized in table headers, or definition of terms. The second element is an integer, set to "1," which is the number that will be assigned to the next defined requirement of this type. Clearly, then, whatever routine is defined to define a requirement, it must update this integer after each definition. Also note, every time *CreateReqType* is used with the same **typename**, any previous instantiation of **typename** is deleted.

Mathematica provides routines to copy, rename, or delete the resulting files, namely:

CopyFile[from_**file**1, to_**file**2]
Rename[from_**file**1, to_**file**2]
DeleteFile[**file**]

4.5.2 Create an End Item Hierarchy

An easy way to depict a hierarchy in *Mathematica* is to create a list consisting of the items numbered as an outline, which is stored in a file given an appropriate name. Here is an example for program **P** with end item structure **Penditems**, saved in a file called **Penditemsfile**:

```
In[148]:= Penditems = {"1 System",
                "1.1 Segment 1",
                    "1.1.1 Element 11",
                    "1.1.2 Element 12",
                    "1.1.3 Element 13",
                "1.2 Segment 2",
                    "1.2.1 Element 21",
                        "1.2.1.1 Subsystem 211",
                        "1.2.1.2 Subsystem 212",
                        "1.2.1.3 Subsystem 213",
                            "1.2.1.3.1 Component 2131",
                            "1.2.1.3.2 Component 2131",
                            "1.2.1.3.3 Component 2133",
                            "1.2.1.3.4 Component 2134",
                    "1.2.2 Element 22",
                        "1.2.2.1 Subsystem 221",
                        "1.2.2.2 Subsystem 222",
                "1.3 Segment 3"
            };
    Save["Penditemsfile", Penditems]
```

4.5.3 Create an Implementation Organization Hierarchy

The implementing organization hierarchy can be depicted the same way as the end items. To keep the example simple, **Porganization** is defined to be exactly the same as the end item organization, with the result saved in a file given the arbitrary name of **Porganizationfile**.

```
In[150]:= Porganization = Penditems;
    Save["Porganizationfile", Penditems]
```

4.5.4 DEFINE A REQUIREMENT OF A PARTICULAR TYPE

DefReq defines a single new requirement of a particular type.

What follows is an example to define a specific functional requirement:

```
In[152]:= DefReq["Pfunctional",
          "P shall have the function to lift a chair.",
          {"draft", {Penditems[[1]]}, "Primary function of system",
          Porganization[[1]], "Name_author", {"self"} , "tbd"}]
```

By displaying the contents of the file **Pfunctinal** we see how the file is constructed:

```
In[153]:= << Pfunctional

Out[153]= {{{Identification, unique number}, {Status, 1 of: draft, approved, discontinued},
          {End item allocation, array of end item name(s) which will implement},
          {Rationale, brief explanation of why need and correct}, {Accepted by,
          name of person/organizational entity accepting responsibility to implement},
          {Author, name and contact information of definer},
          {Parent, array of identifications of same type, that if change, could require this
          requirement change}, {Compliance method, state how will implement}}, 2,
          {P shall have the function to lift a chair., {Pfunctional-1, draft, {1 System},
          Primary function of system, 1 System, Name_author, {self}, tbd}}}
```

Simply, the requirement record is a two-element array: the first element is the natural language statement, and the second element is an array of type attribute values, with the unique identifier the first element of this array. To help distinguish types, this implementation joins the requirement type name with "-" followed by a unique integer. When defining a second requirement, it is appended to the **Pfunctional** database:

```
In[154]:= DefReq["Pfunctional", "P shall have the function to maintain chair height above floor.",
          {"draft", {Penditems[[1]]}, "Primary function of system",
          Porganization[[1]],  "Name_author", {"self"} , "tbd"}]
          << Pfunctional

Out[155]= {{{Identification, unique number}, {Status, 1 of: draft, approved, discontinued},
          {End item allocation, array of end item name(s) which will implement},
          {Rationale, brief explanation of why need and correct}, {Accepted by,
          name of person/organizational entity accepting responsibility to implement},
          {Author, name and contact information of definer},
          {Parent, array of identifications of same type, that if change, could require this
          requirement change}, {Compliance method, state how will implement}}, 3,
          {P shall have the function to lift a chair., {Pfunctional-1, draft, {1 System},
          Primary function of system, 1 System, Name_author, {self}, tbd}},
          {P shall have the function to maintain chair height above floor., {Pfunctional-2,
          draft, {1 System}, Primary function of system, 1 System, Name_author, {self}, tbd}}}
```

4.5.5 EDIT A REQUIREMENT

To change a requirement statement or an attribute value, we must find it in the database, redefine the part that is to change, and return the new information to the database, without changing the ID number. *EditReqAttribute* enables this for attribute information.

For example, if we want to change the status of **Pfunctional-2** from draft to approved and change author from **Name_author** to **Name_author2**, we do the following:

```
In[156]:= EditReqAttribute["Pfunctional", "Pfunctional-2", {"approved", {Penditems[[1]]},
            "Primary function of system", Porganization[[1]], "Name_author2", {"self"} , "tbd"}]
         << Pfunctional

Out[157]= {{{Identification, unique number}, {Status, 1 of: draft, approved, discontinued},
           {End item allocation, array of end item name(s) which will implement},
           {Rationale, brief explanation of why need and correct}, {Accepted by,
             name of person/organizational entity accepting responsibility to implement},
           {Author, name and contact information of definer},
           {Parent, array of identifications of same type, that if change, could require this
             requirement change}, {Compliance method, state how will implement}}, 3,
          {P shall have the function to lift a chair., {Pfunctional-1, draft, {1 System},
            Primary function of system, 1 System, Name_author, {self}, tbd}},
          {P shall have the function to maintain chair height above floor.,
            {Pfunctional-2, approved, {1 System},
            Primary function of system, 1 System, Name_author2, {self}, tbd}}}
```

Similarly, *EditReqStatement* changes a requirement statement. Here is an example:

```
In[158]:= EditReqStatement["Pfunctional", "Pfunctional-1",
            "P shall have the function to raise a chair above the ground."]
         << Pfunctional

Out[159]= {{{Identification, unique number}, {Status, 1 of: draft, approved, discontinued},
           {End item allocation, array of end item name(s) which will implement},
           {Rationale, brief explanation of why need and correct}, {Accepted by,
             name of person/organizational entity accepting responsibility to implement},
           {Author, name and contact information of definer},
           {Parent, array of identifications of same type, that if change, could require this
             requirement change}, {Compliance method, state how will implement}}, 3,
          {P shall have the function to raise a chair above the ground., {Pfunctional-1,
            draft, {1 System}, Primary function of system, 1 System, Name_author, {self}, tbd}},
          {P shall have the function to maintain chair height above floor.,
            {Pfunctional-2, approved, {1 System},
            Primary function of system, 1 System, Name_author2, {self}, tbd}}}
```

To edit requirements in a large database, it's convenient to be able to search the database. *ShowStatement* finds any statements that contain the **test** string. Here is an example:

```
In[160]:= ShowStatement["Pfunctional", "maintain", True]

Out[160]= {{P shall have the function to maintain chair height above floor., {Pfunctional-2,
            approved, {1 System}, Primary function of system, 1 System, Name_author2, {self}, tbd}}}
```

ShowAttributes finds requirements with attributes that match a specified value. Here is an example:

```
In[161]:= ShowAttributes["Pfunctional", {"Author"}, {"Name_author2"}, True]

Out[161]= {{P shall have the function to maintain chair height above floor., {Pfunctional-2,
            approved, {1 System}, Primary function of system, 1 System, Name_author2, {self}, tbd}}}
```

ShowAttributes can also find requirements that match multiple attribute values:

```
In[162]:= ShowAttributes["Pfunctional",
            {"Status", "End item allocation"}, {"draft", {"1 System"}}, True]

Out[162]= {{P shall have the function to raise a chair above the ground., {Pfunctional-1, draft,
            {1 System}, Primary function of system, 1 System, Name_author, {self}, tbd}}}
```

4.5.6 CREATE A DEPENDENCY REPORT

A dependency report identifies all requirements that are either parents or children of an identified requirement.

ReqParentReport finds all parents of a requirement, and recursively calls itself to find parents of the parents.

ReqChildReport finds all children of a requirement, and recursively calls itself to find children of the children.

To demonstrate, first create a sample network of dependent requirements:

```
In[165]:= DefReq["Pfunctional",
      "P shall have the function  to sub level 2-1.",
      {"draft", {Penditems[[2]]}, "Derived subfunction",
      Porganization[[2]], "Name_author", {"Pfunctional-1"} , "tbd"}]
    DefReq["Pfunctional",
      "P shall have the function to sub level 2-2.",
      {"draft", {Penditems[[3]]}, "Derived subfunction", Porganization[[3]],
      "Name_author", {"Pfunctional-1", "Pfunctional-2"} , "B part" }]
    DefReq["Pfunctional",
      "P shall have the function to sub level 2-3.",
      {"draft", {Penditems[[4]]}, "Derived subfunction",
      Porganization[[4]], "Name_author", {"Pfunctional-2"} , "C part"}]
    DefReq["Pfunctional",
      "P shall have the function to sub level 2-4.",
      {"draft", {Penditems[[5]]}, "Derived subfunction",
      Porganization[[5]], "Name_author", {"Pfunctional-2" }, "D part"}]
    DefReq["Pfunctional",
      "P shall have the function to sub level 3-1 ",
      {"draft", {Penditems[[6]]}, "Derived subfunction",
      Porganization[[6]], "Name_author", {"Pfunctional-3"} , "E part"}]
    DefReq["Pfunctional",
      "P shall have the function to sub level 3-2.",
      {"draft", { Penditems[[7]]}, "Derived subfunction", Porganization[[7]],
      "Name_author", {"Pfunctional-3", "Pfunctional-4"} , "F part"}]
    DefReq["Pfunctional",
      "P shall have the function to sub level 3-3.",
      {"draft", {Penditems[[8]], Penditems[[9]]}, "Derived subfunction",
      Porganization[[8]], "Name_author", {"Pfunctional-5"} , "G part"}]
    DefReq["Pfunctional",
      "P shall have the function to sub level 4-1.",
      {"draft", {"tbd"}, "Derived subfunction",
      Porganization[[9]], "Name_author", {"Pfunctional-7"} , "I part"}]
    DefReq["Pfunctional",
      "P shall have the function to sub level 4-2.",
      {"draft", {Penditems[[10]]}, "Derived subfunction",
      Porganization[[10]], "Name_author", {"Pfunctional-7"} , "J part"}]
    DefReq["Pfunctional",
      "P shall have the function to sub level 4-3.",
      {"draft", {"tbd"}, "Derived subfunction",
      Porganization[[11]], "Name_author", {"Pfunctional-9"} , "K part"}]
    DefReq["Pfunctional",
      "P shall have the function to sub level 4-4.",
      {"draft", {Penditems[[12]]}, "Derived subfunction",
      Porganization[[12]], "Name_author", {"Pfunctional-9"} , "tbd"}]
```

Here are several examples of determining parents:

```
In[176]:= ReqParentReport["Pfunctional", "Pfunctional-1", False]

        Parent of Pfunctional-1 is: {self}

In[177]:= ReqParentReport["Pfunctional", "Pfunctional-7", False]

        Parent of Pfunctional-7 is: {Pfunctional-3}

        Parent of Pfunctional-3 is: {Pfunctional-1}

        Parent of Pfunctional-1 is: {self}
```

In[178]:= **ReqParentReport["Pfunctional", "Pfunctional-8", False]**

Parent of Pfunctional-8 is: {Pfunctional-3, Pfunctional-4}

Parent of Pfunctional-3 is: {Pfunctional-1}

Parent of Pfunctional-1 is: {self}

Parent of Pfunctional-4 is: {Pfunctional-1, Pfunctional-2}

Parent of Pfunctional-1 is: {self}

Parent of Pfunctional-2 is: {self}

In[179]:= **ReqParentReport["Pfunctional", "Pfunctional-12", False]**

Parent of Pfunctional-12 is: {Pfunctional-9}

Parent of Pfunctional-9 is: {Pfunctional-5}

Parent of Pfunctional-5 is: {Pfunctional-2}

Parent of Pfunctional-2 is: {self}

Here are several examples of determining children:

In[180]:= **ReqChildReport["Pfunctional", "Pfunctional-2", False]**

Child of Pfunctional-2 is: Pfunctional-4

Child of Pfunctional-4 is: Pfunctional-8

Pfunctional-8 has no children

Child of Pfunctional-2 is: Pfunctional-5

Child of Pfunctional-5 is: Pfunctional-9

Child of Pfunctional-9 is: Pfunctional-12

Pfunctional-12 has no children

Child of Pfunctional-9 is: Pfunctional-13

Pfunctional-13 has no children

Child of Pfunctional-2 is: Pfunctional-6

Pfunctional-6 has no children

In[181]:= **ReqChildReport["Pfunctional", "Pfunctional-1", False]**

Child of Pfunctional-1 is: Pfunctional-3

Child of Pfunctional-3 is: Pfunctional-7

Child of Pfunctional-7 is: Pfunctional-10

Pfunctional-10 has no children

Child of Pfunctional-7 is: Pfunctional-11

Pfunctional-11 has no children

Child of Pfunctional-3 is: Pfunctional-8

Pfunctional-8 has no children

Child of Pfunctional-1 is: Pfunctional-4

Child of Pfunctional-4 is: Pfunctional-8

Pfunctional-8 has no children

4.5.7   CREATE A FUNCTIONAL ALLOCATION REPORT

A useful report is to determine if all functional requirements were allocated to end items. *AllocatedTo* does this check.

The following creates an array, each element of which shows what functional requirement was allocated to the corresponding end item. An "{}" indicates no function has yet been assigned to those end items.

```
In[182]:= Table[
        AllocatedTo["Pfunctional", Penditems[[i]], False],
        {i, 1, Dimensions[Penditems][[1]]}]
Out[182]= {{Pfunctional-1, Pfunctional-2}, {Pfunctional-3}, {Pfunctional-4}, {Pfunctional-5},
        {Pfunctional-6}, {Pfunctional-7}, {Pfunctional-8}, {Pfunctional-9},
        {Pfunctional-9}, {Pfunctional-11}, {}, {Pfunctional-13}, {}, {}, {}, {}, {}, {}}
```

NotAllocatedTo finds any functional requirements not allocated to the approved end items. Here is the output for the database **Pfunctional** and the end items defined by **Penditems**:

```
In[183]:= NotAllocatedTo["Pfunctional", Penditems, False]
```

Requirement Pfunctional-10 is not allocated to an approved item.

Requirement Pfunctional-12 is not allocated to an approved item.

This is correct, since we recorded "{tbd}" in the end item allocated attribute location for each of these requirements.

4.5.8   CREATE A PERFORMANCE ALLOCATION REPORT

To check for any functional requirements for which no performance requirement is allocated, use *AllocatedTo* with **oftype** set equal to the name given the performance requirement type file, and **totype** set equal to an approved functional requirement identification.

Use *NotAllocatedTo* to check for functional requirements that have not been allocated performance requirements by using *NotAllocatedTo* with **oftype** set to the file name for the performance requirements, and **totype** set to an array of approved or draft and approved functional requirement identifications.

4.5.9   CREATE END ITEM SPECIFICATIONS

A specification is a subset of the system or architecture requirements database that are applicable for a stipulated end item. Before the convenience of modern database programs, a specification traditionally was a document, organized per some standard outline, to help ensure all the necessary requirements types were captured and prior requirements could be easily reused. Now, for an end item at any level in a system or architecture, a specification is best visualized as a table that shows requirement statements and all their attributes.

Systems engineering literature is full of similar but different, very well thought out, templates for specifications. Perhaps the granddaddy of all specification standards is Military Standard 490A.[4] This document defined potential formats for a system specification as well as what were called "development specifications," "product specifications," "process specifications," and "material specifications." The recommended outline for a system specification is as follows:

1. Scope
2. Applicable documents
3. Requirements
    3.1. System definition
        3.1.1. General description
        3.1.2. Missions
        3.1.3. Threat
        3.1.4. System diagrams
        3.1.5. Interface definition
        3.1.6. Government-furnished property
        3.1.7. Operational and organizational concepts
    3.2. Characteristics
        3.2.1. Performance characteristics
        3.2.2. Physical characteristics
        3.2.3. Reliability
        3.2.4. Maintainability
        3.2.5. Availability
        3.2.6. System effectiveness models
        3.2.7. Environmental conditions
        3.2.8. Nuclear control requirements
        3.2.9. Transportability
    3.3. Design and construction
        3.3.1. Materials, processes, and parts
        3.3.2. Electromagnetic radiation
        3.3.3. Nameplates and product markings
        3.3.4. Workmanship
        3.3.5. Interchangeability
        3.3.6. Safety
        3.3.7. Human performance and human engineering
    3.4. Documentation
    3.5. Logistics
        3.5.1. Maintenance
        3.5.2. Supply
        3.5.3. Facilities and facility equipment
    3.6. Personnel and training
    3.7. Functional area characteristics
    3.8. Precedence
4. Quality assurance provisions

4.1. General
    4.1.1. Responsibility of tests
    4.1.2. Special tests and examinations
4.2. Quality performance inspections
5. Preparation for delivery
6. Notes
    Appendix

Sections 1 and 2 are preamble and descriptive. Section 3 is called "Requirements," but section 3.1 is explanatory material only. The actual requirements are contained in 3.2 to 3.6, which are roughly grouped by the primary functions, which were not acknowledged as such in the day of the standard. Sections 4, 5, and 6 were also descriptive, perhaps instructional, but did not contain requirements. Note, software and computers are not mentioned in the outline. The Standard's "development specification" template for "computer program," is as follows:

1. Scope
2. Applicable documents
3. Requirements
    3.1. Program definition
    3.2. Detailed functional requirements
        3.2.1. Inputs
        3.2.2. Processing
        3.2.3. Outputs
        3.2.4. Special requirements
    3.3. Adaptation
        3.3.1. General environment
        3.3.2. System parameters
        3.3.3. System capacities
4. Quality assurance provisions
    4.1. Introduction
    4.2. Test requirements
    4.3. Acceptance test requirements
5. Preparation for delivery
6. Notes
    Appendix

Again, the first two and final three sections do not contain requirements. Here, "function" is explicitly used, but it refers to only the operational functionality of the computer program.

Based on this standard, the Aerospace Corporation published a guide,[5] which noted inconsistencies between published standards at the time, and recommend a resolution for system specifications for military space systems. The template for Aerospace's system specification is as follows:

1. Scope
    1.1. Identification
    1.2. System overview

Other than the obvious explicit mention of space vehicles, the Aerospace guide also uses "characteristics," to address that a complicated system may be built in increments, and each increment may have different requirements, as well as "states and modes" as a way to name different ways a system may be configured to be operated with different requirements applicable to the different states and modes. Also note the Aerospace expanded section "3.8 Precedence" to explicitly indicate how to react to conflicting requirements. Finally, there is a new section, "3.9 Qualification," to address that when more than one item is to be made, how should the first and subsequent be qualified. But perhaps Aerospace's most salient difference is the introduction of section "3.7 Characteristics of subordinate items." In this section, the template is recursively applied in whole or part for the items that constitute the system.

Perhaps the most annoying argument that bedevils systems engineering leads on programs for the Department of Defense is whether or not a specification needs to identify both the requirements for the end item and for the subordinate items.

Many system engineers insist there must be a "3.7" in a specification. Many designers and producers find a "3.7" useless at best, and a load of unnecessary work at worst. The system engineer's reasoning is as follows:

"The specification, particularly the system specification, is my product. To produce my product, I and my colleagues spent a great deal of time very carefully analyzing how the requirements should be allocated and divided up between all the segments, elements or subsystems, in such a manner, that the result is the 'best' balance for the overall system. That allocation is very important, and it needs to be recorded, and its record needs to be carefully reviewed and approved at the same time the system spec is approved."

The recipients of the "3.7" requirements have an entirely different perspective. The designers, producers and testers who have the job to make something real, see the "3.7" entries as unnecessary duplication, their reasoning is as follows:

"The requirements in section 3.7 of the higher level specification duplicate the requirements that are in the lower specification section 3.2. Why do we have to define and keep track of all these extra requirements that say exactly the same thing? Why must we define the 3.7 requirements for our subordinate items, when they are recorded perfectly well in the 3.2 section of their respective specifications? This repeating of requirement is exponentially growing duplicate requirements. We have to keep track of our compliance with respect to all these duplicate requirements and we have to prepare and execute verification plans for all these duplicate requirements, creating literally tons of paper work that adds no value at great expense."

Clearly, both sides have a point. One could even suggest to the anti-3.7 folks that any requirements recorded in that section can be exactly the same (that is, have the same identification numbers) as any requirements that will show up in the corresponding lower level documents, so the requirements are not really duplicate, they are simply repeated. But this does not address their primary concern—please provide me a record of what my end item must provide, no more, no less.

I must confess I once was an advocate for having subordinate requirements stipulated in system specifications. My primary motivation was very similar to the argument above, as I saw value in recording the explicit allocations made between subordinate items, with the rationale that this allocation was critical to success. And I thought if the end item explicitly states key requirements imposed on subordinate items, the odds of those requirements being complied with went up.

But I now strongly recommend that one produce a specification that contains only those requirements that apply to that end item. Furthermore, should a requirement apply to multiple end items, it should be recorded as the same requirement allocated to each, not different requirements.

Chapman, Bahill, and Wymore[6] recommend a set of documents be generated:

Document 1. Problem Situation
Document 2. Operational Need
Document 3. System Requirements
Document 4. System Requirements Validation

Document 5. Concept Exploration
Document 6. System Functional Analysis
Document 7. System Physical Synthesis

They also use the terms *input trajectories* and *output trajectories* both to remind and to enforce definition of the variability of each.

The software community was never well served by natural language requirements, so they have devoted considerable attention on the subject and have generally become more and more enamored with graphical representations. Still, many in the field see a need for words to go along with the pictures. Robertson and Robertson[7] offer a large template for software intensive systems, which is actually applicable to many other types of systems. The authors purposely try to address every possible aspect of a software project, so the range of topics is substantially beyond just requirements.

I strongly endorse a multiple document approach. I see both simplicity and value in separately documenting:

1. The need for the system
2. What constitutes solution goodness
3. An identification of the alternatives assessed
4. The rationale for selecting the chosen approach
5. The performance and functional requirements for end items
6. The design of the end item
7. Verification plans
8. Verification evidence

As for the requirements, "all" that is needed is to communicate the functions and associated performance requirements to the end item designers, producers, and supporters. The structure of a specification then is as follows:

Specification for: <end item name>
    Functional requirement: <identification> <statement>
    Accepted by: <value>
    Rationale: <value>
    Allocated performance requirement: <identification> <statement>
    Accepted by: <value>
    Rationale: <value>
    Verification method: <value>
    Verification status: <value>
    Verification evidence: <value>
    [Repeated for each allocated performance requirement.]
    [Repeated for each allocated functional requirement.]
    Sought requirements:
    <none> or <id> <statement>
    Accepted by: <value>
Rationale: <value>

To illustrate, first let's define a few more functional requirements to allocate to the end item **1 System**:

```
In[184]:= DefReq["Pfunctional",
        "P shall have made up function 1.",
        {"draft", {Penditems[[1]]}, "Rationale 1",
        Porganization[[2]],  "Name_1", {"Pfunctional-1"} , "tbr"}]
      DefReq["Pfunctional",
        "P shall have made up function 2.",
        {"draft", {Penditems[[1]]}, "Rationale 2", Porganization[[3]],
        "Name_2", {"Pfunctional-1", "Pfunctional-2"} , "tbr"}]
      DefReq["Pfunctional",
        "P shall have made up function 3.",
        {"draft", {Penditems[[1]]}, "Rationale 3",
        Porganization[[1]], "Name_3", {"Pfunctional-2"} , "tbd" }]
```

Next, we make up a few more that are "inactive" or "approved":

```
In[187]:= DefReq["Pfunctional",
        "P shall have made up function 4.",
        {"approved", {Penditems[[1]]}, "Rationale 4",
        Porganization[[2]], "Name_1", {"Pfunctional-1"} , "A part"}]
      DefReq["Pfunctional",
        "P shall have made up function 5.",
        {"inactive", {Penditems[[1]]}, "Rationale 5", Porganization[[3]],
        "Name_2", {"Pfunctional-1", "Pfunctional-2"} , "B part"}]
      DefReq["Pfunctional",
        "P shall have made up function 6.",
        {"approved", {Penditems[[1]]}, "Rationale 6",
        Porganization[[1]], "Name_3", {"Pfunctional-2"} , "C part"}]
```

To create the sample specification, one needs performance requirements too, so first define a performance requirement type:

```
In[190]:= CreateReqType["Pperformance",
        {{"identification", "unique number"},
        {"Status", "1 of: draft, approved, inactive"},
        {"Functional allocation", "array identifications to which allocated"},
        {"Rationale", "brief explanation of why need and correct"},
        {"Verification responsibility",
          "name or organizational entity acknowleding responsibility
          to show end item achieves"},
        {"Author", "name and contact information of creater"},
        {"Parent",
          "array of  requirement identifications of same type, that if change, could
          require this requirement to change"},
        {"Verification_method", "1 of: record, inspection, analysis,
          test all, test sample, demonstration"},
        {"Verification_status", "1 of: not verified, inwork, verified"},
        {"Verification_evidence", "evidence, or pointer too,
          or file name containing proof end item meets requirement"},
        {"Compliance", "current prediction or margin or how satisfy"}
        }];
      << Pperformance

Out[191]= {{{identification, unique number}, {Status, 1 of: draft, approved, inactive},
        {Functional allocation, array identifications to which allocated},
        {Rationale, brief explanation of why need and correct}, {Verification responsibility,
        name or organizational entity acknowleding responsibility to show end item achieves},
        {Author, name and contact information of creater},
        {Parent, array of  requirement identifications of same type, that if change,
          could require this requirement to change}, {Verification_method,
          1 of: record, inspection, analysis, test all, test sample, demonstration},
        {Verification_status, 1 of: not verified, inwork, verified}, {Verification_evidence,
          evidence, or pointer too, or file name containing proof end item meets requirement},
        {Compliance, current prediction or margin or how satisfy}}, 1}
```

For illustration purposes only, we quickly make up 20 performance requirements, which we allocate to some of the defined functional requirements:

```
In[192]:= Do[DefReq["Pperformance",
           "P shall " <> ToString[i] <> ".",
           {"approved", {"Pfunctional-1"}, "Rationale1", "1 System",
            "Author1", {"self"}, "inspection", "not verified", "tbp", "10% margin"}]
         , {i, 1, 2}];
       Do[DefReq["Pperformance",
           "P shall " <> ToString[i] <> ".",
           {"approved", {"Pfunctional-2"}, "Rationale2", "1 System", "Author2",
            {"Pperformance-1"}, "analysis", "not verified", "tbp", "15% margin"}]
         , {i, 3, 5}];
       Do[DefReq["Pperformance",
           "P shall " <> ToString[i] <> ".",
           {"approved", {"Pfunctional-1"}, "Rationale3", "1 System", "Author3",
            {"Pperformance-3"}, "demonstration", "not verified", "tbp", "tbd" }]
         , {i, 6, 7}];
       Do[DefReq["Pperformance",
           "P shall " <> ToString[i] <> ".",
           {"draft", {"Pfunctional-2"}, "Rationale4", "1 System", "Author4",
            {"Pperformance-4"}, "test all", "not verified", "tbp", "-5% margin"}]
         , {i, 8, 10}];
       Do[DefReq["Pperformance",
           "P shall " <> ToString[i] <> ".",
           {"inactive", {"Pfunctional-2"}, "Rationale5", "1 System", "Author5",
            {"Pperformance-4"}, "inspection", "not verified", "tbp", "20% margin"}]
         , {i, 11, 12}];
       Do[DefReq["Pperformance",
           "P shall " <> ToString[i] <> ".",
           {"approved", {"Pfunctional-3"}, "Rationale6", "1 System", "Author6",
            {"Pperformance-7"}, "test sample", "not verified", "tbp", "1% margin"}]
         , {i, 13, 15}];
       DefReq["Pperformance",
           "P shall 16.",
           {"approved", {"Pfunctional-4"}, "Rationale7", "1 System", "Author7",
            {"Pperformance-9"}, "analysis", "not verified", "tbp", "-10% margin"}];
       DefReq["Pperformance",
           "P shall 17.",
           {"approved", {"Pfunctional-5"}, "Rationale8", "1 System", "Author8",
            {"Pperformance-9"}, "analysis", "not verified", "tbp", "tbd"}];
       DefReq["Pperformance",
           "P shall 18.",
           {"approved", {"Pfunctional-4"}, "Rationale7", "1.1 Segment 1", "Author7",
            {"Pperformance-9"}, "inspection", "not verified", "tbp", "tbd"}];
       DefReq["Pperformance",
           "P shall 19.",
           {"approved", {"Pfunctional-5"}, "Rationale7", "1.1 Segment 1", "Author7",
            {"Pperformance-9"}, "demonstration", "not verified", "tbp", "tbd"}];
       DefReq["Pperformance",
           "P shall 20.",
           {"approved", {"Pfunctional-4"}, "Rationale7", "1.2 Segment 2",
            "Author7", {"Pperformance-9"}, "test all", "not verified", "tbp", "tbd"}];
       << Pperformance
```

```
Out[203]= {{{identification, unique number}, {Status, 1 of: draft, approved, inactive},
          {Functional allocation, array identifications to which allocated},
          {Rationale, brief explanation of why need and correct}, {Verification responsibility,
           name or organizational entity acknowleding responsibility to show end item achieves},
          {Author, name and contact information of creater},
          {Parent, array of  requirement identifications of same type, that if change,
            could require this requirement to change}, {Verification_method,
           1 of: record, inspection, analysis, test all, test sample, demonstration},
          {Verification_status, 1 of: not verified, inwork, verified}, {Verification_evidence,
           evidence, or pointer too, or file name containing proof end item meets requirement},
          {Compliance, current prediction or margin or how satisfy}}, 21,
         {P shall 1., {Pperformance-1, approved, {Pfunctional-1}, Rationale1,
           1 System, Author1, {self}, inspection, not verified, tbp, 10% margin}},
         {P shall 2., {Pperformance-2, approved, {Pfunctional-1}, Rationale1,
           1 System, Author1, {self}, inspection, not verified, tbp, 10% margin}},
```

```
{P shall 3., {Pperformance-3, approved, {Pfunctional-2}, Rationale2, 1 System,
 Author2, {Pperformance-1}, analysis, not verified, tbp, 15% margin}},
{P shall 4., {Pperformance-4, approved, {Pfunctional-2}, Rationale2, 1 System,
 Author2, {Pperformance-1}, analysis, not verified, tbp, 15% margin}},
{P shall 5., {Pperformance-5, approved, {Pfunctional-2}, Rationale2, 1 System,
 Author2, {Pperformance-1}, analysis, not verified, tbp, 15% margin}},
{P shall 6., {Pperformance-6, approved, {Pfunctional-1}, Rationale3, 1 System,
 Author3, {Pperformance-3}, demonstration, not verified, tbp, tbd}},
{P shall 7., {Pperformance-7, approved, {Pfunctional-1}, Rationale3, 1 System,
 Author3, {Pperformance-3}, demonstration, not verified, tbp, tbd}},
{P shall 8., {Pperformance-8, draft, {Pfunctional-2}, Rationale4, 1 System,
 Author4, {Pperformance-4}, test all, not verified, tbp, -5% margin}},
{P shall 9., {Pperformance-9, draft, {Pfunctional-2}, Rationale4, 1 System,
 Author4, {Pperformance-4}, test all, not verified, tbp, -5% margin}},
{P shall 10., {Pperformance-10, draft, {Pfunctional-2}, Rationale4, 1 System,
 Author4, {Pperformance-4}, test all, not verified, tbp, -5% margin}},
{P shall 11., {Pperformance-11, inactive, {Pfunctional-2}, Rationale5, 1 System,
 Author5, {Pperformance-4}, inspection, not verified, tbp, 20% margin}},
{P shall 12., {Pperformance-12, inactive, {Pfunctional-2}, Rationale5, 1 System,
 Author5, {Pperformance-4}, inspection, not verified, tbp, 20% margin}},
{P shall 13., {Pperformance-13, approved, {Pfunctional-3}, Rationale6, 1 System,
 Author6, {Pperformance-7}, test sample, not verified, tbp, 1% margin}},
{P shall 14., {Pperformance-14, approved, {Pfunctional-3}, Rationale6, 1 System,
 Author6, {Pperformance-7}, test sample, not verified, tbp, 1% margin}},
{P shall 15., {Pperformance-15, approved, {Pfunctional-3}, Rationale6, 1 System,
 Author6, {Pperformance-7}, test sample, not verified, tbp, 1% margin}},
{P shall 16., {Pperformance-16, approved, {Pfunctional-4}, Rationale7, 1 System,
 Author7, {Pperformance-9}, analysis, not verified, tbp, -10% margin}},
{P shall 17., {Pperformance-17, approved, {Pfunctional-5}, Rationale8,
 1 System, Author8, {Pperformance-9}, analysis, not verified, tbp, tbd}},
{P shall 18., {Pperformance-18, approved, {Pfunctional-4}, Rationale7,
 1.1 Segment 1, Author7, {Pperformance-9}, inspection, not verified, tbp, tbd}},
{P shall 19., {Pperformance-19, approved, {Pfunctional-5}, Rationale7,
 1.1 Segment 1, Author7, {Pperformance-9}, demonstration, not verified, tbp, tbd}},
{P shall 20., {Pperformance-20, approved, {Pfunctional-4}, Rationale7,
 1.2 Segment 2, Author7, {Pperformance-9}, test all, not verified, tbp, tbd}}}
```

Finally, we make a few arbitrary sought requirements. First, one needs to create the type:

```
In[204]:= CreateReqType["Psought",
    {{"id", "unique number"},
     {"Status", "1 of: draft, approved, inactive"},
     {"End item affected",
      "array of approved end item names for which requirement is needed"},
     {"Rationale", "brief explanation of why need and correct"},
     {"Author", "name and contact information of creater"},
     {"Acceptance record",
      "organizational entity acknowledging responsibility to provide"}
    }]
    << Psought

Out[205]= {{{id, unique number}, {Status, 1 of: draft, approved, inactive},
    {End item affected, array of approved end item names for which requirement is needed},
    {Rationale, brief explanation of why need and correct},
    {Author, name and contact information of creater},
    {Acceptance record, organizational entity acknowledging responsibility to provide}}, 1}
```

Then we make two: one applicable to **1System**, one not:

```
In[206]:= DefReq["Psought",
    "Need to know X.",
    {"approved", {"1 System"}, "So know mass limit if any", "Last, First", "SEIT"}];
    DefReq["Psought",
    "P shall Y.",
    {"approved", {"1.2 Segment 2"},
     "So know power limit if any", "Last2, First2", "SEIT"}];
    <<
    Psought

Out[208]= {{{id, unique number}, {Status, 1 of: draft, approved, inactive},
    {End item affected, array of approved end item names for which requirement is needed},
    {Rationale, brief explanation of why need and correct},
    {Author, name and contact information of creater},
    {Acceptance record, organizational entity acknowledging responsibility to provide}},
    3, {Need to know X., {Psought-1, approved, {1 System},
    So know mass limit if any, Last, First, SEIT}}, {P shall Y.,
    {Psought-2, approved, {1.2 Segment 2}, So know power limit if any, Last2, First2, SEIT}}}
```

SpecWriter prepares the specification. Here is sample output for the make believe requirements generated for end item called **1System**:

In[209]:= **SpecWriter["1 System", "Pfunctional", "Pperformance", "Psought"]**

Specification: 1 System

Functional Requirement:	Pfunctional-1	P shall have the function to
Rationale:	Primary function of system	raise a chair above the ground.

Pperformance-1   P shall 1.
Rationale: Rationale1
Verification Reponsibility: 1 System
Verification Method: inspection

Pperformance-2   P shall 2.
Rationale: Rationale1
Verification Reponsibility: 1 System
Verification Method: inspection

Pperformance-6   P shall 6.
Rationale: Rationale3
Verification Reponsibility: 1 System
Verification Method: demonstration

Pperformance-7   P shall 7.
Rationale: Rationale3
Verification Reponsibility: 1 System
Verification Method: demonstration

Functional Requirement:	Pfunctional-2	P shall have the function to
Rationale:	Primary function of system	maintain chair height above floor.

Pperformance-4   P shall 4.
Rationale: Rationale2
Verification Reponsibility: 1 System
Verification Method: analysis

Pperformance-5   P shall 5.
Rationale: Rationale2
Verification Reponsibility: 1 System
Verification Method: analysis

Pperformance-8   P shall 8.
Rationale: Rationale4
Verification Reponsibility: 1 System
Verification Method: test all

Pperformance-9   P shall 9.
Rationale: Rationale4
Verification Reponsibility: 1 System
Verification Method: test all

Pperformance-10   P shall 10.
Rationale: Rationale4
Verification Reponsibility: 1 System
Verification Method: test all

Functional Requirement:	Pfunctional-14	P shall have made up function 1.
Rationale:	Rationale 1	

Functional Requirement:	Pfunctional-15	P shall have made up function 2.
Rationale:	Rationale 2	

Functional Requirement:	Pfunctional-16	P shall have made up function 3.
Rationale:	Rationale 3	

Functional Requirement:	Pfunctional-17	P shall have made up function 4.
Rationale:	Rationale 4	

Functional Requirement:	Pfunctional-19	P shall have made up function 6.
Rationale:	Rationale 6	

Sought requirements:		
Psought-1	Need to know X.	So know mass limit if any

4.5.10 CREATE A REQUIREMENT COMPLIANCE METHOD REPORT

ComplianceReport shows to what extent each performance requirement is achieved. Here is sample compliance report for **1System**:

In[210]:= **ComplianceReport["1 System", "Pfunctional", "Pperformance"]**

Compliance Report for: 1 System

Identification	Statement	Compliance
Pfunctional-1	P shall have the function to raise a chair above the ground.	tbd
Pperformance-1	P shall 1.	10% margin
Pperformance-2	P shall 2.	10% margin
Pperformance-6	P shall 6.	tbd
Pperformance-7	P shall 7.	tbd
Pfunctional-2	P shall have the function to maintain chair height above floor.	tbd
Pperformance-4	P shall 4.	15% margin
Pperformance-5	P shall 5.	15% margin
Pperformance-8	P shall 8.	-5% margin
Pperformance-9	P shall 9.	-5% margin
Pperformance-10	P shall 10.	-5% margin
Pfunctional-14	P shall have made up function 1.	tbr
Pfunctional-15	P shall have made up function 2.	tbr
Pfunctional-16	P shall have made up function 3.	tbd
Pfunctional-17	P shall have made up function 4.	A part
Pfunctional-19	P shall have made up function 6.	C part

4.5.11 CREATE A VERIFICATION STATUS REPORT

Our chosen specification format reports both the verification responsibility and method. If seeking to report only the verification attribute information, a simple modification to *ComplianceReport*, called *VerificationReport*, does the trick. An example application follows:

In[211]:= **VerificationReport["1 System", "Pfunctional", "Pperformance"]**

Verification Report for: 1 System

Identification	Statement	Method	Status	Evidence
Pfunctional-1	P shall have the function to raise a chair above the ground.	tbd		
Pperformance-1	P shall 1.	inspection	not verified	tbp
Pperformance-2	P shall 2.	inspection	not verified	tbp
Pperformance-6	P shall 6.	inspection	not verified	tbp
Pperformance-7	P shall 7.	inspection	not verified	tbp
Pfunctional-2	P shall have the function to maintain chair height above floor.	tbd		
Pperformance-4	P shall 4.	analysis	not verified	tbp
Pperformance-5	P shall 5.	analysis	not verified	tbp
Pperformance-8	P shall 8.	analysis	not verified	tbp
Pperformance-9	P shall 9.	analysis	not verified	tbp
Pperformance-10	P shall 10.	analysis	not verified	tbp
Pfunctional-14	P shall have made up function 1.	tbr		
Pfunctional-15	P shall have made up function 2.	tbr		
Pfunctional-16	P shall have made up function 3.	tbd		
Pfunctional-17	P shall have made up function 4.	A part		
Pfunctional-19	P shall have made up function 6.	C part		

4.5.12   CREATE AN ARCHITECTURE SPECIFICATION

The DoDAF[8] provides exceptional guidance as to what to describe about an architecture. All that is needed is a little adjustment to acknowledge that the architecture comprises more than the information flow between its constituents. An architecture specification is obtained by slightly modifying the DoDAF All View 1 (AV1) as follows:

1. Architecture identification
    1.1. Name
    1.2. Organization developing the architecture
    1.3. Approval authority
    1.4. Participating organizations
2. Mission, constraints, and assumptions
    2.1. Reference missions
    2.2. Constraints
    2.3. Assumptions
3. As-is architecture
    3.1. As-is system constituents
    3.2. As-is reference mission performance
    3.3. As-is interorganizational relationships
    3.4. As-is communications network
    3.5. As-is information exchange
4. To-be architecture
    4.1. Rationale for to-be architecture selection and corresponding reference missions performance
    4.2. To-be system constituents
    4.3. To-be interorganizational relationships
    4.4. To-be communications network
    4.5. To-be information exchange
    4.6. As-is to To-be transition plan
5. Technology development recommendations
6. Tools used to assess candidate architectures

Notice unlike the system specification, the architecture specification is a document that tells a story; it is not an output of a requirements database indicating which performance requirements correspond to which functional requirements. This is another reason architecting is a different activity than systems engineering.

4.6   VERIFY REQUIREMENTS ARE COMPLIED WITH

There are numerous types of "verifications" that need to be accomplished to bring a system into existence in compliance with its requirements. The word *verification* is often used to denote the effort to determine the real-world solution meets the documented requirements, while the word *validation* is used to denote the effort to show the real-world solution meets the customers' and users' needs.

Though these two efforts are very important, it is equally important to present evidence the documented requirements are complete and correct, that the simulations used to make decisions provide necessarily accurate results, that both the design (instructions to build end items) and the end items themselves comply with requirements, and that the system (collection of end items) or architecture (collection of systems) meets requirements. So there are at least six kinds of "verification efforts." I can't think of six different single-word names for each of these verification efforts, so I will use the word *verify* but add a phrase to explain what is being verified.

This section provides process descriptions and heuristics to perform the six types of verification. Please note that the name given to each of the verification types is purely arbitrary and at the discretion of the traditions of an organization or industry. Based on my experience, it is critical to perform each of the applicable types for a program to succeed.

4.6.1 REQUIREMENTS ARE COMPLETE AND CORRECT

Completeness means all the necessary functions to be implemented were named and only those functions are being implemented. *Correctness* means all the performance requirements needed for all the functional requirements to have meaning are defined, and, the values chosen are the best set in terms of resolving contradictions and ease of implementation.

Since we lack of a specification for the requirements, how do you know when you have found and accurately documented the complete and correct requirements? Most people presented a real-world "thing" to fulfill their needs can tell you what they like or don't like about it, but many have a harder time telling you how they want it different. Many people have trouble relating to abstractions in a manner that enables them to communicate their reaction, and requirements are the ultimate abstraction. So we first need to find ways to ensure requirement comprehension, or the opposite, the extent of ambiguity.

Gause and Weinberg[9] identify the following sources of ambiguity:

1. Observation
2. Recall
3. Interpretation or problem solving

Observation ambiguity is variation in what people see or hear. *Recall ambiguity* is variation in what people remember. *Interpretation ambiguity* is variation in how people in their own minds formulate what they have read or witnessed. Evidence of an interpretation or problem-solving ambiguity is clusters of people interpreting the requirements in fundamentally different ways. Observation and recall ambiguities tend to cause variation amongst people within a given cluster.

Gause and Weinberg suggest three methods for uncovering ambiguity:

1. Definition variants
2. An ambiguity poll
3. A memory poll

We already implemented a means to do the definition variant test, which calls for substituting synonyms for words in the requirement. A caution: it is highly unlikely a customer will in reality consent to read and react to hundreds of variants of the same requirement. So this method is best used by the system engineer or architect to identify the possible misinterpretations they may have regarding the customer's communicated requirements, to then have a direct conversation with the customer on a few specific topics.

To implement the ambiguity poll, select a metric (examples: time to complete, or cost to do) that requires understanding the requirements to estimate. Ask people who must implement the requirements to independently estimate the chosen metric. Then the individuals are brought together to compare and discuss the tabulated results. Large differences in the metric typically are the result of interpretation ambiguity, smaller differences usually the result of observation and recall ambiguities. This is an extremely powerful method to use with the designers, producers and supporters.

To implement the memorization poll, ask individuals to recall requirements from memory. Those requirements remembered well are likely to be unambiguous; the requirements that are not remembered well will likely be ambiguous. Though this method is powerful, the customer is unlikely to care to participate, nor the designers, producers, and supporters. So, a useful variant is to show how you intend to comply with the requirements to the designers, producers, and supporters; their reaction will be doubly helpful to determine if you understood their requirements and are providing a solution they might like. Similarly, by asking the designers, producers, and supporters to write down how they plan to comply with each requirement allocated to them, you will indirectly determine if the requirement was interpreted as you intended.

All these methods share a very basic human activity—conversation. By talking about the requirements, the huge number of potential interpretations is reduced to the desired one. The requirements can be then be reworded to aid maintaining that agreed to interpretation in the future, but once the common understanding is achieved, it really doesn't matter what words are used, provided the people involved remain the same and remember the agreements, which cannot be guaranteed. New participants must be taught how to interpret the requirements to retain the agreed to meaning.

The following explains how to verify requirements are complete and correct:

4.6.1.1   Process to Verify Requirements Are Complete and Correct

Step 1.   Perform the ambiguity checks on each requirement.

Step 2.   Ensure the system or architecture structure closely resembles the functional and physical structure.

Step 3.   Look for and eliminate aggregation of items with conflicts of interest, and partition them into separate end items to ensure checks and balance.

Step 4.   Check to ensure that all functions have parents; if not, delete or establish the appropriate parent.

Step 5.   Check that all subfunctions are independent and collectively name the activities of the parent function that need to be implemented; if not, change the subfunction names and definitions.

Step 6.   Check that all functions have been allocated to at least one end item; if not, delete, or allocate to at least one end item.

Step 7.   Check that all end items have at least one function allocated; if not, delete the end item, or identify the functions the end item performs.

Step 8.   Check all functions have at least one performance requirement allocated to them; if not, delete the function, otherwise record performance requirements for the function.

Step 9.   Check all performance requirements have been allocated to at least one function; if not, delete the performance requirements, otherwise, define the functions for which the performance requirements apply.

Step 10. Check all attributes, except the acceptance attributes, are defined for all requirements.

Step 11: Verify requirements with customer and users.

Step a.   Create a model, either physical or virtual, that simulates the functions to be provided along with performance that will be achieved. Where customer and user desires result in contradictory features, explicitly construct the model so the contradiction can be varied so the customer and user can visualize the impact and express preferences as to what solution to the contradiction is best for them.

Step b.   Demonstrate the model to the customer and users and note what they want as is and what they want changed.

Step c.   Explicitly identify any item that needs technology development to the customers and users, and note their reaction. Find solutions that do not require the development of immature technology, unless there is no alternative for a function and its associated performance. In all such instances, ensure the customer is willing to provide the resources to provide the technology maturation. If the customer is not, but you are certain these features are necessary for success, then you must self-fund the development of the technology.

Step d.   Explain the estimated development and annual operating cost to the customer and users, and note their reaction. If not acceptable, itemize what needs to change. It is possible at this point that what the customer wants and what you can deliver are found to be dramatically at odds. If so, you then have a tough choice, either determine how to get the customer what they want, by perhaps involving others, or undertaking a focused development, or realize that you cannot satisfy this customer and walk away from the program.

Step 12: Verify requirements with those who will implement and support the system.

Step a.   Provide the applicable functional and performance requirements and allow them to study for a time period.

Step b. Hold a requirements familiarization review, asking those attending to individually provide one or more of the following: estimate of time to design and produce the item, nonrecurring and recurring cost of the item, likely reliability of the item, likely availability of the item, and worst-case performance values that will be achieved (for this, do not specify the performance terms, let the presenters identify both the parameter and their prediction). The presentation will most likely identify requirements that were misunderstood, adjust as needed.

Step c. After the designers, producers, and supporters have had some time to react to the result of the requirements familiarization review, hold a requirements compliance review. Ask for indication of how will comply with each allocated requirement, and document these statements in a table that shows each requirement and its associated compliance approach. In so doing, the designers, producers, and supporters may indicate they do not know what the requirement means, so further dialogue and potential requirement editorial is needed before can provide a compliance methodology. Review the compliance methodology and report to the designers, producers, and supporters any that appear contradictory to the intended requirement. Dialogue until agreement is reached and acceptable compliance method is found.

Step d. Explicitly ask the design team what else they need to know to design the end item.

Step e. Explicitly ask the producers if they have the information they need to know how to make the item.

Step f. Explicitly ask the supporters if they have the information they need to know to support the item.

Step g. Complete the acceptance attribute entry for at least the functional requirements.

4.6.2 SIMULATIONS

There are four related but distinct aspects of simulations that need verification:

1. The equations used adequately represent the real-world phenomenon of interest.
2. The input data are correct.
3. The means used to evaluate the equations is correct.
4. The means used to depict the output is correct.

The best general purpose method to verify simulations is to have them independently developed and operated by different teams of people, ideally using different tools. The more significant the potential negative consequence of the simulation, the more independent teams should be formed. Two is the absolute minimum, and five is probably the maximum that will ever be needed. Most organizations are loath to do this, due to the perceived extra cost, but considering the potential negative consequences of wrong simulation data being used to make decisions, the insurance is well worth the cost. The process is the independent teams produce their own simulations and compare outputs. Should the outputs agree, the independence is presumed to provide confidence the agreed too outputs are correct. Should the outputs disagree, the teams work together to find the cause by checking and comparing their input data and equations used. Obviously, completely independently developed and operated simulations are unlikely to agree exactly for any simulation, so a tolerance is required, which typically is set at the threshold that the simulation could be in error before bad consequences would occur.

For equations that predict physical events, verification is best done by using the equations to predict outcomes that can be compared to physical tests. For equations that are fits to data, then the best verification is to be independently derived by at least two separate teams.

Sometimes input data are self-checked, when the equations utilizing the data blow up or produce physically impossible results. Often, input data checking can be semiautomated, as each input may have a likely range or precision, so each input can be tested against their respective test values, and any nonconformance reported. Probably the most common input data errors are as follows:

1. The numbers are not in the units expected.
2. The numbers are in the wrong "sequence" (e.g., yesterday's information rather than today's; decreasing rather than increasing, or vice versa; or location specific with missing location identifiers).

The possibility of this can be minimized by enforcing labels are part of the input data.

Coding of the equations is verified by comparing output to other codes that have been shown to match real-world results. This may be a sequential process, where more and more complicated checks are performed. For example, start by comparing the computed values to those obtained by hand calculations. Then compare more complicated results to programs built independently, or by a different method, or using a different program that is more trusted because it was used for a while and errors have been found and weeded out.

Most complicated equations are that way because of the need for iteration or recursion to obtain the solution. Typical concerns are as follows:

1. How many iterations should be executed?
2. How "small" or "large" can a numerical "step" be and still obtain correct results?

As computer power has increased, these issues have tended to be minimized. But as computer power grows, the problems we seek to solve also grow in complexity as well, so inevitably some system engineers and architects will be addressing simulations that are at the edge of what computing technology enables at the moment. These simulations are very difficult to verify as there is likely to be literally no alternative computer or software combination that can possibly perform the same calculations. That is why the results of these simulations are viewed so skeptically by those not engaged in the effort. For all practical purposes, these leading-edge simulations are not fully verified until sometime after computing capability increases. In the meantime, one can try to logically break down the process into parts that can be independently verified, ideally by people with no vested interest in the result, with the conjecture that since these parts have been independently shown to be performing correctly, the entire result can be presumed to be correct as well.

Verifying the output depictions are correct is roughly the same as verifying the input data was correct. Output data often self-check, in that errors show up in nonsensical plots. Legitimate output ranges are often predictable, so output obtained can be checked against these ranges, with outliers flagged. The labels on plots, titles, and axes need to be checked to ensure the correct units are noted and the data are what they are purported to be. Finally, several points on the plot or table should be selected at random, and an effort can be made to ascertain that the input and equations actually produced those points.

4.6.2.1   Process for Verifying Simulations

Step 1. Formulate at least two independently derived and executed simulations.
Step 2. Compare results, if results match to a publicly declared tolerance, presume
        the simulation is correct, otherwise determine cause and rerun until do.

4.6.3   End Item Design

This verification is to attempt to predict with certainty that the instructions to build the item will build the desired items prior to it being built. *Desired* means it will perform its allocated functions, achieve its performance, and be manufactured and supportable. Different processes are needed for hardware and software.

4.6.3.1   Verify Hardware End Item Design

For a hardware item that needs technology to be matured, the first task is to check the credibility of the technology maturation plan. These plans are typically depicted as a stair step from whatever level the technology is currently, to an acceptable level by a specified date, or milestone. On the maturation plan, each technology maturation improvement must require some explicit criteria to be demonstrated that are independently verifiable. On the plan, the step stays at that maturity level for the predicted time the declared activity will require. So one needs to be examine "proof" the duration of the step is likely, and the maturation jump is the right amount. Of course, how can one possibly know this? The effort, by its very nature, is essentially or completely new and almost completely conjecture! Nevertheless, progress can be made. Of the two factors, duration and maturation increase, the second is the easier to verify. By carefully reviewing the planned activity, one can almost always articulate to what level additional technical maturity is achieved, provided the maturity level definitions are reasonable, such as those shown in Chapter 2. Duration of the activities are much more difficult to predict. The best that can be done is to insist the proposed activities be broken down into comprehensible tasks that consist of things that have been done before and those that have not. Remarkably, many of the tasks will be quite ordinary: material to order, plans drawn up, various facilities and machines acquired and installed, and permissions from agencies or test centers requested and received. These all have historical precedence, and unless something is explicitly done, they are going to take as long as they usually take. Inevitably, there will remain tasks that have never been done before. Some of these "new" activities may be very similar to a prior activity, for which historical actual durations are available. Again, unless something is significantly different, these similar activities will take as long as they have taken in the past. Ultimately, there will be completely new tasks, for which the duration is frankly a guess. Whoever made the guess had some rationale. Ask what it was; you may discover the estimate is more of a wish than a conjecture, or perhaps that was all the time that was left! For these truly new activities, one needs to give as much time as possible. Ideally, one should try to impose cycles, requiring something tangible as quickly as humanly possible, even a bit faster than seems humanly possible. So, then perform at least a second cycle, and ideally a third. At the end of each cycle, have an exit option. That is, if the originally perceived effort is now found to be much harder than anticipated, or even impossible, have a backup alternative that will be invoked to keep the overall program on track. From my experience however, this is usually more than one can ask, as it is exactly those technologies the system is most dependent on for which there is no acceptable alternative, and one has to keep slogging away until the technology is working. This is a royal pain for that program, but once the technology is matured, it's literally a breakthrough, and a blessing for related programs to follow.

4.6.3.1.1   Process to Verify Technology Maturation

Step 1. Document a stair step plan from the current technology level to the desired technology level.
Step 2. Check for explicit criteria to check results against.
Step 3. Check the duration to achieve each step is appropriate.
Step 4. Check the resources to achieve each step are appropriate.
Step 5. Have a backup alternative to invoke if reality stalls the plan.

For hardware, ideally, the end item requirements can be met by buying an existing item, for which at least a specification is available. If the end item must be produced for the first time, then either its description or a manifestation of it is examined for compliance with the requirements.

4.6.3.1.2   Process to Verify Hardware Meets Design Requirements

Step 1.   Check that predicted dimensions will be less than or equal to any allocation.
Step 2.   Check that only acceptable or specified materials are used.
Step 3.   Check that predicted mass will be less than or equal to any allocation (included moments of inertia).
Step 4.   Check that hardware accommodates potential loads to tolerance level specified.

Step 5. Check that it will accommodate potential temperatures, humidity, or any other environmental factors to tolerances specified.

Step 6. Check that any power input will be accommodated, including potential variations in current or voltage.

Step 7. Check how much heat and the peak heat rate the item will produce and make sure as specified or within tolerance.

Step 8. Should radiation or electromagnetic interference tolerances be specified, check for likely compliance.

Step 9. If a new item, have the producers examine and determine how it will be produced, and recommend means to simply its production.

Step 10. Have supporters examine, both as an end item, and where it is in the system surrounded by other items, and confirm it can be supported, and recommend changes to make support time and cost less.

Step 11. Predict reliability and compare to expectation.

Step 12. Predict availability and compare to expectation.

Step 13. If a new item, examine the schedule for development, looking for evidence, usually based on analogy to similar items produced before, that the estimated spans are likely to be achieved.

Step 14. If a new item, examine the cost prediction, looking for evidence, usually based on analogy to similar items produced before, that the material and labor estimates are likely to be achieved.

Step 15. Check that each of the allocated functions will indeed be performed.

Step 16. Predict performance achieved and compare to requirements.

Step 17. Confirm that a plan exists to verify the produced item will meet its requirements and ascertain both the completeness and cost of executing the plan, and make sure acceptable.

4.6.3.2   Verify Software End Item Design

For software, the verification effort is very dependent on the software design process utilized.

4.6.3.2.1   *Process to Verify Software Meets Requirements*

Step 1. Check functions are indeed implemented.

Step 2. Check inputs are as intended, and especially check types and units.

Step 3. Check how will handle spurious inputs, in value, quantity, and rate.

Step 4. Check outputs will be as intended (regardless as to how obtained), and especially check units.

Step 5. Simulate the internal activities to get as much evidence as possible the transformation from input to output will occur as intended (regardless of memory or timing requirements).

Step 6. Check the proposed code is suitable for the application.

Step 7. Check memory required will be within limits.

Step 8. Check processing time required will be within limits.

Step 9. Review the proposed verification method to assess it will indeed verify the final software will indeed meet its requirements.

4.6.4 END ITEM

This verification is to show the produced end item meets its requirements. Each end item is allocated one or more functional requirements with associated performance requirements. Each performance requirement came with attributes stipulating the verification method. Nearly every end item will interface with either another end item or an item external to system. Each of these interface requirements is also a documented function with performance requirements. So the first job is to formulate a verification plan for the end item that honors the verification method stipulations and obtains independently reviewed evidence the requirements are met. Then the planned verification efforts are conducted, and the evidence produced examined to see if met the stipulated requirements. For new end items, it may be necessary to create a facility to mimic environments the end item will face, and this may be as challenging as producing the end item itself. Clearly, though the process is easy to state, end item requirement verification is a complicated task that needs an effort in proportion to the complexity of the system itself, for which deep domain knowledge is critical to success.

The impression may be that it is the requirements that need to be verified. In reality, when planning these efforts, the primary focus is how to show the end item meets the requirements allocated to it, a subtle but critical distinction.

4.6.4.1 Heuristics for End Item Verification

1. Verify interface functions and performance first, simulate the worst thing one side could do to another and make sure to handle appropriately.
2. Inspect the inspectors.
3. When verifying by analysis, first verify the analysis is of the as-built end item, not the could-have-been-built end item.
4. When verifying by test, know how to discern test equipment and process failures and end item failures.
5. When verifying by test, test the test equipment first.
6. When verifying by test, start at the lowest level end items and do not break configuration as test higher level end items.
7. When verifying by test, know how to discern test equipment and process failures and end item failures.

4.6.5 SYSTEM

Systems are verified by first showing their end items have achieved their requirements, then by showing the collection of the end items meets requirements. Recall higher level functions with performance requirements were allocated to higher level parts of the system. As with verifying the end items, the first job is to create a plan to verify these higher level elements achieve their stipulated performance. Some systems are asked to achieve missions that are hoped never to occur. In such instances, it is not possible to fully test the system to its end objective, and we must make do with verifying aspects of the systems with the presumption the aspects will work in unison when called upon.

Though so simple to state, the effort can be enormous and expensive.

4.6.5.1 Heuristics for System Verification

1. If anything can go wrong, it will, so identify everything that can or will go wrong, and put in place something to prevent it, or if it cannot be prevented, recovery from it quickly.
2. If it isn't broken, do not fix it (alternatively, if you have achieved a goal node, stay there).
3. Unless everyone who needs to know does know, there will be a screw-up.
4. Mistakes are inevitable; failure to report is inexcusable.
5. The number of software defects remaining is proportional to those found.
6. Ensure that explicit permission with public acknowledgment is the only way a system undergoing verification can be changed.
7. When searching for a cause, change only one thing at a time, and trace defect to cause by asking, "Why?" at least five times.

4.6.6 ARCHITECTURE

Architectures are verified by first showing their systems have achieved their requirements, then by showing collection of the systems meet requirements. Architectures are often so complex, with elements under so disparate control, that it is virtually impossible to fully verify all requirements desired of them. When establishing architecture requirements, one must keep this in mind, for it is virtually useless to write architecture requirements for which verification is impossible.

4.6.6.1 Heuristics for Architecture Verification

1. Assume previous studies and approaches are flawed.
2. Models are not reality.
3. For architectures that impact society, perceptions are more important than facts, so a coherent constituency must be found and maintained.
4. Unless the politics are a go, the architecture will not go, which means the best engineering solution is not necessarily the best solution.
5. No architecture can be optimum with respect to all desires, so cost rules; so know and appease who benefits, who pays, and who loses.

4.7 MEASURE REQUIREMENTS VOLATILITY

Requirements volatility is defined as

> **(number_new_requirements + number_deleted_requirements + number_revised_requirements) / (Total_number_of_approved_requirements)**

Design efforts should begin after all key requirements are established and volatility is less than 5%. *SRRDatePrediction* takes historical volatility data and projects the date when a specified level will be achieved.

Here is an example:

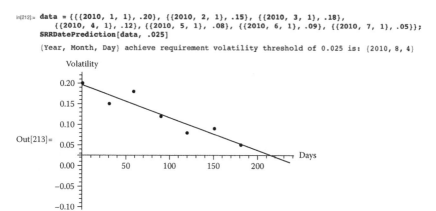

```
In[212]:= data = {{{2010, 1, 1}, .20}, {{2010, 2, 1}, .15}, {{2010, 3, 1}, .18},
        {{2010, 4, 1}, .12}, {{2010, 5, 1}, .08}, {{2010, 6, 1}, .09}, {{2010, 7, 1}, .05}};
      SRRDatePrediction[data, .025]

      {Year, Month, Day} achieve requirement volatility threshold of 0.025 is: {2010, 8, 4}
```

4.8 REQUIREMENTS HEURISTICS

- Requirement allocation requires roles and responsibility clarity.
- Put only the requirements needed for the end item in the specification for the end item.
- Validate requirements first, not last.
- Regardless of the program implementation construct used, all systems have eight primary functions (develop, design, produce, deploy, train, operate, support, and dispose) for which requirements may be specified.
- Different requirement types have different associated attributes.
- Three very useful types of requirements are functional, performance, and sought.
- Convenient attributes for functional requirements are unique identification number, status, parents, end item allocated too, acceptance record, rationale, and author.
- Convenient attributes for performance requirement are unique identification number, status, parents, function allocated too, verification method, verification status, verification evidence, rationale, and author.
- A requirement with TBD or TBR in it communicates no useful information, so why create? Create a "sought" requirement instead.
- Convenient attributes for sought requirements are unique identification number, status, requesting entity, acceptance record, rationale, and author.
- Allocate interface requirements to an end item as any other requirement.
- Treat compliance requirements as any other requirement.
- Requirement statement may be codified with words such as *shall*, *should*, or *will* to denote mandatory, preferred, or deferred compliance, respectively. Any words or phrases accepted by the group with the responsibility

to provide the system or architecture that denotes level of desirability are acceptable to use.
- Good functional requirements minimize the reader's uncertainty as to what activities are required to be implemented.
- The usual functional requirement errors are as follows:
  1. Missing a required activity
  2. Imposing an implementation prematurely
  3. Decomposing functions to lower level than needed to determine how will implement
- Good performance requirements minimize the reader's uncertainty as to how well a function needs to be accomplished.
- The biggest performance requirement error is to ask for more than can be provided with the resources available.
- Good sought requirements make it clear what information is needed and why.
- Requirement statements need to attempt to minimize all the following types of ambiguity:
  1. Compliance level
  2. Completeness
  3. Precision
  4. Comprehension
  5. Referencing
  6. Vague words
  7. Functional requirement
  8. Acronyms
  9. English unit usage
  10. Word emphasis
- Processes to determine requirements are as follows:
  1. Reuse from prior program
  2. Interpreting customer provided documents
  3. Surveys
  4. Witness
  5. Focus groups
  6. Assessing product defects
  7. Deriving concept of operations
  8. Formal diagramming techniques
  9. Quality matrices
  10. Models or prototypes
  11. Incremental build
- The collection of requirements and attribute information is best maintained as a database for which at least the following functions can be performed:
  1. Create a requirement type.
  2. Create an end item hierarchy.
  3. Create an implementation organization hierarchy.
  4. Define a requirement of a particular type.
  5. Edit requirement attributes.

6. Edit requirement statement.
7. Find requirements with specified attribute values.
8. Find requirements with specified content.
9. Identify parents of a requirement.
10. Identify children of a requirement.
11. Identify functional requirements allocated to end items.
12. Identify performance requirements allocated to functions.
13. Identify functional requirements not allocated to an end item.
14. Identify performance requirements not allocated to functional requirements.
15. Create end item specification.
16. Create a requirements compliance report.
17. Create a verification status report.
18. Create an architecture specification.

- The best end item specification format is to simply collect all allocated functional requirements along the corresponding performance requirement and all attribute values.
- An architecture specification format is as follows:
    1. Architecture identification
        1.1. Name
        1.2 Organization developing the architecture
        1.3 Approval authority
        1.4 Participating organizations
    2. Mission, constraints, and assumptions
        2.1 Design reference missions
        2.2 Constraints
        2.3 Assumptions
    3. As-is architecture
        3.1 As-is system constituents
        3.2 As-is design reference mission performance
        3.3 As-is interorganization relationships
        3.4 As-is communications network
        3.5 As-is information exchange
    4. To-be architecture
        4.1 To-be system constituents and rationale for selection
        4.2 As-is design reference mission performance
        4.3 As-is interorganization relationships
        4.4 As-is communications network
        4.5 As-is information exchange
        4.6 Transition plan
    5. Technology development recommendations
    6. Tools used to assess candidate architecture
- To verify requirement are met:
    1. Verify that you have a complete and correct set of requirements using ambiguity checks, showing models of consequences to customer and users, and asking implementers and supporters what it costs and how well it will be accomplished.

2. Verify simulations by comparing results from at least two to five independently developed and operated sources.
3. Verify hardware design by checking physical features and tolerance to achieving performance in off-nominal environment and verify software design screens inputs for suitability, requires fraction of processing and memory available and self-checks output.
4. Verify end item actuality by simulation or test.
5. Verify interface functions and performance first, simulate the worst thing the other side could do to another, and make sure it can handle it appropriately.
6. Inspect the inspectors.
7. When verifying by analysis, first verify the analysis is of the as-built end item, not the could-have-been-built end item.
8. When verifying by test:
   8.1. Know how to discern test equipment and process failures from end item failures.
   8.2. Test the test equipment first.
   8.3. Start at the lowest level end item and do not break the configuration as you test higher level end items.
9. Verify system actuality by simulation or test.
10. If anything can go wrong, it will, so identify everything can think will go wrong, and put in place something to prevent, or if cannot prevent, then recover from quickly.
11. If it isn't broke, do not fix it, or, if you have achieved the goal, stay there.
12. Unless everyone who needs to know does know, there will be a screw up.
13. Mistakes are inevitable; failure to report one is inexcusable.
14. The number of software errors remaining is proportional to those found so far.
15. Ensure that explicit permission with public acknowledgment is the only way a system undergoing verification can be changed.
16. When searching for a cause, change only one thing at a time, and trace defect to cause by asking, "Why?" at least five times.
17. Verify architecture actuality by simulation or test.
18. Assume previous studies and approaches are flawed.
19. Models are not reality.
20. For architectures that impact society, perceptions are more important than facts, so a coherent constituency must be found and maintained.
21. Unless the politics are a go, the architecture will not go, which means the best engineering solutions are not necessarily the best solution.
22. No architecture can be optimum with respect to all desires, so cost rules; so know and appease who benefits, who pays, and who loses.

- Schedule system requirements reviews when the forecast of requirement volatility is below a threshold level such as 5%.

REFERENCES

1. Solomon, Paul J., and Young, Ralph R., *Performance Based Earned Value*, Hoboken, NJ: Wiley Inter-Science, 2007.
2. Lutz, William D., *The Cambridge Thesaurus of American English*, Cambridge: Cambridge University Press, 1994.
3. Akao, Yoji (ed.), *Quality Function Deployment Integrating Customers Requirements into Product Design*, Portland, OR: Productivity Press, 1990.
4. http://www.everyspec.com/MIL-STD/MIL-STD+(00300+-+0499)/MIL_STD_490A_1-378.
5. Aerospace Corporation, Aerospace Report No. TR No. 009 (6904-12), *System Specification Preparation Guide*, El Segundo, CA: The Aerospace Corporation, 1992.
6. Chapman, William L., Bahill, A. Terry, and Wymore, A. Wayne, *Engineering Modeling and Design*, New York: CRC Press, 1992.
7. Robertson, Suzanne, and Robertson, James, *Mastering the Requirements Process*, Reading, MA: Addison-Wesley, 1999.
8. See http://www.architectureframework.com/dodaf/.
9. Gause, Donald C., and Weinberg, Gerald M., *Exploring Requirements: Quality before Design*, New York: Dorset House Publishing, 1989.

5 Improve an Organization's Ability to Do Systems Engineering and Architecting

This chapter recommends an approach to improve both the processes and performance of an organization's systems engineering or architecting. The chapter presents surveys that can be used to gather data to determine where best to concentrate improvement efforts at a point in a program.

The following case studies illustrate issues with attempts to improve systems engineering or architecting.

Case Study 5.1: Reducing Systems Engineering Errors in Competitive Proposals

BACKGROUND

A systems engineering organization had lost several competitive proposals due to inadequacies in the systems engineering section. A systems engineer was asked to examine reported weaknesses and recommend remedial actions.

WHAT HAPPENED

Four of the six losing proposals came with evaluation material available for review. For each, there were a large number of customer-provided feedback briefs, selection letters explaining relative strengths and weaknesses, and "explanation notices," which are written questions or comments from the proposal evaluators asking for a clarification of some kind. Each weakness was characterized as one of the following:

1. Management if it concerned schedule, roles and responsibilities, or configuration management
2. Analysis if it concerned mass claims, estimates or projections, technical performance parameters, trajectories, availability, timeliness, communication link, thermal protection, or guidance, navigation, and control adequacy

3. Requirements if it concerned compliance, derivation, allocation, or the verification method
4. Verification and integration and test if it concerned test plans, facilities, or shipping
5. Risk if it concerned identifying, assessing, or mitigation plans
6. Launch operations if it concerned process and time lines or interfaces
7. Cost if it concerned acceptance and comprehension of the basis of estimates

Each weakness of the above types was given one of four grades, based on the customer's words:

1. Significant weakness or deficiency
2. Weakness
3. Failed to follow proposal instructions
4. Response unclear

Since the proposals were for very different kinds of systems, prepared by mostly different people and for different customers, there were a lot of unique weaknesses. But several types of weaknesses were found on at least 75% of the four proposals. These were the following:

1. Management:
   100% had schedule inconsistencies.
   75% had the following:
   - Government saying they would not provide the named property or inconsistencies in claims as to what property would be provided by government
   - Unclear or inconsistent role and responsibility statements
   - Configuration in consistencies
2. Analysis:
   100% had incomplete trajectory information or unaccepted assumptions
   75% had the following:
   - Mass inconsistencies
   - Key performance parameters not properly calculated
   - Reliability claim errors
3. Requirements: 75% had the following:
   - Failed to comply with some customer stipulated requirement
   - Derived wrong requirements
4. Verification and integration and test: 75% had at least one verification effort not performed properly.
5. Risk: 75% had the following:
   - Significant risks not identified
   - Disagreement with risk assessment
   - Inconsistent statements regarding what risks were or how to mitigate

Finally a root cause analysis was undertaken, and the following was determined:

1. The proposal effort was undertaken while the offering itself was immature or uncertain.
2. The proposal authors lacked the knowledge needed to prepare an appropriate response.
3. There was a flaw in the proposal process.
4. Author or editor error.

Six recommendations were made to counter the root causes:

1. Have a mature offering before start proposal; otherwise, don't bid.
2. Maintain one database of all information for end items.
3. Independently verify analyses. (A process to do so was also recommended.)
4. Conduct a proposal requirements review prior to RFP receipt. (A process to hold the review was also recommended.)
5. Do verification regarding the way the customer wants the proposal; otherwise, don't bid.
6. Find the minimum risk offering. Do risk analyses as the customer wants, be pessimistic, and use the *risk* word only in the risk section of the proposal.

Table 5.1 maps each recommendation to the root causes.

The product was delivered to the head of systems engineering, who was very happy with it and it passed on to the head of proposal operations, who was also very happy with it. However, none of the recommendations was fully implemented on future proposals.

LESSONS LEARNED

Finding process faults is easy; fixing them is difficult. Though at first blush, identifying systemic problems in system engineering proposals may appear to be very difficult, in fact was very easy. Diagnosing potential root causes for the faults and even recommending approaches to prevent the root causes were not too difficult. The recommendations to address the root cause are often obvious. What is very difficult is for people who were not associated with the analysis and recommendation formulation to adopt the recommendations.

As much as practical, have the performers find their faults, recommend the fixes, and implement the fixes. This is the best way to get adults to buy into what must be done.

The list of things to fix always exceeds what one can afford to fix, so one needs a means to prioritize the improvement efforts. Faults are not equal, and some are much more significant than others, so it is critical to provide a means for people to denote the relative importance of the issues.

TABLE 5.1

Fix Recommendations Mapped to Issues and Root Causes

Issue or Cause	Immature Offering	Lack of Knowledge	Process Flaw	Author or Editor Error
Schedule inconsistencies	X-1			X-1
Item on GFE list will not be provided or is inconsistent	X-1	X-2		X-1
Unclear or inconsistent responsibilities	X-1		X-2	X-1
Configuration management inconsistencies	X-1		X-2	X-1
Mass inconsistencies	X-1			X-1
Incorrect performance metric calculations	X-1	X-3		
Trajectory analysis errors	X-1	X-3		
Reliability analysis errors	X-1	X-3		
Missed requirement to comply with	X-1		X-4	
Derived wrong requirement	X-1	X-3	X-4	
Doing verification or integration and test wrong	X-1	X-5		
Fail to identify risks	X-1	X-6	X-6	
Disagree with risk assessment	X-1	X-6		
Inconsistent risk assessment	X-1		X-6	X-1

Recommendations:

1. Have a mature offering; otherwise, don't bid.
2. Maintain a database of all information for end item.
3. Independently verify analyses.
4. Conduct a proposal requirements review prior to RFP receipt.
5. Do verification the way the customer wants; otherwise, don't bid.
6. Find minimum risk offering.

Case Study 5.2: Improving Architecting Efficiency

BACKGROUND

After presenting the results of an architecting effort that took three years for one mission area, our customer asked us to do two more mission areas in the next two years.

WHAT HAPPENED

Examining why it took three years to do the first architecting effort, the first impression was that this was the time needed to collect all the relevant data. But upon closer examination, though the data collection effort certainly took a lot of man-hours and chronological time, the fundamental cause for three years to pass before having the result was a near-constant debate regarding both what was to be accomplished and how to accomplish it. Once we had established the process and the products expected, a complete product was achieved in about a year. So a commitment was made to produce the architecture for the second mission area in one year. This goal was met. What I noticed during the effort to produce the second architecture was that nearly complete products were available much faster than anyone thought. But that didn't mean everything was completely known quickly. What I envisioned now was that an architecting product could be prepared in as little as six months, provided we had opportunity to do an improvement cycle of approximately the same length. So we promised to do the third mission area in six months. In turned out that, about halfway through the effort, there was a pressing need to make some decisions. So we held an architecture review on what we had at about the three-month point. The people dependent on the information were so happy to get what they had that they started asking us to do things then and there. So what then happened was we iterated the architecture on about one-month cycles for the remaining three months until the product was done.

LESSONS LEARNED

Use the learning curve to your advantage. Nearly everyone gets better at doing everything in a predictable way. If can arrange your work to be a series of repetitive efforts, you will end up doing the work very rapidly with less mistakes in a predictable way.

Ask for a product in about half the time people think they can provide the product. Interestingly, since most people get in trouble for delivering things late, you would think most people would pad their time estimates to give themselves as much time as possible. But the pressure to provide immediate good news leads most people, particularly people new at doing something, to promise the result as quickly as possible. Inevitably, something unforeseen happens, or the effort is simply much more difficult than anticipated, and the product is late. So you would think a good heuristic would be to increase the time estimate provided to you. But another heuristic is that work expands to fill the time provided. So, a good method to get things done is by incremental builds on tight schedules. That way, at least the schedule is certain, and to a high degree the cost is as well, since a known number of people will work for a known amount of time. All that is uncertain is exactly how much product will be available. This is also an extremely useful approach if the requirements for the product are very uncertain, as they too can be adjusted with each increment.

5.1 MEASURE SYSTEMS ENGINEERING OR ARCHITECTING PROGRESS

Before I address how to improve doing systems engineering or architecting, we first must agree on how to measure work accomplished. Solomon and Young[1] provide sound guidance in how to use performance-based measures, which very briefly is an enhancement of earned value-based progress monitoring to explicitly include the extent the requirements for the entity are being met is included in the measurement.

For those who may be new to the topic, there are basically two types of work, measurable (sometimes called *discrete*) and level of effort. Measurable work can be monitored in proportion to the extent a product is complete. For example, suppose half the budget is spent, and half the house is painted, in half the anticipated time to complete the job, and the quality is deemed fine. The performance-based earned value is the product or ratios:

(actual days) / (planned days)

(actual cost) / (planned cost)

(actual quality) / (required quality)

(actual quantity) / (planned quantity)

which in this case are all 1's, so the product is 1. But if instead of one half of the house being painted, only one fourth was, the status would be as follows:

$$(1) * (1) * (1/4) / (1/2) * (1) = 0.5$$

That is, the work is substantially behind and it can be estimated that it will take two times the current cost or two times the time spent to complete the house.

Level-of-effort work is necessary, but not attempting to achieve a product with a definite specification. Today there is a great deal of prejudice against level-of-effort work, as it is perceived as non value added because it does not produce a definitive product. Yet, without level-of-effort work, nothing would be accomplished. Management of the project is level-of-effort work. All level-of-effort work can and should be tied to milestones. That is, by a specified date, something is to be done and checked to ensure that it is satisfactory. This could be a first draft, the second draft, or the final product. With the number and even identity of the performing staff fixed for a set time, cost and schedule are met by default. All that is of issue is whether the product is acceptable. Let's examine the four types of systems engineering and architecting work discussed previously to determine which are measurable and which are level of effort:

1. Make decisions. The time and cost to make an ad hoc decision can be allocated budget, but the fraction of the decision accomplished and the quality of decision are very difficult to discern. So recommend efforts to make ad hoc decisions be level of effort. However, once the process for making a routine decision is in place, the time, cost, quantity, and quality of the decision-making process are measurable.
2. Model the system or architecture. If there is a specification for the model to be made, the effort is measurable. Modeling is level of effort, if the model is changed in an ad hoc manner as the need for improvements are noticed, such as when users identify desired changes.

3. Determine natural language requirements. Solomon and Young show how this work can be measurable. To establish the budget baseline, one must know the number of requirements and the budget time allocated for each requirement effort, and presume the fraction of the budget time that will be spent in each of the following activities:
- Define the requirements.
- Validate the requirements.
- Determine the verification method.
- Allocate the requirement.
- Document the verification procedure.
- Verify the requirement is met.

By then monitoring the actual accomplishment of the above activities for each requirement, the performance earned value is determined.

ReqManPerfEval reports the schedule variance for requirements determination efforts.

For example, suppose the total number of requirements is three, with a budget of 80 days to get all requirements verified. Suppose 15% of the effort is expended on each of defining, validating, determining verification method, and preparing verification plan, while 20% of the effort is expended on each of allocating the requirements and verifying the requirements are met. Then the input variables are as follows:

```
In[146]:= numberrequirements = 3;
          totalbudgetdays = 80;
          percenteffortbyphase = {.15, .15, .15, .20 , .15, .20};
```

Suppose the effort is stretched out over seven months, and the number of requirements that achieve each state per month is as follows:

```
In[149]:= schedule = Table[0, {6}, {7}];
          schedule[[1, 1]] = 3;
          schedule[[2, 2]] = 2;
          schedule[[2, 3]] = 1;
          schedule[[3, 3]] = 1;
          schedule[[3, 4]] = 2;
          schedule[[4, 5]] = 3;
          schedule[[5, 6]] = 3;
          schedule[[6, 7]] = 3;
          schedule
```

Out[158]= {{3, 0, 0, 0, 0, 0, 0}, {0, 2, 1, 0, 0, 0, 0}, {0, 0, 1, 2, 0, 0, 0},
          {0, 0, 0, 0, 3, 0, 0}, {0, 0, 0, 0, 0, 3, 0}, {0, 0, 0, 0, 0, 0, 3}}

Finally, suppose after five months, the actual count of requirements that achieve each state is as follows:

```
In[159]:= progress = Table[0, {6}, {7}];
          progress[[1, 2]] = 3;
          progress[[2, 4]] = 1;
          progress[[2, 5]] = 1;
          progress[[3, 4]] = 1;
          progress
```

Out[164]= {{0, 3, 0, 0, 0, 0, 0}, {0, 0, 0, 1, 1, 0, 0}, {0, 0, 0, 1, 0, 0, 0},
          {0, 0, 0, 0, 0, 0, 0}, {0, 0, 0, 0, 0, 0, 0}, {0, 0, 0, 0, 0, 0, 0}}

The result is as follows:

In[165]:= `ReqManPerfEval[totalbudgetdays, percenteffortbyphase, schedule, progress]`

`Current requirements schedule variance by reporting period:`

Out[165]= `{-36., -24., -48., -48., -84., -120., -168.}`

5.2 IMPROVE PROCESSES USED

Deming[2] definitively demonstrated that production and standardized service processes are either statistically "in control" or "out of control." A process is in control when the measured process output does not exceed control limits. The control limits are established mathematically based on the nature of the process output measurement and the sampling amount. A process is out of control when control limits cannot be determined, or an output occurs that exceeds previously determined control limits, or a sequence of outputs occur that show recognizable nonrandom behavior while still bounded by the control limits. Deming's first great lesson is that only in control can processes be improved. The first job for anyone attempting to improve a process is to achieve statistical control. Once achieved, there are two types of improvement possible: first, to reduce the variability in the output; and, second, to nudge the mean of the output in a more favorable direction. Neither of these improvements can be achieved for an out-of-control process because it is impossible to determine if the results witnessed after a process change are due to the process change, or the result of inherent variation of the out-of-control process. Deming's second great lesson is that process improvement is achieved incrementally by a series of experiments that are planned, implemented, assessed, and acted upon. If the experiment results in a controlled process with less variability or a better mean result, then continue to do what the experiment attempted; if not, then don't.

Deming's insights apply to processes that are measurable and repetitive and that occur often, approximately at least 30 times. System engineers and architects may believe their activities are so unique, or creative, that a process cannot be defined for their efforts, and therefore process improvement techniques do not apply to them. Process zealots, on the other hand, may claim that process improvement efforts are applicable to all system engineering and architecting activities.

To get a process under control, the activities involved must produce a result that is measurable with minimal ambiguity. The outputs must occur in sufficient number that we can estimate the mean and variance of the product quality, at least by using samples, and can work to improve the output measures using multiple experiments. A process with defined steps that produces only one or a few outputs is not easily improved, as it cannot be established that the process is in control, and there are insufficient opportunities to apply process improvement experiments to achieve controlled improvements. So, for a process improvement effort to be achievable, the following three criteria must be present:

1. The activity can be broken down into describable, repeatable steps.
2. The quality of the product is unambiguously measurable quickly enough to affect the process.
3. A sufficient amount of product is produced to establish if the process is in statistical control.

Describable and repeatable steps simply means the process can be taught to others, or automated. The execution of the process may take considerable judgment or skill. The product quality measure may be binary (e.g., accept or reject), but ideally it has a range of possible values. For many systems engineering and architecting processes, the primary obstacle to implementing process improvements is that the most obvious measures of goodness are not determined until long after the process is completed, which of course makes it impossible to use the process output as a means to improve the process. The best one can do is record "lessons learned" and hope they are learned by the team doing the next similar job. Of course, if the next team is not the team that learned the lessons, they probably won't. The time and effort to produce the product measure the resources expended, which certainly are important, but they are not product quality measures. If the nature of the process is such that an acceptable quality level is essentially guaranteed, then retaining that quality with increasing efficiency could be the process improvement sought.

Let's examine the processes discussed in the previous chapters to determine which satisfy these three conditions.

Make decisions. There are two situations to consider: making specific decisions within an established framework, and making ad hoc decisions to establish the framework. Examples of making decisions in an established framework are acceptance per defined criteria, travel route selection, diagnosis, activities to perform based on a diagnosis, and resource allocations. Making specific decisions in an established framework often satisfies all three criteria—the process can be broken down into identifiable steps, the quality of the product is readily measured either using the decision criteria or (post decision audits) examining the results of the decision, and a large number of decisions may need to be made. Making ad hoc decisions to establish a framework is unlikely to meet all three criteria. The process includes determining what needs to be decided, and when, then selecting the methodology to make the decision and executing that methodology. Activities to determine what to decide, and by when, can be defined only very vaguely, and for any particular issue, happens once. The quality of the decision is measurable against the decision criteria, but it is extremely difficult to know if the "best" decision criteria are being used. Selecting the most appropriate decision-making method to use is a repeatable process that can happen often, and one can usually judge if the method chosen is appropriate or not. The exception is when what has to be decided is so new and unique that none of the currently known decision-making methods apply, and a new method must be invented. Since a big part of the value of system engineers and architects to an enterprise is to help efficiently make good decisions, when we can add methodologies to make decisions, we are potentially increasing our value. So with respect to making ad hoc decisions, at most, only the subprocess of applying the chosen decision-making method can potentially be brought under statistical control.

Model the system or architecture. The process includes determining what is important to model and how to represent it as equations, followed by evaluating the equations. Determining for the first time what to model and how to write the appropriate equations are not describable processes, nor repetitive, nor unambiguously measurable. However, once the key equations are derived and have been shown to be reliably useful, the process to evaluate the equations is well defined, often repeatable, and the product quality can be measured by how well the model mimics reality, or the index of performance value achieved. So with respect to modeling systems or architectures, at most, only the model evaluation process can potentially be brought under statistical control. What we model are consequences of candidate requirements. We model the requirements to predict the performance that will be achieved by the end items as well as the cost and time to deliver. Prediction is an extremely difficult endeavor as it involves simplifications and assumptions, and even the smallest error may totally invalidate the result. As discussed in Chapter 3, humans are reasonably good at forecasting results within a known context, but very weak at predicting the unlikely potential that will radically change the context. If system engineers and architects can widen the context in which we can make useful predictions at lower cost, then the value of our expertise to our enterprise increases.

Determining natural language requirements. The process includes finding the requirements, documenting them unambiguously, identifying the desired end items, allocating the requirements to the end items, establishing verification plans for each end item, and collecting evidence to show if the requirements were satisfied. Let's go through each subprocess in detail.

Finding requirements. In Chapter 4, several different processes to find requirements were presented, so the processes are certainly describable. Do they happen often, and are there useful goodness measures? There are two situations to consider. If the job is to find specific instances of known requirement types (such as to specify options regarding features of a house, automobile, computer, or other system), then the process may occur often. This is an example where a process all but guarantees acceptable quality, if the measure of goodness is that the chosen options can indeed be produced. Alternatively, the quality measure could be the percentage of customers who provide inputs and then actually purchase the selection, as this goodness measure addresses if the option combinations offered are indeed acceptable to customers. So finding specific instances for requirements in an established framework can be brought under statistical control and incrementally improved. The second situation is when the job is to find for the first time the basic requirements for a systems or architecture, even if we also expect to reuse existing requirements. Then, the process is done only once, and the goodness of the requirements is difficult to measure. We can and should ask the designers, producers, users, and supporters, "Do you have the requirements you need to do your job?" And we can try to determine if the requirements are sufficiently unambiguous by "testing" for comprehension. The quality of the requirement set could be measured by how well it balances the customer's conflicting needs, or how easily in terms of cost and time it will take to implement the solution. We can and should ask the customer if the proposed requirements best balance their needs, and hopefully they will tell us. But, unfortunately, if they are seeking the solution by competitive bidding, they may wait to tell us until they selected the provider with the explanation that after seeing their options, they actually prefer more of feature A than feature B. So, finding requirements for the first time cannot be brought under statistical control.

Documenting requirements. Regardless of the requirement-finding process used, once declared, maintaining and communicating the results comprise a repeatable process that happens often. A possible product quality measure is how accurately the retained requirements data match the initially determined requirements. This could be measured by auditing retained data with respect to initial data. If the initial requirements are recorded in any semblance of a database, then perfect accuracy is all but guaranteed. So this is another example of where process quality is essentially guaranteed and efficiency is to be improved. So documenting and communicating the determined requirements can be brought under statistical control and incrementally improved.

Identify the end items and allocate requirements to the end items. Again there are two situations to consider. If the job is to create something by utilizing existing end items, then the process steps are potentially definable, and goodness is measured by the extent to which new end items are avoided, and could be done often, so this situation could be brought under statistical control. But, for new systems and architectures, these activities are very difficult to describe and happen only once, but goodness can be measured by how well the result satisfied the requirements, as well as the estimated cost and time to provide the final system or architecture. So defining end items and allocating for the first time cannot be brought under statistical control. However, the process of literally naming which performance requirements are allocated to which functional requirements and which functional requirements are allocated to which end items is describable, and it happens often, with goodness potentially measurable by how succinctly the allocation is documented, or the efficiency of the process. So this subprocess can be brought under statistical control.

Establishing verification plans for each end item. The two situations previously mentioned exist here as well. If the job is to verify specific instances of known requirement types, then the process is desirable, repeatable, and may happen often, with goodness measured by certainty that the requirements are verified, or efficiency, and therefore it can be brought under statistical control. If the job is to figure out how to verify new requirements for a new end item, it will be describable and measurable, but happens once, and cannot be brought under statistical control.

Collecting evidence for verification. Once again, if we are collecting well-established evidence to verify well-established requirements for well-established end items, the process is describable, repeatable, and potentially happens often, with goodness potentially measured by evidence rejection rate or efficiency, so it can be brought under statistical control. Collecting evidence for new requirements will be describable, and also be able to use goodness measures such as rejection rate or cost or time to accomplish. If the effort is repeated often, we can seek statistical control, but if happens infrequently, we won't be able to.

So we have a very complicated situation. Some aspects of the systems engineering and architecting process remain "an art," subject to the varieties of individual practitioners, for which the activities are yet to be completely describable, or lack repetition, or for which quality is difficult to measure. These activities cannot be brought to statistical control, and therefore cannot be improved by incremental process improvement methods. These areas include the following:

1. Determining what decisions to make, and by when
2. Determining for the first time how best to model a system or architecture and write the appropriate equations

3. Finding "new" requirements, as well as establishing the best end items for implementation
4. The associated verification plans and verification evidence collection

A goal of the profession is to find ways to bring these activities into the realm of processes that can be brought under statistical control. All the other aspects of systems engineering and architecting are likely to be definable processes with measurable goodness, and if they happen repetitively enough, they can and should be brought under statistical control.

To deal with these complications, we separate process improvements into two areas. First and foremost, for processes for which statistical process control can be achieved, apply Deming-inspired processes improvement experiments. For process areas not yet under control, seek to mature the process, aiming eventually to achieve statistical process control. Possible broad levels of process maturity are as follows:

1. Not needed
2. Performed ad hoc (that is, in a manner such that the next activity is determined only after the current activity is completed or there is more trial and error than in adherence to a documented set of repeatable steps)
3. Performed per formal documentation that someone can be taught to follow or that can be automated
4. Performed per formal documentation that someone can be taught to follow or that can be automated with product quality assessed
5. Performed per documentation that someone can be taught to follow or that can be automated with adjustments made based on statistically valid product quality measures

5.2.1 SYSTEMS ENGINEERING PROCESS AREAS

In Chapter 1 we defined a narrow set of systems engineering activities—namely, to make decisions, model systems, and establish natural language and formal requirements. Each of these activities occurs within the context of a number of program activities. We should attempt to improve systems engineering in relation to other activities on a program, just as we seek to improve a design that needs fasteners first by trying to eliminate the need for fasteners. To this end, we identify 16 program process areas for which we can assess process maturity:

1. Planning: identifying, defining, and integrating tasks, responsibilities, resources, and schedules associated with program activities.
2. Tracking and oversight: tracking establishes metrics associated with planned tasks and products. Oversight assesses these metrics and then initiates corrective actions as required to resolve or otherwise mitigate deficiencies.
3. Subcontractor management: selecting, tracking, and overseeing subcontractors engaged in the effort.
4. Intergroup coordination: enabling and achieving effective communication and issue resolution between a program's groups.

5. Configuration management: the planning, identification, change control, status accounting, and auditing of the product elements which include requirements, interfaces, design representations, production procedures, test plans and results, and end items.

6. Quality assurance: evaluate and assess products and processes to reduce noncompliance rates and enhance robustness.

7. Risk management: involves three related activities: risk identification, risk analysis, and risk mitigation. Risk identification determines what might impact technical performance, cost, or schedule. Risk analysis determines the likelihood of the impact and the magnitude of the consequences. Risk mitigation determines and executes actions to deal with the identified risks.

8. Training: ensure program participants are prepared and qualified to perform defined program tasks.

9. Technology management: identifying, evaluating, and selecting appropriate technologies for the program's products and processes.

10. Environment and tool support: determine and make available appropriate facilities, procedures, and tools for the tasks to be performed within cost and schedule constraints.

11. Integrated engineering analysis: create simulations and conduct analyses to assess requirements to determine the best set with respect to development, production, deployment, training, operations, support, disposal effectiveness, cost, and schedule.

12. System concept definition: establish a top-level description of the solution to meet customer and user needs that is consistent with technology, cost or schedule limitations, and risk tolerance level.

13. System requirements: identify, refine, allocate, and communicate the necessary and sufficient information to define performance goals or constraints for the system to achieve for each of the following primary functions: development, design, production, deployment, training, operation, support, and disposal.

14. System design: allocate functions with associated performance requirements and constraints to implementing end items.

15. System integration: define and control internal and external interfaces to ensure that the end items come together as a complete system that satisfies the system-level requirements within the defined operating environment.

16. System verification: ensures that the completed system satisfies the system-level requirements. Successively applied to the requirements set, simulations used, end item design, produced end items, and total system.

5.2.2 Architecting Process Areas

The process areas for architecting are defined as follows:

1. Planning: identifying, defining, and integrating tasks, responsibilities, resources, and schedules associated with the architecting effort.

2. Tracking and oversight: tracking establishes metrics associated with planned tasks and products. Oversight assesses these metrics and then

initiates corrective actions as required to resolve or otherwise mitigate deficiencies.
3. Subcontractor management: selecting, tracking, and overseeing subcontractors engaged in the effort.
4. Intergroup coordination: enabling and achieving effective communication and issue resolution among groups involved in the architecting effort.
5. Configuration management: the planning, identification, change control, status accounting, and auditing of the product elements which include requirements, interfaces, concept definitions, simulation tools, and cost estimates.
6. Training: ensure architecting participants are prepared and qualified to perform defined architecting tasks.
7. Environment and tool support: determine and make available appropriate facilities, procedures, and tools for the tasks to be performed within cost and schedule constraints.
8. Establish as-is architecture and requirements: define the (1) scenarios or reference missions the family of systems are to accomplish, (2) existing family of systems, and (3) current organizational and information exchange interface requirements.
9. Determine the as-is family of systems shortfalls: Conduct campaign, mission, engagement, and engineering analyses to explicitly identify the as-is family of systems performance, capability, sufficiency, or utility shortfalls.
10. Develop concepts to mitigate shortfalls: Define candidate concepts to remove shortfalls in sufficient depth to enable accurate life cycle cost estimates.
11. Identify key enabling technologies: Define quantitatively technological advances that could mitigate shortfalls.
12. Determine the utility of the candidate families of systems: identify and estimate the utility and cost of the candidate families of systems.
13. Establish to-be architecture and requirements: Determine the desired family of system consistent within cost constraints and specify key requirements for each contributing system. Document a time-phased acquisition and deployment plan from the as-is to the to-be family of systems showing the estimated incremental improvement in sufficiency utility achieved and the required funding profile. Establish the to-be organizational and information exchange implementing requirements.

5.3   IMPROVE PROCESS PRODUCT QUALITY

The second area to improve is the quality of the process area outputs. One could argue that since systems engineering and architecting contribute to the creation of the end item, the features of the end item should be used to judge quality. System engineers and architects seek to provide a requirement set for the product that best balances conflicting customer and user needs while requiring the minimum resources to produce and operate. So, the most basic product quality measure could be how well the desired technical performance will be achieved at what expense of resources. Obviously, this has two problems. First, as was shown in Chapter 3, there is no such thing as the single best system or architecture. All we provide are possible features at different price points; it is a customer judgment as to what combination of features and price is most acceptable to them. Second, the system engineering and architects can only predict performance, cost, and schedule. Only after the designers and producers translate the requirements into real-world items do we begin to really know the true resources required and the true performance achieved. And, once created, user expertise can still dramatically affect both demonstrated performance and efficiency of operation. Since the evidence of the real performance, cost, and schedule materialize well after the bulk of the requirement establishment effort is completed, this information is not available when needed to guide process improvements, unless the customer grants a "do over" and adjustments are made to the requirement set post initial production or even operation. Also since many other program activities "cause" the real performance and resource needs, neither the predictions nor the reality can be used to guide changes to the system engineering or architecting process independent of actions to be taken on all the other related activities on the program. We can assess how well we achieve goals associated with the process areas described in the previous section. Rather than attempting to define an absolute scale for goal achievement, we can compare relatively any approach and result A with respect to any approach and result B, which is good enough for us to enable a means to improve process product quality. Simply put, to improve process area product quality, we name goals for each process area, and we periodically assess both how important each goal is to the program and how well we are achieving the goals. Regardless of the grading scheme used, a subset of the goals will always be more important than the mean importance, with goal satisfaction less than the mean goal satisfaction. These most important and least satisfied goals are the ones the program should now concentrate on to improve.

5.3.1 Program Process Quality

Planning process goals are as follows:

- Work tasks are complete, efficient, and well documented.
- Design, production, test, training, support, and operations tasks are well integrated.
- Cost, schedule, and technical objectives are achievable.

Tracking and oversight process goals are as follows:

- Tasks to track and oversee program activities are complete, efficient, and well documented.
- Current cost, schedule, and technical status is accurately known.
- Actions taken to achieve cost, schedule, or technical goals best balance the impact on impacted groups.
- Work effort to track and oversee activities is appropriate and productive.

Subcontractor management process goals are as follows:

- Select fully qualified subcontractors who are contributing to a successful program.
- Subcontractor tasks are complete, efficient, and well documented.
- Subcontractor cost, schedule, and technical goals are achievable and agreed to.
- Subcontractor cost, schedule, and technical accomplishments are tracked, enabling timely issue resolution.
- Actions to resolve subcontractor cost, schedule, or technical issues best balance impact to all groups.

Intergroup coordination process goals are as follows:

- Group roles and responsibilities are defined and are adequately unambiguous.
- Methods for intergroup coordination are documented in a useful manner.
- Appropriate groups are involved in resolving technical or program issues.
- All groups have the program and technical information they need, when they need it.

Configuration management process goals are as follows:

- Tasks to document, control, and communicate products are complete, efficient, and well documented.
- Proposed configuration changes are identified and communicated in a useful manner.
- Configuration changes are evaluated accurately with respect to cost, schedule, and technical impacts.
- Configuration changes are controlled.
- Configuration change decisions are appropriate.

Quality control process goals are as follows:

- Tasks to achieve error-free and robust products are complete, efficient, and well documented.
- Product acceptance standards are complete and well documented.
- Inspectors are utilized in an appropriate manner.
- Noncompliance issues are tracked to closure.

Risk management process goals are as follows:

- Tasks to identify, assess, and mitigate risks are complete, efficient, and well documented.
- Technical, cost, and schedule risks are identified and assessed.
- Actions are taken to mitigate risks.
- Risk mitigation actions are monitored.
- Risk status is communicated and coordinated across affected groups.

Training process goals are as follows:

- Training tasks to benefit program personnel are complete, efficient, and well documented.
- People on the program are fully qualified to do their jobs.
- Training meets the needs of the people on the program.
- Training effectiveness is measured to provide feedback for additional training needs.

Technology management process goals are as follows:

- Tasks to identify and select appropriate technologies to utilize are complete, efficient, and well documented.
- Technologies are selected for use that support the business goals.
- Technology innovations are implemented that improve processes.
- Technology innovations are implemented that improve products.

Environment and tool support process goals are as follows:

- Tasks to acquire and make available facilities, procedures, and tools are complete, efficient, and well documented.
- Facilities, procedures, and tools are provided which enable completing tasks within cost and schedule constraints.
- Facilities, procedures, and tools support intergroup interaction.
- Tools used are compatible and enhance efficiency.

Integrated engineering analysis process goals are as follows:

- An integrated multidisciplinary approach to problem solving is coached and fostered.
- Analysis tasks are complete, efficient, and well documented.
- All relevant groups are involved that need to be involved in making decisions.
- Methodology and criteria for making decisions are clear, complete, appropriate, and consistently applied.
- Analyses are sufficient, accurate, and timely enough to best guide decision making.

System concept development process goals are as follows:

- Tasks to determine best solution to offer customer and users are complete, efficient, and well documented.
- Customer and user needs are accurately documented in a useful manner.
- Derived requirements are traceable to customer and user needs and well documented.
- Chosen end items are allocated appropriate and complete functional and performance requirements.
- Chosen concept features best satisfy customer and user needs.

System requirements process goals are as follows:

- Tasks to determine performance requirements for all eight primary functions are complete, efficient, and well documented.
- All requirement attributes are logical, documented, and easy to find and comprehend.
- The functional and performance requirements enable appropriate implementation options.
- All groups have all the requirements they need to do their jobs.

System design process goals are as follows:

- Tasks to determine instructions to produce the end items are complete, efficient, and well documented.
- Design requirements are traceable to customer and user requirements.
- Production, test, training, and support implications of design are determined and acceptable.
- Design baseline is established and is the best solution.

System integration process goals are as follows:

- Tasks to achieve internal and external interfaces are complete, efficient, and well documented.
- Interface requirements between end items are clear, complete, and unambiguous.
- Interface requirements with external entities are clear, complete, and unambiguous.
- The end items will assemble into a uniform whole that meets system-level requirements.

System verification process goals are as follows:

- Requirement set verification criteria and tasks are complete, efficient, and well documented.
- Simulation verification criteria and tasks are complete, efficient, and well documented.
- End item design verification criteria and tasks are complete, efficient, and well documented.
- Produced end items verification criteria and tasks are complete, efficient, and well documented.
- System verification criteria and tasks are complete, efficient, and well documented.
- Verification evidence obtained is documented in a useful manner.
- All verification evidence obtained so far indicates that requirements are being appropriately met.

5.3.2 Architecting Process Quality

For architecting, the process area goals are as follows.

Planning process goals are as follows:

- Work tasks are complete, efficient, and well documented.
- Cost, schedule, and technical objectives are achievable.

Tracking and oversight process goals are as follows:

- Tasks to track and oversee architecting activities are complete, efficient, and well documented.
- Current cost, schedule, and technical status is accurately known.
- Actions taken to achieve cost, schedule, or technical goals best balance the impact on impacted groups.
- Work effort to track and oversee activities is appropriate and productive.

Subcontractor management process goals are as follows:

- Select fully qualified subcontractors who are contributing to a successful architecting effort.
- Subcontractor tasks are complete, efficient, and well documented.
- Subcontractor cost, schedule, and technical goals are achievable and agreed to.
- Subcontractor cost, schedule, and technical accomplishments are tracked, enabling timely issue resolution.
- Actions to resolve subcontractor cost, schedule, or technical issues best balance impact to all groups.

Intergroup coordination process goals are as follows:

- Group roles and responsibilities are defined and are adequately unambiguous.
- Methods for intergroup coordination are documented in a useful manner.
- Appropriate groups are involved in resolving technical or program issues.
- All groups have the program and technical information they need, when they need it.

Configuration management process goals are as follows:

- Tasks to document, control, and communicate products are complete, efficient, and well documented.
- Proposed configuration changes are identified and communicated in a useful manner.

- Configuration changes are evaluated accurately with respect to cost, schedule, and technical impacts.
- Configuration changes are controlled.
- Configuration change decisions are appropriate.

Training process goals are as follows:

- Training tasks to benefit personnel are complete, efficient, and well documented.
- People on the architecting effort are fully qualified to do their jobs.
- Training meets the needs of the people on the architecting effort.
- Training effectiveness is measured to provide feedback for additional training needs.

Environment and tool support process goals are as follows:

- Tasks to acquire and make available facilities, procedures, and tools are complete, efficient, and well documented.
- Facilities, procedures, and tools are provided which enable completing tasks within cost and schedule constraints.
- Facilities, procedures, and tools support intergroup interaction.
- Tools used are compatible and enhance efficiency.

Establish the as-is architecture and its requirements:

- Architecture reference missions adequately represent the key purpose for the family of systems.
- Have data for the family of systems to assess performance, capability, sufficiency, and utility.
- Organization relationships governing the existing family of systems operations are adequately defined.
- Information needed for the existing family of systems elements to perform mission is adequately defined.

Determine the as-is family of systems shortfalls:

- Accurate and efficient simulations exist to determine the family of system performance.
- Accurate and efficient simulations exist to determine the family of system capability.
- Accurate and efficient simulations exist to determine the family of system sufficiency.
- Accurate and efficient simulations exist to determine the family of system utility.
- People with authority to establish the to-be architecture easily comprehend the nature, cause, and impact of existing shortfalls.

Develop concepts to mitigate shortfalls:

- Adequate number and diversity of concepts are defined to address all key shortfalls.
- All concepts are realistic with respect to capability, availability, and cost.

- Concept definitions enable determining performance, capability, sufficiency, and utility.
- Concept definitions enable accurate cost estimates.
- Concept definitions are sufficiently detailed to serve as requirements for implementation.
- People with authority to establish the to-be architecture easily comprehend the candidate concepts.

Identify key enabling technologies to mitigate shortfalls:

- Advancements from the current state of the art that could mitigate shortfalls are adequately identified.
- The explicit performance level for technology advancements sought is quantitatively defined.
- Technology maturation experts can easily develop technology maturation plans based on the shortfall and technology need documentation.

Determine utility and cost of candidate family of systems:

- Data needed to assess individual systems are available.
- Data needed to assess combinations of systems are available.
- All credible families of systems combinations are assessed.
- Secondary system selection criteria were adequately explored for leading candidate family of system solutions.
- People with the authority to establish the to-be architecture have the best appropriate data to easily make the decision.

Establish the to-be architecture and its requirements:

- The requirements for systems constituents are accurately and completely documented.
- To-be organizational relationships are accurately and completely documented.
- To-be information flow requirements and solutions are accurately and completely documented.

5.4 IMPROVE EFFICIENCY

The third and final area to improve is efficiency in producing the products.

5.4.1 Systems Engineering Efficiency

As the primary job of systems engineering is to produce verified requirements, the fundamental efficiency metric is the number of verified requirements per effort expended. More specifically, the efficiency measure is

(Number of unique verified functional and performance requirements – Number of sought requirements) / (Total cost of requirement establishment effort to date)

The number of verified requirements will be essentially zero at the start of the project, grow as the system definition is matured, and ultimately reach a final count. *Unique* means needed with no duplications. Two requirements with different identifications that have the same content, even if allocated to different end items, are NOT unique; they are duplicates, and would count as one requirement in the numerator. Recall that the word *verified* means different things at different times (see Chapter 4). The number of sought requirements will initially be large, and will probably grow as the system becomes better understood, but will settle to 0 as the needs of the design, production, training, and support and operation communities are satisfied. The total costs of the effort to date are all expenses made to produce the verified requirements. Early in the program these costs are those associated with finding, documenting, and communicating top-level system requirements, but recall, eventually, that *verified* means "show by a means that the allocated functional and performance requirements are achieved by the end item," so the total cost of the verification effort ends up in the denominator. So this measure starts at minus infinity at the start of a program, peaks at a positive value once the requirement set has settled out, then stays positive but decreases as the cost of verification effort grows.

Clearly this measure motivates:

1. Reduce the cost of efforts, as larger costs reduce the measure.
2. Reuse, as that will tend to increase the number of verified requirements, reduce the number of sought requirements, and reduce the cost.
3. Minimize the requirements needed, as any effort that produces nonunique requirements will add to the cost but not add to the numerator count.
4. Get the user organizations the minimum requirements they need, as each user has the right and obligation to declare the requirements they need, but the system engineers will push back to keep the list small, both to maximize the numerator and to keep the cost low.

The efficiency achieved is comparable between programs at similar milestones, such as follows:

1. System requirement review, when fundamental functional and performance requirements that affect the entire system are formally accepted
2. Preliminary design review, when the fundamental functional and performance requirements for system segments, elements, or subsystems are formally accepted
3. Critical design review, when the detailed functional and performance requirements for all end items are formally accepted
4. Functional configuration audit, when the efforts to show the individual end items achieve stipulated functional and performance requirements are completed
5. Physical configuration audit, when the efforts to show the collective end items are assembled into an integrated whole that meets at least its stipulated functional and performance requirements
6. System delivery, when the customer takes possession of the system and hands it to the users

5.4.2 ARCHITECTING EFFICIENCY

The architect is expected to determine what systems to have given the key features of candidate systems, and how those systems are to interact. The architecting effort does not include the development of the systems themselves. Prior to placing an architecture specification under configuration control, the primary architecting efficiency metric is

(Number of unique verified architecture functional and performance requirements – Number of sought requirements) / (Total cost of requirements establishment effort to date)

Once the architecture specification is under configuration control, the job is one of updating the architecture specification in the face of changes to the operating environment or mission goals. The less cost per change to incorporate, the more efficient, so the architecture specification maintenance efficiency metric is

(Number of requirement changes in a period) / (Total man-hours expended to maintain and update the architecture specification in period)

5.5 USE SURVEYS TO DETERMINE WHAT IS MOST URGENT TO IMPROVE AT ANY POINT IN A PROGRAM

To help a program or architecting effort periodically assess and seek to improve process, quality, and efficiency, it is recommended all participants, or at least a sample from organizations involved, as well as the customer if willing, be surveyed. The survey frequency is at least just after each major milestone, including the kickoff, or if the program is very long running with years or more between major milestones, then every three to six months. The survey consists of four parts. The first part defines the process areas. The second part establishes the relative importance of each process areas at that point in the program. The third part establishes the perception of the process maturity. The fourth part establishes the perception of how well each process achieves stated goals and allows for a free-form input from the survey participant.

Appendix 5A contains a survey for systems engineering and Appendix 5B contains one for architecting.

The survey recipient is first asked to rate each process area on a scale from 1 to 9 as to how important they perceive that process to success at this point in the program.

Second, the survey recipient is asked to assess the maturity of the process area on a scale from 1 to 7, with written descriptions for each level of maturity.

Third, the survey recipient is asked to assess how well each goal for each process area is currently being achieved on a scale from 1 to 9.

Finally, the survey recipient is asked to provide any optional comments regarding the process area.

The formal requirements *ProgProcSurResults* were created to automate the assessment of the program's process survey. *ProgProcSurResults* takes the survey inputs, then calculates for each process area: relative importance, relative process maturity, and relative process goal achievement. The routine then prints the geo-metric mean of the input importance and maturity scores for each process area. The routine prints a plot of the process areas with respect to process maturity and impor-tance. The routine calculates the geometric centroid of the points, and lists those process areas that are least mature that are most important, that is, have importance values more than the centroid and maturity values less than the centroid. The use of a centroid turns whatever scores the graders provided into relative scores, so no matter how inflated or deflated they are, there will always be a set of processes above and below the centroid values. The printed list is in order of decreasing "distance" of that process area data point from the centroid. That is, the process areas are listed in order from most important to improve at this time on the program. The routine then outputs the geometric mean of the survey inputs of process importance and goal satisfaction for each process goal. The routine then prints a plot of the process area goals with respect to goal achievement and process area importance (it is assumed the process area importance is the same for all goals associated with that process). The geometric centroid of the points is determined, and the routine prints those goals that have achievement scores below the centroid and importance scores above the centroid. The printed list is in order of decreasing "distance" of the goal from the centroid, again to help facilitate finding the goals most important to improve at this time.

As an example, assume there are three surveys providers, who provided the fol-lowing scores between 1 and 9 for each of the 16 process areas:

```
In[168]:= importance = {{7, 6, 5, 4, 3, 2, 1, 7, 6, 5, 4, 3, 6, 2, 1, 8},
                        {8, 7, 6, 5, 4, 3, 2, 1, 8, 7, 6, 5, 4, 4, 3, 0},
                        {6, 5, 4, 3, 2, 1, 0, 0, 6, 5, 4, 3, 2, 1, 0, 0}};
```

Note, the number zero ("0") is the survey input if the provider indicated don't know or did not apply.

Similarly, the three survey recipients would provide a number between 1 and 7 to indicate the process maturity at this time:

```
In[169]:= maturity = {{1, 2, 3, 4, 5, 6, 7, 1, 2, 3, 4, 5, 6, 7, 0, 0},
                      {2, 2, 3, 3, 4, 4, 5, 5, 6, 6, 7, 7, 6, 6, 5, 5},
                      {0, 0, 3, 4, 5, 6, 7, 1, 2, 3, 4, 5, 6, 7, 1, 2}};
```

Last, the three surveys recipients provided the following scores from 1 to 9 for the 71 process area goals, with their opinions of *don't know* or *not applicable* recorded as 0:

```
In[170]:= ga = {{0, 8, 7, 6, 5, 4, 3, 2, 1, 9, 8, 7, 6, 5, 4, 3, 2, 1, 9, 8, 7, 6, 5,
                 4, 3, 2, 1, 9, 8, 7, 6, 5, 4, 3, 2, 1, 9, 8, 7, 6, 5, 4, 3, 2, 1, 9, 8,
                 7, 6, 5, 4, 3, 2, 1, 9, 8, 7, 6, 5, 4, 3, 2, 1, 9, 8, 7, 6, 5, 4, 3, 2},
                {9, 0, 7, 6, 5, 4, 3, 2, 1, 9, 8, 7, 6, 5, 4, 3, 2, 1, 9, 8, 7, 6, 5, 4,
                 3, 2, 1, 9, 8, 7, 6, 5, 4, 3, 2, 1, 9, 8, 7, 6, 5, 4, 3, 2, 1, 9, 8, 7,
                 6, 5, 4, 3, 2, 1, 9, 8, 7, 6, 5, 4, 3, 2, 1, 9, 8, 7, 6, 5, 4, 3, 2},
                {9, 8, 0, 6, 5, 4, 3, 2, 1, 9, 8, 7, 6, 5, 4, 3, 2, 1, 9, 8, 7, 6, 5, 4,
                 3, 2, 1, 9, 8, 7, 6, 5, 4, 3, 2, 1, 9, 8, 7, 6, 5, 4, 3, 2, 1, 9, 8, 7,
                 6, 5, 4, 3, 2, 1, 9, 8, 7, 6, 5, 4, 3, 2, 1, 9, 8, 7, 6, 5, 4, 3, 2}}};
```

Given these inputs, *ProgProcSurResults* provides the following survey tabulation:

In[171]:= **ProgProcSurResults[importance, maturity, ga]**

Program Process Area, Importance(1-9), Maturity(1-7)

01 Planning	6.95	1.41
02 Tracking and Oversight	5.94	2.
03 Subcontractor Management	4.93	3.
04 Intergroup Coordination	3.91	3.63
05 Configuration Management	2.88	4.64
06 Quality Assurance	1.82	5.24
07 Risk Management	1.41	6.26
08 Training	2.65	1.71
09 Technology Management	6.6	2.88
10 Environment and Tool Support	5.59	3.78
11 Integrated Engineering Analysis	4.58	4.82
12 System Concept Definition	3.56	5.59
13 System Requirements	3.63	6.
14 System Design	2.	6.65
15 System Integration	1.73	2.24
16 System Verification	8.	3.16

Program Process Areas with Respect to Importance to Program and Maturity at this Time

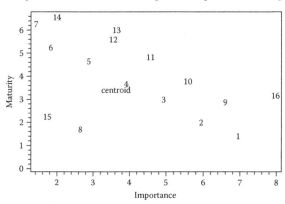

Program Process Areas that are Most Important to Improve Are:

[Distance to Centroid, Process ID and Name]

4.378	16 System Verification
3.947	01 Planning
3.040	09 Technology Management
2.783	02 Tracking and Oversight
1.409	03 Subcontractor Management

Program Process Area Goals, Importance(1-9), Satisfaction(1-9)

#	Goal	Importance	Satisfaction
01.1	Work tasks are complete, efficient, and well documented	6.95	9.
01.2	Design, production, test, training, support and operations tasks are well integrated	6.95	8.
01.3	Cost, schedule and technical objectives are achievable	6.95	7.
02.1	Tasks to track and oversee program activities are complete, efficient and well documented	5.94	6.
02.2	Current cost, schedule and technical status is accurately known	5.94	5.
02.3	Actions taken to achieve cost, schedule, or technical goals best balance impact to all groups	5.94	4.
02.4	Work effort to track and oversee activities is appropriate and productive	5.94	3.
03.1	Selected fully qualified subcontractors who are contributing to a successful program	4.93	2.
03.2	Subcontractor tasks are complete, efficient, and well documented	4.93	1.
03.3	Subcontractor cost, schedule and technical goals are achievable and agreed too	4.93	9.
03.4	Subcontractor cost, schedule and technical accomplishments are tracked, enabling timely issue resolution	4.93	8.
03.5	Actions to resolve subcontractor cost, schedule or technical issues best balance impact to all groups	4.93	7.
04.1	Group roles and responsibilities are defined and are adequately unambiguous	3.91	6.
04.2	Methods for inter-group coordination are documented in a useful manner	3.91	5.
04.3	Appropriate groups are involved in resolving technical or program issues	3.91	4.
04.4	All groups have the program and technical information they need, when they need it	3.91	3.
05.1	Tasks to document, control and communicate products are complete, efficient, and well documented	2.88	2.
05.2	Proposed configuration changes are identified and communicated in a useful manner	2.88	1.
05.3	Configuration changes are evaluated accurately with respect to cost, schedule and technical impacts	2.88	9.
05.4	Configuration changes are controlled	2.88	8.
05.5	Configuration change decisions are appropriate	2.88	7.
06.1	Tasks to achieve error free and robust products are complete, efficient and well documented	1.82	6.
06.2	Product acceptance standards are well documented	1.82	5.
06.3	Inspectors are utilized in an appropriate manner	1.82	4.
06.4	Noncompliance issues are tracked to closure	1.82	3.
07.1	Tasks to identify, assess and mitigate risks are complete, efficient, and well documented	1.41	2.
07.2	Technical, cost, and schedule risks are identified and assessed	1.41	1.
07.3	Actions are taken to mitigate risks	1.41	9.
07.4	Risks mitigation actions are monitored	1.41	8.
07.5	Risk status is communicated and coordinated across affected groups	1.41	7.
08.1	Training tasks to benefit program personnel are complete, efficient, and well documented	2.65	6.
08.2	People on the program are fully qualified to do their jobs	2.65	5.
08.3	Training meets the needs of the people on the program	2.65	4.
08.4	Training effectiveness is measured to provide feedback for additional training needs	2.65	3.

Item	Description		
09.1	Tasks to identify and select appropriate technologies to utilize are complete, efficient, and well documented	6.6	2.
09.2	Technologies are selected for use that support the business goals	6.6	1.
09.3	Technology innovations are implemented that improve processes	6.6	9.
09.4	Technology innovations are implemented that improve products	6.6	8.
10.1	Tasks to make available facilities, procedures and tools are complete, efficient, and well documented	5.59	7.
10.2	Facilities, procedures and tools provided enable completing tasks within cost and schedule constraints	5.59	6.
10.3	Facilities, procedures and tools support inter-group interaction	5.59	5.
10.4	Tools used are compatible and enhance efficiency	5.59	4.
11.1	An integrated multi-discipline approach to problem solving is coached and fostered	4.58	3.
11.2	Analysis tasks are complete, efficient and well documented	4.58	2.
11.3	All relevant groups are involved that need to be involved in making decisions	4.58	1.
11.4	Methodology and criteria for making decisions are clear, appropriate and consistently applied	4.58	9.
11.5	Analyses are sufficient, accurate and timely enough to best guide decision making	4.58	8.
12.1	Tasks to determine best solution to offer care complete, efficient, and well documented	3.56	7.
12.2	Customer and user needs are accurately documented in a useful manner	3.56	6.
12.3	Derived requirements are traceable to customer and user needs and are well documented	3.56	5.
12.4	Chosen end items are allocated appropriate and complete functional and performance requirements	3.56	4.
12.5	Chosen concept features best satisfy customer and user needs	3.56	3.
13.1	Tasks to determine performance requirements for all 8 primary functions are complete, efficient, and well documented	3.63	2.
13.2	All requirement attributes are logical, documented, and easy to find and comprehend	3.63	1.
13.3	The functional and performance requirements enable appropriate implementation options	3.63	9.
13.4	All groups have all the requirements they need to do their jobs	3.63	8.
14.1	Tasks to determine instructions to produce the end items are complete, efficient, and well documented	2.	7.
14.2	Design requirements are traceable to customer and user requirements	2.	6.
14.3	Production, test, training, and support implications of design are determined and acceptable	2.	5.
14.4	Design baseline is established and is best solution	2.	4.
15.1	Tasks to achieve internal an external interfaces are complete, efficient, and well documented	1.73	3.
15.2	Interface requirements between end items are clear, complete and unambiguous	1.73	2.
15.3	Interface requirements with external entities are clear, complete and unambiguous	1.73	1.
15.4	The end items will assemble into a uniform whole that meets system level requirements	1.73	9.
16.1	Requirement set verification criteria and tasks are complete, efficient, and well documented	8.	8.
16.2	Simulation verification criteria and tasks are complete, efficient, and well documented	8.	7.
16.3	End item design verification criteria and tasks are complete, efficient, and well documented	8.	6.
16.4	Produced end items verification criteria and tasks are complete, efficient and well documented	8.	5.
16.5	System verification criteria and tasks are complete, efficient and well documented	8.	4.
16.6	Verification evidence obtained is documented in a useful manner	8.	3.
16.7	All verification evidence obtained so far indicates requirements are being appropriately met	8.	2.

Program Process Areas Goals with Respect
to Importance to Program and Goal Satisfaction at this Time

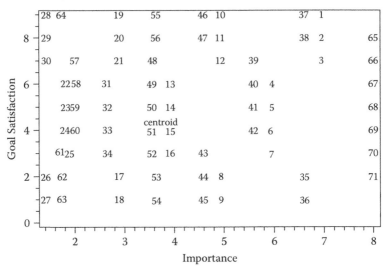

Program Process Areas Goals that are Most Important to Improve Are:

[Distance to Centroid, Goal ID and Statement]

4.850	16.7 All verification evidence obtained so far indicates requirements are being appropriately met
4.479	16.6 Verification evidence obtained is documented in a useful manner
4.348	09.2 Technologies are selected for use that support the business goals
4.312	16.5 System verification criteria and tasks are complete, efficient and well documented
3.667	09.1 Tasks to identify and select appropriate technologies to utilize are complete, efficient, and well documented
3.460	03.2 Subcontractor tasks are complete, efficient, and well documented
3.350	11.3 All relevant groups are involved that need to be involved in making decisions
2.565	02.4 Work effort to track and oversee activities is appropriate and productive
2.552	03.1 Selected fully qualified subcontractors who are contributing to a successful program
2.400	11.2 Analysis tasks are complete, efficient and well documented
2.262	02.3 Actions taken to achieve cost, schedule, or technical goals best balance impact to all groups
1.914	10.4 Tools used are compatible and enhance efficiency
1.516	11.1 An integrated multi-discipline approach to problem solving is coached and fostered
1.251	04.4 All groups have the program and technical information they need, when they need it
0.320	04.3 Appropriate groups are involved in resolving technical or program issues

ArchProcSurResults provides architecting process survey results.

As an example, presume there are four survey inputs, and the process importance scores are as follows:

```
In[172]:=  api = {{3, 3, 1, 8, 9, 1, 3, 9, 9, 8, 5, 4, 4},
              {2, 3, 2, 8, 8, 2, 3, 8, 9, 8, 5, 3, 2},
              {3, 0, 1, 9, 8, 2, 3, 9, 6, 8, 5, 4, 5},
              {3, 3, 0, 9, 8, 1, 2, 9, 9, 6, 5, 5, 4}};
```

And the process area maturities are

```
In[173]:=  apm = {{3, 3, 1, 7, 7, 7, 7, 7, 7, 7, 5, 4, 4},
              {2, 3, 2, 6, 6, 2, 3, 6, 6, 6, 5, 3, 2},
              {3, 0, 1, 6, 6, 2, 3, 6, 6, 6, 5, 4, 5},
              {3, 3, 0, 7, 7, 1, 2, 7, 7, 6, 5, 5, 4}};
```

And goal satisfaction is as follows:

```
In[174]:= apga = {{0, 8, 7, 6, 5, 4, 3, 2, 1, 9, 8, 7, 6, 5, 4, 3, 2, 1, 9, 8, 7, 6, 5, 4, 3, 2, 1,
         9, 8, 7, 6, 5, 4, 3, 2, 1, 9, 8, 7, 6, 5, 4, 3, 2, 1, 9, 8, 7, 6, 5, 4, 3, 2, 1},
         {9, 0, 7, 6, 5, 4, 3, 2, 1, 9, 8, 7, 6, 5, 4, 3, 2, 1, 9, 8, 7, 6, 5, 4, 3, 2,
         1, 9, 8, 7, 6, 5, 4, 3, 2, 1, 9, 8, 7, 6, 5, 4, 3, 2, 1, 9, 8, 7, 6, 5, 4, 3, 2, 1},
         {9, 8, 0, 6, 5, 4, 3, 2, 1, 9, 8, 7, 6, 5, 4, 3, 2, 1, 9, 8, 7, 6, 5, 4, 3, 2,
         1, 9, 8, 7, 6, 5, 4, 3, 2, 1, 9, 8, 7, 6, 5, 4, 3, 2, 1},
         {0, 0, 0, 0, 0, 4, 4, 4, 4, 4, 0, 0, 0, 0, 0, 8, 8, 8, 8, 0, 0, 0, 0, 0, 3, 3,
         3, 3, 3, 0, 0, 0, 0, 1, 1, 1, 1, 1, 0, 0, 0, 0, 0, 0, 0, 0, 0, 0, 0, 0, 0, 0, 0}
         };
```

Then *ArchProcSurResults* provides the survey results as follows:

In[175]:= **ArchProcSurResults[api, apm, apga]**

Architecting Process Area, Importance(1-9), Maturity(1-7)

01	Planning	2.71	2.71
02	Tracking and Oversight	3.	3.
03	Subcontractor Management	1.26	1.26
04	Intergroup Coordination	8.49	6.48
05	Configuration Management	8.24	6.48
06	Training	1.41	2.3
07	Environment and Tool Support	2.71	3.35
08	Establish As-is Architecture and Requirements	8.74	6.48
09	Determine As-is Family of Systems Shortfals	8.13	6.48
10	Deveop Concepts to Mitigate Shortdalls	7.44	6.24
11	Identify Key Enabling Technologies	5.	5.
12	Determine Utility of Candidate Family of Systems	3.94	3.94
13	Establish To-be Architecture and Requirements	3.56	3.56

Architecting Process Areas with
    Respect to Importance to Program and Maturity at this Time

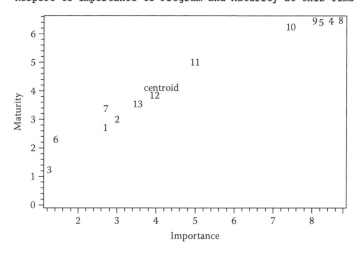

Process Areas that are Most Important to Improve Are:

[Distance to Centroid, Process ID and Name]

{}

Architecting Process Area Goals, Importance(1-9), Satisfaction(1-9)

01.1	Work tasks are complete,efficient,and well documented	2.71	9.
01.2	Cost,schedule and technical objectives are achievable	2.71	8.
02.1	Tasks to track and oversee architecting activities are complete,efficient and well documented	3.	7.
02.2	Current cost,schedule and technical status is accurately know	3.	6.
02.3	Actions taken to achieve cost,schedule,or technical goals best balance impact on impacted groups	3.	5.
02.4	Work effort to track and oversee activities is appropriate and productive	3.	4.
03.1	Selected fully qualified subcontractors who are contributing to a successful architecting effort	1.26	3.22
03.2	Subcontractor tasks are complete, efficient, and well documented	1.26	2.38
03.3	Subcontractor cost, schedule and technical goals are achievable and agreed too	1.26	1.41
03.4	Subcontractor cost, schedule and technical accomplishments are tracked, enabling timely issue resolution	1.26	7.35
03.5	Actions to resolve subcontractor cost, schedule or technical issues best balance impact to all groups	1.26	8.
04.1	Group roles and responsibilities are defined and are adequately unambiguous	8.49	7.
04.2	Methods for inter-group coordination are documented in a useful manner	8.49	6.
04.3	Appropriate groups are involved in resolving technical or program issues	8.49	5.
04.4	All groups have the program and technical information they need, when they need it	8.49	4.
05.1	Tasks to document,control and communicate products are complete,efficient,and well documented	8.24	3.83
05.2	Proposed configuration changes are identified and communicated in a useful manner	8.24	2.83
05.3	Configuration changes are evaluated accurately with respect to cost,schedule and technical impacts	8.24	1.68
05.4	Configuration changes are controlled	8.24	8.74
05.5	Configuration change decisions are appropriate	8.24	8.
06.1	Training tasks to benefit personnel are complete,efficient,and well documented	1.41	7.
06.2	People on the architecting effort are fully qualified to do their jobs	1.41	6.
06.3	Training meets the needs of the people on the architecting effort	1.41	5.
06.4	Training effectiveness is measured to provide feedback for additional training needs	1.41	4.
07.1	Tasks to acquire and make available facilities,procedures and tools are complete,efficient,and well documented	2.71	3.
07.2	Facilities,procedures and tools provided which enable completing tasks within cost and schedule constraints	2.71	2.21
07.3	Facilities,procedures and tools support inter-group interaction	2.71	1.32
07.4	Tools used are compatible and enhance efficiency	2.71	6.84
08.1	Scenarios or design reference missions adequately represent the key purpose for the family of systems	8.74	6.26
08.2	Have data for family of systems to assess performance,capability,sufficiency and utility	8.74	7.
08.3	Organization relationships governing existing family of systems operations are adequately defined	8.74	6.
08.4	Information needed for existing family of systems elements to perform mission are adequately defined	8.74	5.
09.1	Accurate and efficient simulations exist to determine family of system performance	8.13	4.
09.2	Accurate and efficient simulations exist to determine family of system capability	8.13	3.
09.3	Accurate and efficient simulations exist to determine family of system sufficiency	8.13	1.68
09.4	Accurate and efficient simulations exist to determine family of system utility	8.13	1.
09.5	People with authority to establish the to-be architecture easily comprehend the nature,cause and impact of existing shortfall	8.13	5.2
10.1	Adequate number and diversity of concepts are defined to address all key shortfalls	7.44	4.76
10.2	All concepts are realistic with respect to capability,availability and cost	7.44	4.3
10.3	Concept definitions enables determining performance,capability,sufficiency and utility	7.44	6.
10.4	Concept definitions enable accurate cost estimates	7.44	5.
10.5	Concept definition are sufficiently detailed to serve as requirements for implementation	7.44	4.
10.6	People with authority to establish the to-be architecture easily comprehend the candidate concepts	7.44	4.
11.1	Advancements from the current state of the art that could mitigate shortfalls are adequately identified	5.	2.
11.2	The explicit performance level for technology advancements sought are quantitatively defined	5.	1.
11.3	Technology maturation experts can easily develop technology maturation plans based on the shortfall and technology need documentation	5.	9.
12.1	Data needed to assess individual systems is available	3.94	8.
12.2	Data needed to assess combinations of systems is available	3.94	6.
12.3	All credible family of systems combinations are assessed	3.94	6.
12.4	Secondary system selection criteria were adequately explored for leading candidate family of system solutions	3.94	4.
12.5	People with the authority to establish the to-be architecture have the best appropriate data to easily make the decision	3.94	4.
13.1	The requirements for systems constituents are accurately and completely documented	3.56	3.
13.2	To-be organizational relationships are accurately and completely documented	3.56	2.
13.3	To-be information flow requirements and solutions are accurately and completely documented	3.56	1.

Architecting Process Areas Goals with Respect
to Importance to Program and Goal Satisfaction at this Time

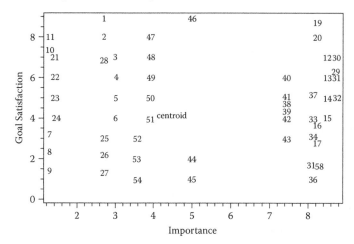

Architecting Process Areas Goals that are Most Important to Improve Are:

[Distance to Centroid, Goal ID and Statement]

```
4.944   09.4 Accurate and efficient simulations exist to determine family of system utility
4.635   05.3 Configuration changes are evaluated accurately with respect to cost,schedule and technical impacts
4.545   09.3 Accurate and efficient simulations exist to determine family of system sufficiency
4.192   04.4 All groups have the program and technical information they need, when they need it
4.149   05.2 Proposed configuration changes are identified and communicated in a useful manner
3.997   09.2 Accurate and efficient simulations exist to determine family of system capability
3.955   05.1 Tasks to document,control and communicate products are complete,efficient,and well documented
3.840   09.1 Accurate and efficient simulations exist to determine family of system performance
3.342   10.6 People with authority to establish the to-be architecture easily comprehend the candidate concepts
3.195   11.2 The explicit performance level for technology advancements sought are quantitatively defined
3.152   10.5 Concept definition are sufficiently detailed to serve as requirements for implementation
2.231   11.1 Advancements from the current state of the art that could mitigate shortfalls are adequately identified
```

APPENDIX 5A: SYSTEMS ENGINEERING EFFECTIVENESS SURVEY

PLEASE READ FIRST

Please complete and return this survey to [insert name] by [insert time and date].

Your response is confidential.

Your participation is critical to determine how best to improve our systems engineering efforts.

The survey has four sections:

Section 1 defines the 16 process areas.

Section 2 obtains your assessment of the relative importance of improving each process area at this time.

Section 3 obtains your assessment of the maturity of each process area at this time.

Section 4 obtains your assessment of the product quality of each process area at this time.

Thank you for taking the time to provide this information.

If you have any questions, please contact [insert name, phone number, and email address].

To improve the survey process, please answer the following questions after completing the survey:

1. Approximately how many minutes were needed for you to complete this survey? _____

2. How would you improve this survey?

SECTION 1: PROCESS AREA DEFINITIONS

Planning: Identify, define, and integrate tasks, responsibilities, resources, and schedules associated with program activities.

Tracking and oversight: Tracking establishes metrics associated with planned tasks and products. Oversight assesses these metrics and then initiates corrective actions as required to resolve or otherwise mitigate deficiencies.

Subcontractor management: Select, track, and oversee subcontractors engaged in the effort.

Intergroup coordination: Enable and achieve effective communication and issue resolution between a program's groups.

Configuration management: The planning, identification, change control, status accounting, and auditing of the product elements, which include requirements, interfaces, design representations, production procedures, test plans and results, and end items.

Quality assurance: Evaluate and assess products and processes to reduce noncompliance rates and enhance robustness.

Risk management: Involves three related activities: risk identification, risk analysis, and risk mitigation. Risk identification determines what might impact technical performance, cost, or schedule. Risk analysis determines the likelihood of the impact and the magnitude of the consequences. Risk mitigation determines and executes actions to deal with the identified risks.

Training: Ensure that program participants are prepared and qualified to perform defined program tasks.

Technology management: Identify, evaluate, and select appropriate technologies for the program's products and processes.

Environment and tool support: Determine and make available appropriate facilities, procedures, and tools for the tasks to be performed within cost and schedule constraints.

Integrated engineering analysis: Create simulations and conduct analyses to assess requirements to determine the best set with respect to development, production, deployment, training, operations, support, and disposal effectiveness, cost, and schedule.

System concept definition: Establish a top-level description of the solution to meet customer and user needs that is consistent with technological, cost, or schedule limitations and risk tolerance level.

System requirements: Identify, refine, allocate, and communicate the necessary and sufficient information to define performance goals or constraints for the system to achieve for each of the following primary functions: development, design, production, deployment, training, operation, support, and disposal.

System design: Allocate functions with associated performance requirements and constraints to implementing end items.

System integration: Define and control internal and external interfaces to ensure that the end items come together as a complete system that satisfies the system-level requirements within the defined operating environment.

System verification: Ensure that the completed system satisfies the system-level requirements. Successively applied to the requirements set, simulations used, end-item design, produced end items, and the total system.

SECTION 2: ASSESS THE RELATIVE IMPORTANCE OF IMPROVING EACH PROCESS AREA

For each listed process area defined in Section 1, please mark an X or checkmark in the column indicating your perception of the relative importance of the process area at this time for the program, from 1 for *least important* to 9 for *most important*.

If you perceive some subprocess areas to be equally important at some level between 1 and 9, then mark all those subprocess areas at that same level.

Process Area Defined in Section 1	Process Area Importance at This Time Least...Most								
	1	2	3	4	5	6	7	8	9
Planning									
Tracking and oversight									
Subcontractor management									
Intergroup coordination									
Configuration management									
Quality assurance									
Risk management									
Training									
Technology management									
Environment and tool support									
Integrated engineering analysis									
System concept definition									
System requirements									
System design									
System integration									
System verification									

SECTION 3: ASSESS THE MATURITY OF EACH PROCESS AREA

For each listed process area defined in Section 1, please mark an X or checkmark in the column that you believe best describes the maturity level of the process at this time.

If you do not know a maturity level, please mark the last column.

Process Area Defined in Section 1	Has Process Maturity Best Described By						
	1 Not Needed	2 Not Performed	3 Performed Ad Hoc	4 Performed per Informal Definition	5 Performed per Formal Definition	6 Performed per Formal Definition with Product Quality Assessed	7 Performed with Adjustments Made Based on Statistically Valid Product Quality Measures	0 I Do Not Know
Planning								
Tracking and oversight								
Subcontractor management								
Intergroup coordination								
Configuration management								
Quality assurance								
Risk management								
Training								
Technology management								
Environment and tool support								
Integrated engineering analysis								
System concept definition								
System requirements								
System design								
System integration								
System verification								

SECTION 4: ASSESS THE EFFECTIVENESS OF EACH PROCESS AREA

PLEASE READ FIRST

What follows is a chart for each of the process areas defined in Section 1.

For each process area, goals are stated.

Please mark the box to indicate how well each goal is currently being achieved, from 1 for *not at all* to 9 for *perfectly*.

If you believe the goal is not applicable, or you do not know how well the goal is being achieved, then mark the far right box.

Each sheet provides space for you to record any input you care to provide.

PROCESS AREA: PLANNING

Check the box to the right of each goal statement that indicates how effectively the stated goal is being achieved.

The Goal	Is Being Achieved: Not at All......Perfectly									0 Not Applicable or Do Not Know
	1	2	3	4	5	6	7	8	9	
01.1 Work tasks are complete, efficient, well documented, and achievable.										
01.2 Design, production, test, training, support, and operations tasks are well integrated.										
01.3 Cost, schedule, and technical objectives are achievable.										

Optional comments:

Process Area: Tracking and Oversight

Check the box to the right of each goal statement that indicates how effectively the stated goal is being achieved.

The Goal	Is Being Achieved: Not at All......Perfectly									0 Not Applicable or Do Not Know
	1	2	3	4	5	6	7	8	9	
02.1 Tasks to track and oversee program activities are complete, efficient, and well documented.										
02.2 Current cost, schedule, and technical status are accurately known.										
02.3 Actions taken to achieve cost, schedule, or technical goals best balance impact to all groups.										
02.4 Work effort to track and oversee activities is appropriate and productive.										

Optional comments:

Process Area: Subcontractor Management

Check the box to the right of each goal statement that best indicates how effectively the stated goal is being achieved.

The Goal	Is Being Achieved: Not at All......Perfectly									0 Not Applicable or Do Not Know
	1	2	3	4	5	6	7	8	9	
03.1 Selected fully qualified subcontractors who are contributing to a successful program.										
03.2 Subcontractor tasks are complete, efficient, and well documented.										
03.3 Subcontractor cost, schedule, and technical goals are achievable and agreed to.										

The Goal	Is Being Achieved: Not at All......Perfectly									0 Not Applicable or Do Not Know
	1	2	3	4	5	6	7	8	9	
03.4 Subcontractor cost, schedule, and technical accomplishments are tracked, enabling timely issue resolution.										
03.5 Actions to resolve subcontractor cost, schedule, and technical issues best balance impact to all groups.										

Optional comments:

PROCESS AREA: INTERGROUP COORDINATION

Check the box to the right of each goal statement that best indicates how effectively the stated goal is being achieved.

The Goal	Is Being Achieved: Not at All......Perfectly									0 Not Applicable or Do Not Know
	1	2	3	4	5	6	7	8	9	
04.1 Group roles and responsibilities are defined and are adequately unambiguous.										
04.2 Methods for intergroup coordination are documented in a useful manner.										
04.3 Appropriate groups are involved in resolving technical or program issues.										
04.4 All groups have the program and technical information they need.										

Optional comments:

PROCESS AREA: CONFIGURATION MANAGEMENT

Check the box to the right of each goal statement that best indicates how effectively the stated goal is being achieved.

The Goal	Is Being Achieved: Not at All......Perfectly									0 Not Applicable or Do Not Know
	1	2	3	4	5	6	7	8	9	
05.1 Tasks to document, control, and communicate products are complete, efficient, and well documented.										
05.2 Proposed configuration changes are identified and communicated in a useful manner.										
05.3 Configuration changes are evaluated accurately with respect to cost, schedule, and technical impacts.										
05.4 Configuration changes are controlled.										
05.5 Configuration change decisions are appropriate.										

Optional comments:

PROCESS AREA: QUALITY ASSURANCE

Check the box to the right of each goal statement that best indicates how effectively the stated goal is being achieved.

The Goal	Is Being Achieved: Not at All......Perfectly									0 Not Applicable or Do Not Know
	1	2	3	4	5	6	7	8	9	
06.1 Tasks to achieve error-free and robust products are complete, efficient, and well documented.										
06.2 Product acceptance standards are well documented.										
06.3 Inspectors are utilized in an appropriate manner.										
06.4 Noncompliance issues are tracked to closure.										

Optional comments:

Process Area: Risk Management

Check the box to the right of each goal statement that indicates how effectively the stated goal is being achieved.

The Goal	Is Being Achieved: Not at All......Perfectly									0 Not Applicable or Do Not Know
	1	2	3	4	5	6	7	8	9	
07.1 Tasks to identify, assess, and mitigate risks are complete, efficient, and well documented.										
07.2 Technical, cost, and schedule risks are identified and assessed.										
07.3 Actions are taken to mitigate risks.										
07.4 Risks mitigation actions are monitored.										
07.5 Risk status is communicated and coordinated across affected groups.										

Optional comments:

Process Area: Training

Check the box to the right of each goal statement that indicates how effectively the stated goal is being achieved.

The Goal	Is Being Achieved: Not Applicable									0 Not Applicable or Do Not Know
	1	2	3	4	5	6	7	8	9	
08.1 Training tasks are complete, efficient, and well documented.										
08.2 People on the program are fully qualified to do their jobs.										
08.3 Training meets the needs of the people on the program.										
08.4 Training effectiveness is measured to provide feedback for additional training needs.										

Optional comments:

PROCESS AREA: TECHNOLOGY MANAGEMENT

Check the box to the right of each goal statement that indicates how effectively the stated goal is being achieved.

The Goal	Is Being Achieved: Not at All......Perfectly									0 Not Applicable or Do Not Know
	1	2	3	4	5	6	7	8	9	
09.1 Tasks to identify and select appropriate technologies to utilize are complete, efficient, and well documented.										
09.2 Technologies are selected for uses that support the business goals.										
09.3 Technology innovations are implemented that improve processes.										
09.4 Technology innovations are implemented that improve products.										

Optional comments:

PROCESS AREA: ENVIRONMENT AND TOOL SUPPORT

Check the box to the right of each goal statement that indicates how effectively the stated goal is being achieved.

The Goal	Is Being Achieved: Not at All......Perfectly									0 Not Applicable or Do Not Know
	1	2	3	4	5	6	7	8	9	
10.1 Tasks to acquire and make available facilities, procedures, and tools are complete, efficient, and well documented.										
10.2 Facilities, procedures, and tools provided enable completing tasks within cost and schedule constraints.										
10.3 Facilities, procedures, and tools support intergroup interaction.										
10.4 Tools used are compatible and enhance efficiency.										

Optional comments:

PROCESS AREA: INTEGRATED ENGINEERING ANALYSIS

Check the box to the right of each goal statement that indicates how effectively the stated goal is being achieved.

The Goal	Is Being Achieved: Not at All......Perfectly									0 Not Applicable or Do Not Know
	1	2	3	4	5	6	7	8	9	
11.1 An integrated, multidisciplinary approach to problem solving is coached and fostered.										
11.2 Analysis tasks are complete, efficient, and well documented.										
11.3 All relevant groups are involved that need to be involved in making decisions.										
11.4 Methodology and criteria for making decisions are clear, appropriate, and consistently applied.										
11.5 Analyses are sufficient, accurate, and timely enough to best guide decision making.										

Optional comments:

PROCESS AREA: SYSTEM CONCEPT DEVELOPMENT

Check the box to the right of each goal statement that indicates how effectively the stated goal is being achieved.

The Goal	Is Being Achieved: Not at All......Perfectly									0 Not Applicable or Do Not Know
	1	2	3	4	5	6	7	8	9	
12.1 Tasks to determine best solutions to offer customers and users are complete, efficient, and well documented.										
12.2 Customer and user needs are accurately documented in a useful manner.										

The Goal	Is Being Achieved: Not at All......Perfectly									0 Not Applicable or Do Not Know
	1	2	3	4	5	6	7	8	9	
12.3 Derived requirements are traceable to customer and user needs and are well documented.										
12.4 Chosen end items are allocated appropriate and complete functional and performance requirements.										
12.5 Chosen concept features best satisfy customer and user needs.										

Optional comments:

PROCESS AREA: SYSTEM REQUIREMENTS

Check the box to the right of each goal statement that indicates how effectively the stated goal is being achieved.

The Goal	Is Being Achieved: Not at All......Perfectly									0 Not Applicable or Do Not Know
	1	2	3	4	5	6	7	8	9	
13.1 Tasks to determine performance requirements for all eight primary functions are complete, efficient, and well documented.										
13.2 All requirement attributes are logical, documented, and easy to find and comprehend.										
13.3 The functional and performance requirements enable appropriate implementation options.										
13.4 All groups have all the requirements they need to do their jobs.										

Optional comments:

PROCESS AREA: SYSTEM DESIGN

Check the box to the right of each goal statement that indicates how effectively the stated goal is being achieved.

The Goal	Is Being Achieved: Not at All......Perfectly									0 Not Applicable or Do Not Know
	1	2	3	4	5	6	7	8	9	
14.1 Tasks to determine instructions to produce the end items are complete, efficient, and well documented.										
14.2 Design requirements are traceable to customer and user requirements.										
14.3 Production, test, training, and support implications of design are determined and acceptable.										
14.4 Design baseline is established and is the best solution.										

Optional comments:

PROCESS AREA: SYSTEM INTEGRATION

Check the box to the right of each goal statement that indicates how effectively the stated goal is being achieved.

The Goal	Is Being Achieved: Not at All......Perfectly									0 Not Applicable or Do Not Know
	1	2	3	4	5	6	7	8	9	
15.1 Tasks to achieve internal and external interfaces are complete, efficient, and well documented.										
15.2 Interface requirements between end items are clear, complete, and unambiguous.										
15.3 Interface requirements with external entities are clear, complete, and unambiguous.										
15.4 The end items will assemble into a uniform whole that meets system-level requirements.										

Optional comments:

Process Area: System Verification

Check the box to the right of each goal statement that indicates how effectively the stated goal is being achieved.

The Goal	Is Being Achieved: Not at All......Perfectly									0 Not Applicable or Do Not Know
	1	2	3	4	5	6	7	8	9	
16.1 Requirement set, verification criteria, and tasks are complete, efficient, and well documented.										
16.2 Simulation verification criteria and tasks are complete, efficient, and well documented.										
16.3 End-item design verification criteria and tasks are complete, efficient, and well documented.										
16.4 Produced end-item verification criteria and tasks are complete, efficient, and well documented.										
16.5 System verification criteria and tasks are complete, efficient, and well documented.										
16.6 Verification evidence obtained is documented in a useful manner.										
16.7 All verification evidence obtained so far indicates requirements are being appropriately met.										

Optional comments:

APPENDIX 5B: ARCHITECTING EFFECTIVENESS SURVEY

PLEASE READ FIRST

Please complete and return this survey to [insert name] by [insert time and date].

Your response is confidential.

Your participation is critical to determine how best to improve our systems engineering efforts.

The survey has four sections:

Section 1 defines the 13 process areas.

Section 2 obtains your assessment of the relative importance of improving each process area at this time.

Section 3 obtains your assessment of the maturity of each process area at this time.

Section 4 obtains your assessment of the product quality of each process area at this time.

Thank you for taking the time to provide this information.

If you have any questions, please contact [insert name, phone number, and email address].

To improve the survey process, please answer the following questions after completing the survey:

1. Approximately how many minutes were needed for you to complete this survey? _____
2. How would you improve this survey?

SECTION 1: PROCESS AREA DEFINITIONS

Planning: Identify, define, and integrate tasks, responsibilities, resources, and schedules associated with activities underway.

Tracking and oversight: Tracking establishes metrics associated with planned tasks and products. Oversight assesses these metrics and then initiates corrective actions as required to resolve or otherwise mitigate deficiencies.

Subcontractor management: Select, track, and oversee subcontractors engaged in the effort.

Intergroup coordination: Enable and achieve effective communication and issue resolution among groups.

Configuration management: The planning, identification, change control, status accounting, and auditing of the product elements, which include requirements, interfaces, design representations, production procedures, test plans and results, and end items.

Training: Ensure that program participants are prepared and qualified to perform defined program tasks.

Environment and tool support: Determine and make available appropriate facilities, procedures, and tools for the tasks to be performed within cost and schedule constraints.

Establish as-is architecture and requirements: Define the (1) scenarios or reference missions the family of systems are to accomplish; (2) existing family of systems; and (3) current organizational and information exchange interface requirements.

Determine the as-is family of systems shortfalls: Conduct campaign, mission, engagement, and engineering analyses to explicitly identify the as-is family of systems performance, capability, sufficiency, and/or utility shortfalls.

Develop concepts to mitigate shortfalls: Define candidate concepts to remove shortfalls in sufficient depth to enable accurate life cycle cost estimates.

Identify key enabling technologies: Define quantitatively technological advances that could mitigate shortfalls.

Determine the utility of candidate families of systems: Identify and estimate the utility and cost of candidate families of systems.

Establish to-be architecture and requirements: Determine a desired family of systems consistent with cost constraints, and specify key requirements for each contributing system. Document time-phased acquisition and deployment plan from the as-is to the to-be family of systems showing the estimated incremental improvement in sufficiency utility achieved and the required funding profile. Establish the to-be organizational and information exchange-implementing requirements.

SECTION 2: ASSESS THE RELATIVE IMPORTANCE OF IMPROVING EACH PROCESS AREA

For each listed process area defined in Section 1, please mark an X or checkmark in the column indicating your perception of the relative importance of the process area at this time for the program, from 1 for *least important* to 9 for *most important*.

If you perceive some subprocess areas to be equally important at some level between 1 and 9, then mark all those subprocess areas at that same level.

Process Area Defined in Section 1	Process Area Importance at This Time Least.....................................Most								
	1	2	3	4	5	6	7	8	9
Planning									
Tracking and oversight									
Subcontractor management									
Intergroup coordination									
Configuration management									
Training									
Environment and tool support									
Establish as-is architecture and requirements									
Determine the as-is family of systems shortfalls									
Develop concepts to mitigate shortfalls									
Identify key enabling technologies									
Determine the utility of candidate families of systems									
Establish to-be architecture and requirements									

SECTION 3: ASSESS THE MATURITY OF EACH PROCESS AREA

For each listed process area defined in Section 1, please mark an X or checkmark in the column that you believe best describes the maturity level of the process at this time.

If you do not know a maturity level, please mark the last column.

Process Area Defined in Section 1	Has Process Maturity Best Described By							
	1 Not Needed	2 Not Performed	3 Performed Ad Hoc	4 Performed per Informal Definition	5 Performed per Formal Definition	6 Performed per Formal Definition with Product Quality Assessed	7 Performed with Adjustments Made Based on Statistically Valid Product Quality Measures	0 I Do Not Know
Planning								
Tracking and oversight								
Subcontractor management								
Intergroup coordination								
Configuration management								
Training								
Environment and tool support								
Establish as-is architecture and requirements								
Determine the as-is family of systems shortfalls								
Develop concepts to mitigate shortfalls								
Identify key enabling technologies								
Determine the utility of candidate families of systems								
Establish to-be architecture and requirements								

SECTION 4: ASSESS THE EFFECTIVENESS OF EACH PROCESS AREA

PLEASE READ FIRST

What follows is a chart for each of the process areas defined in Section 1.

For each process area, goals are stated.

Please mark the box to indicate how well each goal is currently being achieved, from 1 for *not at all* to 9 for *perfectly*.

If you believe the goal is not applicable, or you do not know how well the goal is being achieved, then mark the far right box.

Each chart provides space for you to record any input you care to provide.

PROCESS AREA: PLANNING

Check the box to the right of each goal statement that indicates how effectively the stated goal is being achieved.

The Goal	Is Being Achieved Not at All..........Perfectly									0 Not Applicable or Do Not Know
	1	2	3	4	5	6	7	8	9	
01.1 Work tasks are complete, efficient, and well documented.										
01.2 Cost, schedule, and technical objectives are achievable.										

Optional comments:

PROCESS AREA: TRACKING AND OVERSIGHT

Check the box to the right of each goal statement that indicates how effectively the stated goal is being achieved.

The Goal	Is Being Achieved Not at All..........Perfectly									0 Not Applicable or Do Not Know
	1	2	3	4	5	6	7	8	9	
02.1 Tasks to track and oversee architecting activities are complete, efficient, and well documented.										
02.2 Current cost, schedule, and technical status are accurately known.										
02.3 Actions taken to achieve cost, schedule, or technical goals best balance impact on all affected groups.										
02.4 Work effort to track and oversee activities is appropriate and productive.										

Optional comments:

Process Area: Subcontractor Management

Check the box to the right of each goal statement that best indicates how effectively the stated goal is being achieved.

The Goal	Is Being Achieved Not at All..........Perfectly									0 Not Applicable or Do Not Know
	1	2	3	4	5	6	7	8	9	
03.1 Selected fully qualified subcontractors who are contributing to a successful architecting effort.										
03.2 Subcontractor tasks are complete, efficient, and well documented.										
03.3 Subcontractor cost, schedule, and technical goals are achievable and agreed to.										
03.4 Subcontractor cost, schedule, and technical accomplishments are tracked, enabling timely issue resolution.										
03.5 Actions to resolve subcontractor cost, schedule, or technical issues best balance impact to all groups.										

Optional comments:

Process Area: Intergroup Coordination

Check the box to the right of each goal statement that indicates how effectively the stated goal is being achieved.

The Goal	Is Being Achieved Not at All..........Perfectly									0 Not Applicable or Do Not Know
	1	2	3	4	5	6	7	8	9	
04.1 Group roles and responsibilities are defined and are adequately unambiguous.										
04.2 Methods for intergroup coordination are documented in a useful manner.										
04.3 Appropriate groups are involved in resolving technical or program issues.										
04.4 All groups have the program and technical information they need, when they need it.										

Optional comments:

PROCESS AREA: CONFIGURATION MANAGEMENT

Check the box to the right of each goal statement that best indicates how effectively the stated goal is being achieved.

The Goal	Is Being Achieved Not at All..........Perfectly									0 Not Applicable or Do Not Know
	1	2	3	4	5	6	7	8	9	
05.1 Tasks to document, control, and communicate products are complete, efficient, and well documented.										
05.2 Proposed configuration changes are identified and communicated in a useful manner.										
05.3 Configuration changes are evaluated accurately with respect to cost, schedule, and technical impacts.										
05.4 Configuration changes are controlled.										
05.5 Configuration change decisions are appropriate.										

Optional comments:

PROCESS AREA: TRAINING

Check the box to the right of each goal statement that indicates how effectively the stated goal is being achieved.

The Goal	Is Being Achieved Not at All..........Perfectly									0 Not Applicable or Do Not Know
	1	2	3	4	5	6	7	8	9	
06.1 Training tasks to benefit personnel are complete, efficient, and well documented.										
06.2 People on the architecting effort are fully qualified to do their jobs.										
06.3 Training meets the needs of the people on the architecting effort.										
06.4 Training effectiveness is measured to provide feedback for additional training needs.										

Optional comments:

Process Area: Environment and Tool Support

Check the box to the right of each goal statement that indicates how effectively the stated goal is being achieved.

The Goal	Is Being Achieved Not at All.........Perfectly									0 Not Applicable or Do Not Know
	1	2	3	4	5	6	7	8	9	
07.1 Tasks to acquire and make available facilities, procedures, and tools are complete, efficient, and well documented.										
07.2 Facilities, procedures, and tools are provided that enable completing tasks within cost and schedule constraints.										
07.3 Facilities, procedures, and tools support intergroup interaction.										
07.4 Tools used are compatible and enhance efficiency.										

Optional comments:

Process Area: Establish As-Is Architecture and Requirements

Check the box to the right of each goal statement that indicates how effectively the stated goal is being achieved.

The Goal	Is Being Achieved Not at All.........Perfectly									0 Not Applicable or Do Not Know
	1	2	3	4	5	6	7	8	9	
08.1 Scenarios or design reference missions adequately represent the key purpose for the family of systems.										
08.2 Have data for family of systems to assess performance, capability, sufficiency, and utility.										
08.3 Organization relationships governing the existing family of systems operations are adequately defined.										
08.4 Information needed for the existing family of systems elements to perform mission is adequately defined.										

Optional comments:

Process Area: Determine the As-Is Family of Systems Shortfalls

Check the box to the right of each goal statement that indicates how effectively the stated goal is being achieved.

The Goal	Is Being Achieved Not at All..........Perfectly									0 Not Applicable or Do Not Know
	1	2	3	4	5	6	7	8	9	
09.1 Accurate and efficient simulations exist to determine the family of system performance.										
09.2 Accurate and efficient simulations exist to determine the family of system capability.										
09.3 Accurate and efficient simulations exist to determine the family of system sufficiency.										
09.4 Accurate and efficient simulations exist to determine the family of system utility.										
09.5 People with authority to establish the to-be architecture easily comprehend the nature, cause, and impact of existing shortfalls.										

Optional comments:

Process Area: Develop Concepts to Mitigate Shortfalls

Check the box to the right of each goal statement that indicates how effectively the stated goal is being achieved.

The Goal	Is Being Achieved Not at All..........Perfectly									0 Not Applicable or Do Not Know
	1	2	3	4	5	6	7	8	9	
10.1 Adequate number and diversity of concepts are defined to address all key shortfalls.										
10.2 All concepts are realistic with respect to capability, availability, and cost.										
10.3 Concept definitions enable determining performance, capability, sufficiency, and utility.										

The Goal	Is Being Achieved Not at All..........Perfectly									0 Not Applicable or Do Not Know
	1	2	3	4	5	6	7	8	9	
10.4 Concept definitions enable accurate cost estimates.										
10.5 Concept definitions are sufficiently detailed to serve as requirements for implementation.										
10.6 People with authority to establish the to-be architecture easily comprehend the candidate concepts.										

Optional comments:

Process Area: Identify Key Enabling Technologies to Mitigate Shortfalls

Check the box to the right of each goal statement that indicates how effectively the stated goal is being achieved.

The Goal	Is Being Achieved Not at All..........Perfectly									0 Not Applicable or Do Not Know
	1	2	3	4	5	6	7	8	9	
11.1 Advancements from the current state of the art that could mitigate shortfalls are adequately identified.										
11.2 The explicit performance levels for technology advancements sought are quantitatively defined.										
11.3 Technology maturation experts can easily develop technology maturation plans based on the shortfall and technology need documentation.										

Optional comments:

Process Area: Determine the Utility and Cost of Candidate Families of Systems

Check the box to the right of each goal statement that indicates how effectively the stated goal is being achieved.

The Goal				Is Being Achieved Not at All..........Perfectly						0 Not Applicable or Do Not Know
	1	2	3	4	5	6	7	8	9	
12.1 Data needed to assess individual systems are available.										
12.2 Data needed to assess combinations of systems are available.										
12.3 All credible family-of-system combinations are assessed.										
12.4 Secondary system selection criteria were adequately explored for leading candidate family-of-system solutions.										
12.5 People with the authority to establish the to-be architecture have the best appropriate data to easily make the decision.										

Optional comments

PROCESS AREA: ESTABLISH TO-BE ARCHITECTURE AND REQUIREMENTS

Check the box to the right of each goal statement that indicates how effectively the stated goal is being achieved.

The Goal	Is Being Achieved Not at All..........Perfectly									0 Not Applicable or Do Not Know
	1	2	3	4	5	6	7	8	9	
13.1 The requirements for systems constituents are accurately and completely documented.										
13.2 To-be organizational relationships are accurately and completely documented.										
13.3 To-be information flow requirements and solutions are accurately and completely documented.										

Optional comments:

REFERENCES

1. Solomon, Paul J. and Young, Ralph R., *Performance-Based Earned Value*, New York: John Wiley & Sons, 2007.
2. Deming, W. Edwards, *Out of the Crisis*, Cambridge, MA: Massachusetts Institute of Technology, 1989.

Index

Printed and bound by CPI Group (UK) Ltd, Croydon, CR0 4YY

18/10/2024

01776261-0015